T0220458

# Materials Reliability in Microelectronics II

MATERIALS RESEARCH SOCIETY SYMPOSIUM PROCEEDINGS VOLUME 265

# Materials Reliability in Microelectronics II

Symposium held April 27-May 1, 1992, San Francisco, California, U.S.A.

EDITORS:

## C.V. Thompson

Massachusetts Institute of Technology
Cambridge, Massachusetts, U.S.A.

## J.R. Lloyd

Digital Equipment Corporation
Hudson, Massachusetts, U.S.A.

MATERIALS RESEARCH SOCIETY
Pittsburgh, Pennsylvania

CAMBRIDGE UNIVERSITY PRESS
Cambridge, New York, Melbourne, Madrid, Cape Town,
Singapore, São Paulo, Delhi, Mexico City

Cambridge University Press
32 Avenue of the Americas, New York NY 10013-2473, USA

Published in the United States of America by Cambridge University Press, New York

www.cambridge.org
Information on this title: www.cambridge.org/9781107409682

Materials Research Society
506 Keystone Drive, Warrendale, PA 15086
http://www.mrs.org

This publication has been registered with Copyright Clearance Center, Inc.
For further information please contact the Copyright Clearance Center,
Salem, Massachusetts.

First published 1992
First paperback edition 2012

Single article reprints from this publication are available through
University Microfilms Inc., 300 North Zeeb Road, Ann Arbor, MI 48106

CODEN: MRSPDH

ISBN 978-1-107-40968-2 Paperback

# Contents

*Invited Paper

## PART III: METALLIZATION SCHEMES FOR
## INTERCONNECTS, WIRING, AND PACKAGING

*Invited Paper

## PART IV: OXIDE AND DEVICE RELIABILITY

## PART V: ANALYTICAL TECHNIQUES AND OTHER TOPICS

*Invited Paper

# Preface

Design constraints imposed by concerns with device and interconnect reliability currently limit further evolution to faster and more complex integrated circuits. While the reliability of materials in integrated circuits is related to their performance in electronic systems, and has therefore been dealt with primarily by electrical engineers, the study and understanding of the physical mechanisms of failure, and the design of processes and structures to avoid failure, present classical challenges in materials science and engineering. For example, the mechanical and chemical stability of metals in integrated circuits is affected by the relationships in the way they have been processed, their microstructures, and their properties, just as in large-scale structures, such as bridges and automobiles. However, the interesting challenges come in when the macrostructures and microstructures are shrunk to micrometer and submicrometer sizes, and also when metals are encapsulated in rigid oxides and then subjected to stresses considerably higher than their bulk yield stresses, and when they are subjected to current densities several orders of magnitude higher than those allowed in house wiring. These unusual dimensions and physical conditions lead to new behavior, which can be studied and understood using the methods developed for bulk materials in other applications.

This symposium and its predecessor have provided forums for materials scientists and engineers to meet with other scientists and engineers to discuss, in detail, the physical phenomena that lead to failure of materials in microelectronics. From these symposia new perspectives on reliability issues have emerged. In this volume, for example, considerable progress is seen in the merging of the understanding of the mechanical deformation and electromigration-induced damage of metallic interconnects. Also, new tools for characterizing thin film microstructure have been described, along with the insights into failure processes that they have lead to. Progress in modelling and design for circuit reliability has also been reported.

We thank the participants for lively discussions, and the speakers, especially the invited speakers, for thoughtful and informative presentations. We also thank Sienna Technologies, Inc. and Digital Equipment Corporation for financial support of the symposium, and Peter Heron for his efforts in preparing these proceedings for publication.

Carl V. Thompson
James R. Lloyd

June 1992

Volume 239—Thin Films: Stresses and Mechanical Properties III, W.D. Nix, J.C. Bravman, E. Arzt, L.B. Freund, 1992, ISBN: 1-55899-133-6

Volume 240—Advanced III-V Compound Semiconductor Growth, Processing and Devices, S.J. Pearton, D.K. Sadana, J.M. Zavada, 1992, ISBN: 1-55899-134-4

Volume 241—Low Temperature (LT) GaAs and Related Materials, G.L. Witt, R. Calawa, U. Mishra, E. Weber, 1992, ISBN: 1-55899-135-2

Volume 242—Wide Band Gap Semiconductors, T.D. Moustakas, J.I. Pankove, Y. Hamakawa, 1992, ISBN: 1-55899-136-0

Volume 243—Ferroelectric Thin Films II, A.I. Kingon, E.R. Myers, B. Tuttle, 1992, ISBN: 1-55899-137-9

Volume 244—Optical Waveguide Materials, M.M. Broer, G.H. Sigel, Jr., R.Th. Kersten, H. Kawazoe, 1992, ISBN: 1-55899-138-7

Volume 245—Advanced Cementitious Systems: Mechanisms and Properties, F.P. Glasser, G.J. McCarthy, J.F. Young, T.O. Mason, P.L. Pratt, 1992, ISBN: 1-55899-139-5

Volume 246—Shape-Memory Materials and Phenomena—Fundamental Aspects and Applications, C.T. Liu, H. Kunsmann, K. Otsuka, M. Wuttig, 1992, ISBN: 1-55899-140-9

Volume 247—Electrical, Optical, and Magnetic Properties of Organic Solid State Materials, L.Y. Chiang, A.F. Garito, D.J. Sandman, 1992, ISBN: 1-55899-141-7

Volume 248—Complex Fluids, E.B. Sirota, D. Weitz, T. Witten, J. Israelachvili, 1992, ISBN: 1-55899-142-5

Volume 249—Synthesis and Processing of Ceramics: Scientific Issues, W.E. Rhine, T.M. Shaw, R.J. Gottschall, Y. Chen, 1992, ISBN: 1-55899-143-3

Volume 250—Chemical Vapor Deposition of Refractory Metals and Ceramics II, T.M. Besmann, B.M. Gallois, J.W. Warren, 1992, ISBN: 1-55899-144-1

Volume 251—Pressure Effects on Materials Processing and Design, K. Ishizaki, E. Hodge, M. Concannon, 1992, ISBN: 1-55899-145-X

Volume 252—Tissue-Inducing Biomaterials, L.G. Cima, E.S. Ron, 1992, ISBN: 1-55899-146-8

Volume 253—Applications of Multiple Scattering Theory to Materials Science, W.H. Butler, P.H. Dederichs, A. Gonis, R.L. Weaver, 1992, ISBN: 1-55899-147-6

Volume 254—Specimen Preparation for Transmission Electron Microscopy of Materials-III, R. Anderson, B. Tracy, J. Bravman, 1992, ISBN: 1-55899-148-4

Volume 255—Hierarchically Structured Materials, I.A. Aksay, E. Baer, M. Sarikaya, D.A. Tirrell, 1992, ISBN: 1-55899-149-2

Volume 256—Light Emission from Silicon, S.S. Iyer, R.T. Collins, L.T. Canham, 1992, ISBN: 1-55899-150-6

Volume 257—Scientific Basis for Nuclear Waste Management XV, C.G. Sombret, 1992, ISBN: 1-55899-151-4

Volume 258—Amorphous Silicon Technology—1992, M.J. Thompson,
Y. Hamakawa, P.G. LeComber, A. Madan, E. Schiff, 1992,
ISBN: 1-55899-153-0

Volume 259—Chemical Surface Preparation, Passivation and Cleaning for
Semiconductor Growth and Processing, R.J. Nemanich, C.R. Helms,
M. Hirose, G.W. Rubloff, 1992, ISBN: 1-55899-154-9

Volume 260—Advanced Metallization and Processing for Semiconductor Devices and
Circuits II, A. Katz, Y.I. Nissim, S.P. Murarka, J.M.E. Harper, 1992,
ISBN: 1-55899-155-7

Volume 261—Photo-Induced Space Charge Effects in Semiconductors: Electro-optics,
Photoconductivity, and the Photorefractive Effect, D.D. Nolte,
N.M. Haegel, K.W. Goossen, 1992, ISBN: 1-55899-156-5

Volume 262—Defect Engineering in Semiconductor Growth, Processing and Device
Technology, S. Ashok, J. Chevallier, K. Sumino, E. Weber, 1992,
ISBN: 1-55899-157-3

Volume 263—Mechanisms of Heteroepitaxial Growth, M.F. Chisholm, B.J. Garrison,
R. Hull, L.J. Schowalter, 1992, ISBN: 1-55899-158-1

Volume 264—Electronic Packaging Materials Science VI, P.S. Ho, K.A. Jackson,
C-Y. Li, G.F. Lipscomb, 1992, ISBN: 1-55899-159-X

Volume 265—Materials Reliability in Microelectronics II, C.V. Thompson,
J.R. Lloyd, 1992, ISBN: 1-55899-160-3

Volume 266—Materials Interactions Relevant to Recycling of Wood-Based Materials,
R.M. Rowell, T.L. Laufenberg, J.K. Rowell, 1992,
ISBN: 1-55899-161-1

Volume 267—Materials Issues in Art and Archaeology III, J.R. Druzik,
P.B. Vandiver, G.S. Wheeler, I. Freestone, 1992, ISBN: 1-55899-162-X

Volume 268—Materials Modification by Energetic Atoms and Ions, K.S. Grabowski,
S.A. Barnett, S.M. Rossnagel, K. Wasa, 1992, ISBN: 1-55899-163-8

Volume 269—Microwave Processing of Materials III, R.L. Beatty, W.H. Sutton,
M.F. Iskander, 1992, ISBN: 1-55899-164-6

Volume 270—Novel Forms of Carbon, C.L. Renschler, J. Pouch, D. Cox, 1992,
ISBN: 1-55899-165-4

Volume 271—Better Ceramics Through Chemistry V, M.J. Hampden-Smith,
W.G. Klemperer, C.J. Brinker, 1992, ISBN: 1-55899-166-2

Volume 272—Chemical Processes in Inorganic Materials: Metal and Semiconductor
Clusters and Colloids, P.D. Persans, J.S. Bradley, R.R. Chianelli,
G. Schmid, 1992, ISBN: 1-55899-167-0

Volume 273—Intermetallic Matrix Composites II, D. Miracle, J. Graves, D. Anton,
1992, ISBN: 1-55899-168-9

Volume 274—Submicron Multiphase Materials, R. Baney, L. Gilliom, S.-I. Hirano,
H. Schmidt, 1992, ISBN: 1-55899-169-7

Volume 275—Layered Superconductors: Fabrication, Properties and Applications,
D.T. Shaw, C.C. Tsuei, T.R. Schneider, Y. Shiohara, 1992,
ISBN: 1-55899-170-0

Volume 276—Materials for Smart Devices and Micro-Electro-Mechanical Systems,
A.P. Jardine, G.C. Johnson, A. Crowson, M. Allen, 1992,
ISBN: 1-55899-171-9

Volume 277—Macromolecular Host-Guest Complexes: Optical, Optoelectronic, and
Photorefractive Properties and Applications, S.A. Jenekhe, 1992,
ISBN: 1-55899-172-7

Volume 278—Computational Methods in Materials Science, J.E. Mark,
M.E. Glicksman, S.P. Marsh, 1992, ISBN: 1-55899-173-5

Prior Materials Research Society Symposium Proceedings
available by contacting Materials Research Society

## PART I

# Stress and Electromigration/Modelling

# DEFORMATION MECHANISMS IN THIN FILMS.

H.J. Frost
Thayer School of Engineering, Dartmouth College, Hanover, New Hampshire 03755

## ABSTRACT

There are a variety of different mechanisms which may contribute to the plastic deformation of a polycrystalline material, involving such processes as dislocation glide, dislocation climb, grain boundary sliding and diffusion of vacancies. Each mechanism can be characterized by a relationship which gives strain-rate as a function of stress, temperature and microstructural state. A deformation mechanism may actually involve multiple, coupled atomic processes. For example, diffusional creep results from the coupled operation of vacancy diffusion and grain boundary sliding. The total strain-rate of the polycrystal is the sum of the contributions of the various different mechanisms, with different mechanisms dominant in different regimes of stress and temperature. In bulk polycrystals this behavior has been described with deformation-mechanism maps. For thin films the same atomic processes of deformation can operate, but the deformation mechanisms may have different stress---strain-rate---temperature relationships because of the presense of free surfaces or thin-film---substrate interfaces.

## INTRODUCTION

Mechanisms of plastic deformation in polycrystalline thin films has been a topic of increasing interest in recent years. Recent reviews are given by Nix [1] and Alexopoulos and O'Sullivan [2]. This is primarily because of the technological importance of controlling the microstructure, properties and behavior of thin films in electronic and protective coating applications. The topic also involves several scientifically interesting phenomena and provides an excellent setting to apply the modelling developed for bulk materials to a new environment.

It is sometimes useful to make a distinction between the atomic processes by which deformation occurs, and deformation mechanisms which are used to describe how those processes operate. The atomic process include:

> Dislocation Glide
> Dislocation Climb
> Atomic diffusion (usually vacancy migration) through the lattices of the crystals
> Atomic diffusion along grain boundaries
> Grain Boundary Sliding (sometimes accompanied by grain boundary migration)
> Deformation Twinning (glide of partial dislocations, organized in particular
>    arrangements, sometimes accompanied by local atomic reshuffling)
> Cracking and Cavitation.

We might define an atomic deformation process as a process which results in a relative displacement between two reference atoms. A deformation mechanism might be defined as the operation of one or more atomic processes in a manner which accomplishes the strain of the polycrystal as a whole. A deformation mechanism often involves the operation of two or more of the atomic processes, coupled together by rate limiting steps so that their operation may be conveniently described by a single relationship between stress, $\sigma$, temperature, T, and strain-rate, $\dot{\epsilon}$ (uniaxial) or $\dot{\gamma}$ (in shear). More precisely, each mechanism has a relationship between stress, temperature, strain-rate and microstructural state. For some mechanisms the important components of the microstructure (e.g. dislocation density and arrangement) may evolve during the deformation, so that the rate-equation for instaneous deformation depends on the prior history of deformation. The situation is usually simplified by considering only steady-state microstructural states which arise during deformation (e.g. steady-state creep) or particular, specified initial microstructures. A selection of such rate equations is given in Table 1.

A standard list of deformation mechanisms includes the following:

Dislocation Glide: (Figure 1) [3]
  Limited by Discrete Obstacles
  Limited by Lattice Resistance

Dislocation Climb and Glide: (Figure 2)
  Power-law Creep
    Lattice Diffusion Dominated
      (High Temperature Creep [4] )
    Dislocation-Core Diffusion Dominated
      (Low Temperature Creep [5] )

Diffusional Flow: (Figure 3)
  Vacancy Migration / Grain Boundary Sliding
    Lattice Diffusion Dominated:
      (Nabarro-Herring Creep [6,7])
    Boundary Diffusion Dominated:
      (Coble Creep [8])

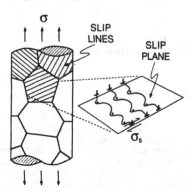

Figure 1. Low-temperature plasticity by dislocation glide limited by discrete obstacles. The strain-rate is determined by the kinetics of obstacle cutting.

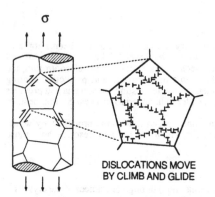

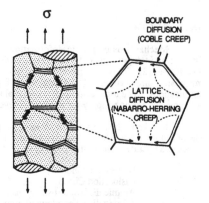

Figure 2. Power-law creep involving cell formation by dislocation climb. Power-law creep limited by glide processes alone is also possible. Grain boundary sliding often contributes to the strain. At higher stresses dislocation glide contributes increasingly to the overall strain rate, and the power-law dependence on stress breaks down.

Figure 3. Diffusional flow by diffusional transport through and around the grains. Geometric compatibility usually requires grain boundary sliding. The strain rate may be limited by the rate of diffusion (along the fastest diffusion path) or by the rate of an interface reaction.

Some additional mechanisms are sometimes important. Deformation twinning is important for HCP metals at low and moderate temperatures, and may also occur in BCC and FCC metals at very low temperatures. At very high temperatures dynamic recrystallization becomes important. At very low stresses, dislocation creep may operate as Harper-Dorn Creep, in which the strain rate is linearly proportional to stress. [9] A more important phenomenon for this paper is the breakdown of the power-law behavior of dislocation creep at high stresses. [10] We have simply included that power-law breakdown within the dislocation creep mechanism.

TABLE 1.
DEFORMATION–MECHANISM RATE EQUATIONS

| Mechanism | Rate Equation | Notation |
|---|---|---|
| **Dislocation Glide:** <br><br> Limited by Discrete Obstacles | $$\dot{\gamma} = \dot{\gamma}_0 \exp\left[-\frac{\Delta F}{kT}\left(1 - \frac{\sigma_s}{\hat{\tau}}\right)\right]$$ | $\dot{\gamma}_0$ = Constant <br> $\Delta F$ = Free Energy required to overcome obstacles <br> $\hat{\tau}$ = Flow Stress at 0 K |
| Limited by Lattice Resistance | $$\dot{\gamma} = \dot{\gamma}_p\left(\frac{\sigma_s}{\mu}\right)^2 \exp\left\{-\frac{\Delta F_p}{kT}\left[1 - \left(\frac{\sigma_s}{\hat{\tau}_p}\right)^{3/4}\right]^{4/3}\right\}$$ | $\dot{\gamma}_p$ = Constant <br> $\Delta F_p$ = Free Energy of an isolated pair of kinks <br> $\hat{\tau}_p$ = Flow Stress at 0 K. |
| **Dislocation Climb and Glide:** <br><br> Powerlaw Creep | $$\dot{\gamma} = \frac{A_2 \, D_{eff} \, \mu b}{kT}\left(\frac{\sigma_s}{\mu}\right)^n$$ | $A_2$ = Dimensionless Constant <br> $n$ = Power-law Exponent typically $n \approx 3 - 7$ |
| Power-law Breakdown | $$\dot{\gamma} = \frac{A'_2 \, D_{eff} \, \mu b}{kT}\left[\sinh\left(\alpha' \frac{\sigma_s}{\mu}\right)\right]^{n'}$$ <br><br> $$D_{eff} = D_v + \frac{10 \, a_c}{b^2}\left(\frac{\sigma_s}{\mu}\right)^2 D_c$$ <br><br> Lattice Diffusion    Dislocation Core <br> Dominated    Diffusion Dominated <br> (High Temperature)   (Low Temperature) | $A'_2 \, \alpha'^{n'} = A_2$ <br> $n' = n$ <br><br> $D_v$ = Lattice Diffusion Coefficient <br> $D_c$ = Dislocation Core Diffusion Coeff. <br> $a_c$ = Cross-sectional Area of Dislocation Core |
| **Diffusional Flow** <br><br> Grain-Boundary Sliding and Vacancy Migration | $$\dot{\gamma} = \frac{42 \, \sigma_s \, \Omega}{kT \, d^2} D_{eff}$$ <br><br> $$D_{eff} = D_v + \frac{\pi \delta D_b}{d}$$ <br><br> Lattice Diffusion    Boundary Diffusion <br> Dominated:    Dominated: <br> Nabarro-Herring Creep   Coble Creep | $d$ = Grain Size <br> $\Omega$ = Atomic Volume <br><br> $D_b$ = Grain Boundary Diffusion Coeff. <br> $\delta$ = Thickness of Grain Boundary |

$\sigma_s$ = Effective Shear Stress      $\mu$ = Shear Modulus
$\dot{\gamma}$ = Effective Shear Strain Rate      $b$ = Burgers Vector
$T$ = Temperature      $k$ = Boltzmann's Constant

Table 1. Rate equations for deformation-mechanisms used to construct the maps shown in this paper. A full discussion of the choice of equations is given in Frost and Ashby [13].

## CONSTRUCTION OF MAPS:

The various mechanisms are dominant in different regimes of temperature and stress. The presentation of the different regions of mechanism dominance on axes of temperature and stress was introduced by Weertman [11] and elaborated by Ashby [12] and Frost and Ashby [13]. Figure 4 shows such a map for pure aluminum. It is convenient to use axes of homologous temperature (normalized by the melting point $T_m$) and logarithmic stress normalized by the shear modulus, $\mu$. Since plastic deformation results from the shear component of the stress tensor (that is, to first approximation, does not depend on the hydrostatic component) we use an effective shear stress, $\sigma_s$, which is given in terms of principal stresses: $\sigma_1, \sigma_2, \sigma_3$, as:

$$\sigma_s = \frac{1}{\sqrt{6}}\left\{(\sigma_1 - \sigma_2)^2 + (\sigma_2 - \sigma_3)^2 + (\sigma_3 - \sigma_1)^2\right\}^{1/2} .$$

Similarly, we use an equivalent shear strain-rate, $\dot{\gamma}$, given by principal strain-rates: $\dot{\varepsilon}_1, \dot{\varepsilon}_2, \dot{\varepsilon}_3$ as:

$$\dot{\gamma}_s = \frac{\sqrt{2}}{\sqrt{3}}\left\{(\dot{\varepsilon}_1 - \dot{\varepsilon}_2)^2 + (\dot{\varepsilon}_2 - \dot{\varepsilon}_3)^2 + (\dot{\varepsilon}_3 - \dot{\varepsilon}_1)^2\right\}^{1/2} .$$

By using equivalent stress and equivalent strain-rate we are essentially adopting a von Mises yield criterion. One important point here for thin film applications is that uniaxial tension and biaxial tension are predicted to have different strain-rates for a given stress:

Uniaxial Tension: $\sigma_2 = \sigma_3 = 0$ : $\quad \sigma_s = \sigma_1/\sqrt{3}$ ; $\quad \dot{\gamma}_s = \sqrt{3}\,\dot{\varepsilon}_1$

Biaxial Tension: $\sigma_2 = \sigma_1$ ; $\sigma_3 = 0$ : $\quad \sigma_s = \sigma_1/\sqrt{3}$ ; $\quad \dot{\gamma}_s = 2\sqrt{3}\,\dot{\varepsilon}_1$

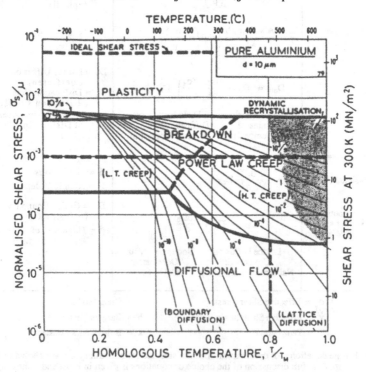

Figure 4. Deformation-mechanism map for pure aluminum of grain size 10 microns.

Since each mechanism has a rate equation, it is possible to plot contours of constant strain rate on the field of stress and temperature. Sometimes the different mechanisms can be considered to be additive (e.g.: Coble creep and Nabarro-Herring creep; diffusional creep and dislocation creep). Sometimes the mechanisms are alternatives: Low temperature plasticity is essentially an alternative mechanism involving the same defects as dislocation creep. Care must also be taken to recognize the restrictions on the stress–temperature–strain-rate relationships that are implicitly assumed on such a map. Figure 5 sketches the two most common alternatives. In a deformation experiment we may choose the independent variables. Typically we specify either temperature and strain-rate, or alternatively, temperature and stress. At low temperatures, holding the strain-rate constant produces the standard stress-strain behavior observed in tension or compression tests. Holding the stress constant may result in a rapidly decreasing strain-rate which vanishes as the material rapidly work-hardens. At high temperatures it is common to hold the stress constant, and observe the resulting strain-rate. If the temperature is of order half the melting point or higher, a steady-state strain-rate may be reached. The typical deformation-mechanism map is based on the relationship between this steady-state strain-rate and the applied stress and temperature. It is possible to construct maps which portray the transient strain-rate behavior (Frost and Ashby [13]), but they are more complicated.

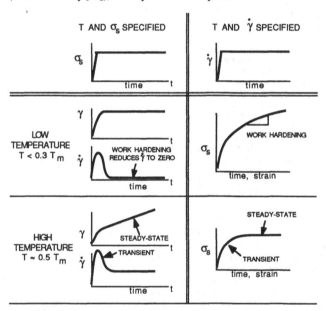

Figure 5. Relationships between $\sigma_s$, T, $\dot{\gamma}$ and $\gamma$ for different choices of the independent variable. For creep tests at high temperature $\sigma_s$ and T are prescribed. For tensions tests at low temperatures $\dot{\gamma}$ and T are prescribed.

## DEFORMATION IN THIN FILMS

There are several ways in which a thin film could be subjected to plastic deformation. There have been several recent studies of the indentation behavior of films; for thin films this involves nanoindenter experiments. This technique measures indentation size as a function of load and loading time. Although the stress and strain fields vary in time and in space within the film, it is possible to extract an estimate of the stress required for plastic deformation as a function of the imposed strain-rate. [1] Another type of thin film mechanical test is the bending deflection of thin film microbeams. [14] A microbeam cantilever is a specimen of thin film which has been removed from its constraining substrate by etching away the substrate under it. In this case also the stresses and strains are not uniform throughout the region of deformation.

One important situation for thin film deformation involves the stresses that are created during thermal cycling due to the mismatch of thermal expansion coefficients between the film and the substrate. A practical example of this is aluminum films on silicon substrates, as used for electronic interconnections in integrated circuits. The thermal expansion coefficient for aluminum far exceeds that of silicon. If adhesion is maintained between the film and substrate, temperature changes will cause stresses to develop. Since the silicon substrate is much thicker than the aluminum layer, the silicon expansion is hardly influenced by the aluminum, and the aluminum is constrained to follow the strain of the silicon. The stresses that develop can result in bending of the entire silicon wafer, and can therefore be easily measured by measuring the resulting curvature. (See Fig. 6.)

Many different authors have reported on the stress relaxation mechanisms which operate during thermal cycling.[15-22] The experiment actually provides an opportunity for a direct comparison between experiment and the predictions of a deformation-mechanism map. Murakami [23] made just such a direct comparison between experimentally observed stress relaxations in thin Pb films and a deformation-mechanism map constructed from the expected rate equations. We will repeat that demonstrational comparison for the case of pure aluminum.

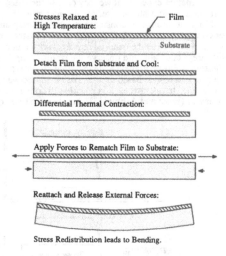

Figure 6. Illustration of the relationship between the biaxial stress in a thin film and the associated bending of the substrate, following Nix [1].

Figure 7 shows a measurement of biaxial stress through a thermal cycle as reported by Doerner, Gardener and Nix [24], showing the heating from room temperature to 445° C and cooling back down. The initial state of biaxial tension resulted from prior thermal history. Because aluminum expands faster than the silicon, the stress is initially relieved by the heating. At about 200° C the experimental stress begins to deviate from the elastic projection by an amount representing the plastic deformation. This experimental stress—temperature history can be plotted directly onto the stress—temperature field of a deformation-mechanism map.(See Figure 7.) From the strain-rate contours we can read off the strain-rates predicted from rate equations for the bulk deformation mechanisms. Given the heating rate used in the experiment, we may deduce the plastic strain rate associated with the plastic relaxation. We may therefore directly compare predicted and experimental strain rates. This comparison is made in Figure 8.

The predicted strain-rates in Figure 8 are several orders of magnitude higher than the observed strain-rates. That is to say, the actual thin aluminum film displays a much greater flow stress than would bulk pure aluminum at the same temperature and strain-rate. The difference is not a negligible; it represents the fundamental differences between the operation of the plastic deformation processes in a thin film and in a bulk polycrystal. We will discuss some of these differences in the next section.

DEFORMATION MECHANISMS IN THIN FILMS:

Deformation in thin films differs from deformation in the bulk in several fundamental ways. There are two basic concerns. One is that the thickness dimension of a thin film may be smaller than the characteristic distance involved in the action of the bulk deformation mechanism.

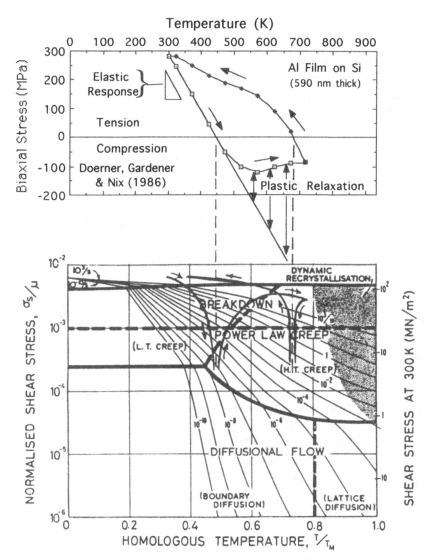

Figure 7. The history of biaxial stress in an aluminum film taken through an thermal cycle, reported by Doerner, Gardener and Nix [24], plotted on the deformation-mechanism map for pure aluminum, from Frost and Ashby [13].

This constrains the mechanisms which involve dislocation glide or grain boundary sliding to operate in a manner in which the thickness direction is fundamentally different from the two in-plane directions. The second basic concern is that thin films are often constrained in two dimensions by direct attachment to their substrates, and sometimes in the third dimension by covering layers. The sketch in Figure 9 shows three typical cases: a free-standing film without any constraint, a film on a substrate in which the deformation in two directions is constrained to

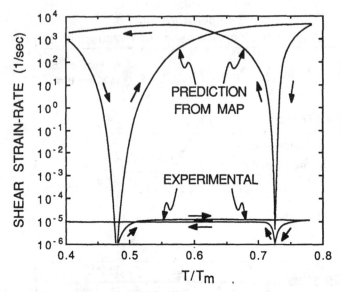

Figure 8. Comparison of the experimental strain-rates of the plastic relaxation during a thermal cycle of a thin film (Figure 7), and the strain-rates predicted theoretically from a bulk deformation-mechanism map for the same stress—temperature history.

match the substrate, and a film on a substrate with an encapsulating layer in which deformation is constrained in all three directions. Let us discuss in turn how these constraints effect each of the classes of deformation mechanism.

Diffusional Creep was first analyzed for bulk polycrystals, or strings of single crystals along a cylindrical wire. [6] Several authors have realized that grain boundary sliding is a necessary part of diffusional creep for a bulk polycrystal in which compatibility between the grains is maintained (e.g. [25,26.27]). Conversely, Raj and Ashby [28] showed that diffusional transport is the likely mechanism to provide the necessary accommodation for sliding of non-planar boundaries.

The rate equation for diffusional creep was adapted for the case of thin foils by Gibbs [29], who proposed that the uniaxial tensile strain-rate, $\dot{\varepsilon}$, would be:

$$\dot{\varepsilon} = \frac{10\,\sigma\,\Omega}{kT\,t\,d}\,D_v + \frac{15\,\sigma\,\Omega}{kT\,t^2\,d}\,\delta D_b$$

where $t$ = foil thickness, $d$ = grain size, and other symbols are defined in Table 1. Gibbs' treatment of Nabarro-Herring creep is simply to replace the $1/d^2$ dependence with $1/(td)$. For Coble creep it replaces $1/d^3$ with $1/(t^2 d)$. This treatment seems appropriate for some conditions,

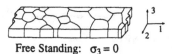

Free Standing: $\sigma_3 = 0$

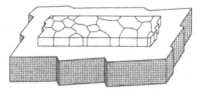

On Substrate: $\varepsilon_1$ and $\varepsilon_2$ constrained.

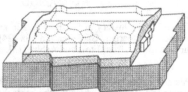

With Capping Confinement: $\sigma_3 \neq 0$

Figure 9. Alternative degrees of constraint for a polycrystalline thin film strip.

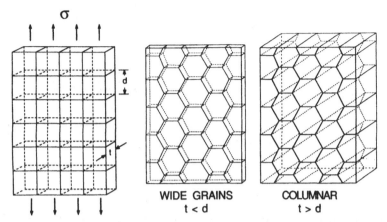

Figure 10. Geometries of polycrystalline thin films used for modelling diffusional creep. Gibbs [29] used the geometry on the left hand side. The other two alternatives have different thickness dependencies.

but perhaps not all conditions. Figure 10 shows three regimes based on the ratio of grain diameter to film thickness. The sketch in Gibbs paper shows equiaxed grains (grain size ≈ thickness). In another alternative, the grains are much wider than thickness. In this case the are of the free surface becomes greater than the area of the boundary. Surface diffusion could then become a significant path for matter transport, replacing in a sense the Coble creep mechanism. The Nabarro-Herring creep begins to depend on the free surfaces as sinks for vacancies.

In the alternative extreme the film is much thicker than the grain diameter. If the grains are equiaxed (approximately spherical) this becomes the same as a bulk polycrystal. If the grains are long, narrow columns which run entirely through the film thickness (Figure 10, t > d), then the diffusional path lengths are much smaller in the plane of the film than in the perpendicular direction. The diffusional fields can approach a two-dimensional configuration and plane strain can operate faster than other strain components. The rate for this plane strain deformation becomes independent of film thickness.

Diffusional creep on a substrate should have an additional constraint. (Figure 11) For a film to follow the deformation of its substrate by diffusional flow, some sort of sliding would be required at the interface between them.

Power-law creep is also fundamentally different in thin films. The steady-state dislocation creep that is characterized by the typical deformation-mechanism map is a result of the cooperative motion of the dislocations tangled up in the cell walls. The diameter of the cells or subgrains is typically proportional to the inverse of the stress. Figure 12 shows a sketch of some typical experimental cell sizes. [30] Note that for typical stresses the cell size is larger than the film thickness. In this case, thin films cannot develop the type of steady-state subgrain structure that leads to steady-state power-law creep. (Figure 12).

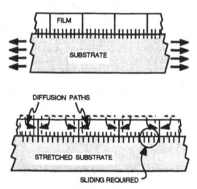

Figure 11. Stretching a film by diffusional creep created to match a stretching of the substrate is constrained by the attachment to the substrate.

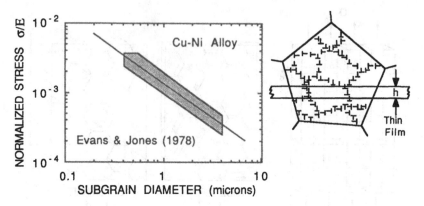

Figure 12. Dislocation cell (subgrain) sizes observed during creep tests, from Evans & Jones [30]. Thin films are often thinner than the expected cell diameter for the applied stresses.

There is an additional problem with applying the power-law creep rate equation to the deformation of thin films. The empirical constant and the map both are intended to describe steady state creep. In typical tensile creep experiments steady-state is not reached until after a primary creep deformation which may involve a strain of 10% or more. Total strains in many thin film cases are 1% or less. The initial strain rate at the beginning of primary creep may be an order of magnitude faster than the eventual steady-state rate.

There has been considerable interest in the mechanisms of dislocation glide in crystalline films. The thickness of the film may well be less than the average spacing between obstacles in a bulk polycrystal. The grain size of a polycrystalline film is likely to be no larger than the thickness, which may be a very fine grain size compared to polycrystals. There is therefore the possibility that grain-size strengthening may be significant. A Hall-Petch relationship (stress proportional to inverse square-root of grain size) has been reported. (e.g. Griffin et al. [31]) A more important cause of very high yield stresses is the constraint on the dislocation motion provided by the constraints on the film. If the film is securely attached to the substrate, then a travelling dislocation must leave behind a misfit dislocation. (See left-hand side of Figure 14.) If the film is also constrained on the top by, for example, an oxide layer, then two trailing dislocations will be left behind. (See right-hand side of Figure 14.) These constraints provide a potent strengthening mechanism. The minimum biaxial stresses need to move the dislocation of Burgers vector b, in a film of thickness h covered by an oxide of thickness t, is given as [1]:

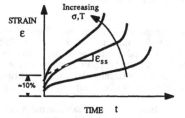

Figure 13. Typical tensile creep curves showing the usual primary creep behavior. The strain required to complete primary creep often exceeds the levels of strain of interest for thin film applications.

$$\sigma = \frac{\sin\phi \, b}{\cos\phi \cos\lambda \, 2\pi(1-\nu)h}\left[\frac{\mu_f\mu_s}{\mu_f+\mu_s}\ln\left(\frac{\beta_s h}{b}\right) + \frac{\mu_f\mu_o}{\mu_f+\mu_o}\ln\left(\frac{\beta_o t}{b}\right)\right]$$

where $\mu_f$, $\mu_s$, and $\mu_o$ are the shear moduli of the film, the substrate, and the oxide, respectively; $\beta_s = 2.6$ and $\beta_o = 17.5$ are numerical constants; and $\phi$ and $\lambda$ are angles defined in Figure 14. (The notation of h for film thickness follows Nix [1], and differs from the rest of this paper.)

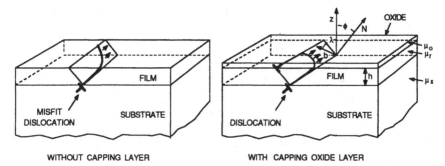

Figure 14. Dislocation glide in thin films, following Nix [1]. Capping the film with a rigid layer increases the required stress.

Although we have not yet explicitly considered cavitation and cracking as deformation mechanisms, they are often important for both thin films and bulk polycrystals. Ghandi and Ashby [32,33] have summarized the variety of fracture behaviors with fracture-mechanism maps, which are directly analogous to deformation-mechanism maps. Particular attention is due for the diffusional growth of cavities on grain boundaries, which has been extensively treated in the metallic creep literature (e.g. [34,35]), because the phenomenon may lead to failures of aluminum interconnect lines. (e.g.[36-40]) Electromigration-induced failure also involves the diffusional transport of atoms with both electron-wind and stress-gradient driving forces. [41-44]

CONCLUSIONS

A direct comparison has been made between the plastic deformation behavior of a thin polycrystalline film and the behavior predicted from rate equations which describe bulk polycrystal deformation. The difference is several orders of magnitude in strain-rate for a given stress, or more than an order of magnitude in stress for a given strain-rate. The common reason for this difference is the constraints imposed upon a thin film by its attachment to a substrate and possibly a capping layer. For a free-standing film the proximity of the free surfaces has a strong influence. In either case, the basic operation of the deformation mechanism is altered, and the associated rate-equation should be amended. A compilation of all deformation mechanisms and rate equations, comparable to that done for bulk polycrystals in deformation-mechanism maps, has yet to be completed, but it would be valuable for understanding certain failure mechanisms.

ACKNOWLEDGEMENTS

The author thanks Profs. W.D. Nix, Che-Yu Li and C.V. Thompson and Dr. P. Nagpal.

REFERENCES AND BIBLIOGRAPHY

1. W.D. Nix, "Mechanical Properties of Thin Films", Met. Trans. **20A**, 2217-2245, 1989.
2. P.S. Alexopoulos and T.C. O'Sullivan, "Mechanical Properties of Thin Films", Annual Review of Materials Science **20**, 391-420, 1990.
3. U.F. Kocks, A.S. Argon and M.F. Ashby, "Thermodynamics and Kinetics of Slip", Progress in Materials Science **19**, 1975
4. A.K. Mukherjee, J.E. Bird & J.E. Dorn. "Experimental Correlations for High Temperature Creep", Trans. ASM **62**, 155-179, 1969
5. S.L. Robinson and O.D. Sherby, "Mechanical Behavior of Polycrystalline Tungsten at Elevated Temperature", Acta Metall. **17**, 109-125, 1969.
6. C. Herring, "Diffusional Viscosity of a Polycryst. Solid", J. Appl. Phys. **21**,437-45,1950
7. F.R.N. Nabarro, Report on a Conf. on the Strength of Metals, Phys. Soc. London, 1948.
8. R.L. Coble, "A Model for Boundary Diffusion Controlled Creep in Polycrystalline Materials", J. Appl. Phys. **34**, 1679-1682, 1963.
9. J.-P. Poirier, Creep of Crystals, Cambridge Univ. Press, Cambridge, 1985, pp. 114-117.
10. F. Garofalo, "An Empirical Relation Defining the Stress Dependence of Minimum Creep Rate in Metals", Trans. TMS-AIME **227**, 351-356, 1963.

11. J. Weertman, "Disloc. Climb Theory of Steady-State Creep", Trans.ASM **61**, 681-94, 1968
12. M.F. Ashby, "A first report on deformation-mechanism maps", Acta Met.**20**, 887-98,1972
13. H.J. Frost and M.F. Ashby, Deformation-Mechanism Maps, Pergamon Press, Oxford, 1982
14. T.P. Weihs, S. Hong, J.C. Bravman and W.D. Nix, "Mechanical deflection of cantilever microbeams: A new technique....", J. Mater. Res. **3**, 931-942, 1988
15. A. Gangulee, "Strain-Relaxation in Thin Films on Substrates", Acta Met.**22**, 177-83, 1974
16. A.K. Sinha and T.T. Sheng, "The Temperature Dependence of Stresses in Aluminum Films on Oxidized Silicon Substrates", Thin Solid Films **48**, 117-126, 1978.
17. P.A. Flinn, D.S. Gardner and W.D. Nix, "Measurement and Interpretation of Stress in Al-Based Metallization ....", IEEE Trans. on Electron Devices **ED-34**, 689-699, 1987.
18. B. L. Draper and T.A. Hill, "Stress and stress relaxation in integrated circuit metals and dielectrics", J. Vac. Sci. Technol. B **9**, pp. 1956-1962, 1991.
19. P.A. Flinn, "Measurement and interpretation of stress in Cu films as a function of thermal history", J. Mater. Res. **6**, 1498-1501, 1991.
20. M. Hershkovitz, I.A. Blech and Y. Komem, "Stress Relaxation in Thin Aluminium Films", Thin Solid Films **130**, 87-93, 1985.
21. M.S. Jackson & C.-Y. Li, "Stress Relax.& Hillock Growth .", Acta Met **30**,p.1993, 1982
22. D.S. Gardner and P.A. Flinn, "Mechanical stress as a function of temperature for aluminum alloy films", J. Appl. Phys. **67**, 1831-1844, 1990.
23. M. Murakami, "Thermal Strain in Lead Thin Films II: Strain Relaxation Mechanisms", Thin Solid Films **55**, 101-111, 1978.
24. M.F. Doerner, D.S. Gardner & W.D. Nix, "Plastic properties of thin films on substrates .. by submicron indentation hardness and substrate curvature..", J. Mat. Res.**1**,845-51,1986.
25. R.N. Stevens, "Grain Boundary Sliding and Diffusion Creep in Polycrystalline Solids", Phil. Mag. **23**, 265-283, 1971.
26. W. Roger Cannon, "The Contribution of Grain Boundary Sliding to Axial Strain during Diffusional Creep", Phil. Mag. **25**, 1489-1497, 1972.
27. M.F. Ashby and R. Verrall, "Diffusion-Accommodated Flow and Superplasticity", Acta Metall. **21**, 149-163, 1973.
28. R. Raj & M.F. Ashby, "On GB Sliding & Diffusional Creep", Met.Trans.**2**,1113-27, 1971
29. G.B. Gibbs, "Diffusion Creep in a Thin Foil", Phil Mag. **13**, 589-593, 1966
30. R.W. Evans & F.L. Jones, "Hot ductility of wrought Cu-Ni alloy", Metals Tech. **5**,1,1978
31. A.J. Griffin, Jr., F.R. Brotzen and C. Dunn, "Hall-Petch Relation in Thin Film Metallizations", Scripta Metall. **20**, 1271-1272, 1986
32. M.F. Ashby, C. Gandhi and D.M.R. Taplin, "Fracture-Mechanism Maps and their construction for F.C.C. Metals and Alloys", Acta Metall. **27**, 699-729, 1979.
33. C. Gandhi and M.F. Ashby, "Fracture-Mechanism Maps for materials which cleave: F.C.C., B.C.C. and H.C.P. Metals and Ceramics", Acta Metall. **27**, 1565-1602, 1979.
34. D. Hull and D.E. Rimmer, "The Growth of Grain-boundary Voids Under Stress", Philos. Mag. **4**, 673-687, 1959.
35. I.-W. Chen & A.S. Argon, "Diffusive Growth of G.B. Cavities", Acta Met.**29**, 1759,1981
36. F.G. Yost, D.E. Amos and A.D. Romig, Jr., "Stress-Driven Diffusive Voiding of Aluminum Conductor Lines", IEEE/IRPS, pp. 193-200, 1989.
37. K. Hinode, N. Owada, T. Nishida and K. Mukai, "Stress-induced grain boundary fractures in Al-Si interconnects", J. Vac. Sci. Technol. B **5**, 518-522, 1987.
38. Anne I. Sauter and W.C. Nix, "A study of stress-driven diffusive growth of voids in encapsulated interconnect lines", J. Mater. Res. **7**, 1133-1143, 1992.
39. C.-Y. Li, R.D. Black & W.R. LaFontaine, "Analysis of thermal stress-induced grain bdry. cavitation and notching in narrow Al-Si metallizations", Appl. Phys. Lett. **53**, 31-33, 1988
40. M.A. Korhonen, C.A. Paszkiet & C.-Y. Li, "Mechanisms of thermal stress relaxation and stress-induced voiding in narrow Al-based metallizations", J.Appl.Phys. **69**,8083-91,1991
41. E. Artz & W.D. Nix, "..the effect of line width and mech. strength on electromigration failure of interconnects with "near-bamboo" grain structures", J.Mater.Res. **6**,731-36, 1991
42. W.D. Nix and E. Artz, "On Void Nucleation and Growth in Metal Interconnect Lines under Electromigration Conditions", Metallurgical Transactions, in press, 1992.
43. M.A. Korhonen, P. Børgensen and Che-Yu Li, "Electromig. in Al-based Interconnects ...., with and without ...Stress-Migration Damage", Mat. Res. Soc. Fall Meeting, 1991.
44. M.A. Korhonen, P. Børgensen, K.N. Tu and Che-Yu Li, "Stress Evolution due to Electromigration in Confined Metal Lines, submitted to Journal of Applied Physics, 1992

# UNDERSTANDING VOID PHENOMENA IN METAL LINES: EFFECTS OF MECHANICAL AND ELECTROMIGRATION STRESS

PAUL A. FLINN
Intel Corporation, 3065 Bowers Avenue, Santa Clara, CA 95052, and
Department of Materials Science and Engineering, Stanford University,
Stanford, CA, 94305.

## ABSTRACT

As the shrinking of VLSI devices continues, the problem of voids in interconnections becomes of steadily increasing concern. Voids can result from the effects of triaxial tensile stresses produced during fabrication; they can also arise from electromigration. The effects can combine: voids arising from mechanical stress can move and grow under electromigration stress. A detailed understanding of the phenomena requires both knowledge of the properties of the metal and dielectric as functions of time and temperature, and direct observations of the void behavior in real time under varying stress conditions. The material property information can be obtained by a combination of wafer curvature, X-ray diffraction and ultramicro indentation measurements. Void behavior can be inferred from high precision resistivity measurements, and observed directly with Scanning Electron Microscopy. With these data it is possible to evaluate various models for the phenomena.

## INTRODUCTION

Void formation in interconnections as a consequence of electromigration has long been known as a potential reliability hazard for VLSI (Very Large Scale Integration) devices[1]. In spite of extensive efforts in the intervening years, however, the interpretation of some aspects of the phenomena remain controversial, and a complete, generally accepted physical model of the process does not yet exist. There have been relatively few reports of detailed direct observation of void formation and motion. Recent observations[2, 3, 4] have not been consistent with widely held views on the subject.

Void formation in interconnections as a consequence of mechanical stress was first reported more recently[5, 6]. This effect also has serious implications for reliability and has also been a subject of extensive investigation. Although significant controversy remains, this problem is somewhat simpler than that of electromigration, and is somewhat better understood.

The appearance of mechanically induced voids is quite similar to that of electromigration induced voids; they may be slit-like, as shown in Figure 1(a) suggesting the opening of a grain boundary, or irregular in shape, roughly resembling a grain as shown in Figure 1(b). Normally they are quite few in number relative to common microstructural features such as grain boundaries, triple points or precipitate particles. It is generally believed that, in both cases, voids are the result of the condensation of excess vacancies. The difference is in the origin of the excess of vacancies: the electron wind in the electromigration case, and hydrostatic tension in the stress induced case. Since the mechanically induced case is somewhat simpler, we will consider it first.

Mat. Res. Soc. Symp. Proc. Vol. 265. ©1992 Materials Research Society

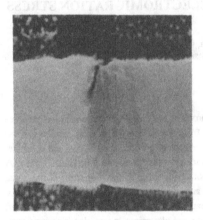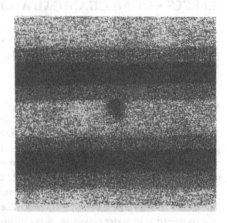

Figure 1: (a) (left) Typical slit-like void observed in the early 1980's. Metal deposition under poor conditions (by today's standards). SEM micrograph taken after removal of passivation.
(b) (right) Typical irregular void observed in metal deposited under good conditions. Photograph taken through passivation with 120 KV SEM[4].

## ORIGINS OF MECHANICAL STRESS IN INTERCONNECTION LINES

Mechanical stress induced voiding occurs only in metal lines encased in dielectric. (We will use the term "dielectric" both for interlayer dielectric, and for the top layer, commonly called "passivation", since the basic physical phenomena are the same, whichever layer is involved.) In the absence of a dielectric coating, the stress normal to the free surface of the metal is zero; there is no hydrostatic tension, and no driving force for void formation. When the metal is encased in, and bonded to, dielectric, however, a decrease in the metal volume relative to its surroundings, will result in a triaxial tensile stress. The hydrostatic component of this stress cannot be relaxed by plastic flow, since plastic flow, to a good approximation, conserves volume. Such a volume decrease, with accompanying hydrostatic tension and potential voiding, can arise from any of several causes.

The simplest, and first reported, cause for voiding in metal lines is the contraction of the metal on cooling after dielectric deposition. The coefficient of thermal expansion for aluminum is much larger than that of silicon or the common dielectric materials. In order to understand the material properties, we consider first the case of a uniform film, as sketched in Figure 2. The $z$ component of stress is zero, and the stretching of the metal in the plane of the film is accompanied by a reduction in thickness. The stress changes and plastic behavior of the metal during a thermal cycle can be determined by wafer curvature measurements on a uniform film[9, 8]. Typical behavior is shown in Figure 3. As is evident in the figure, in the usual temperature range for

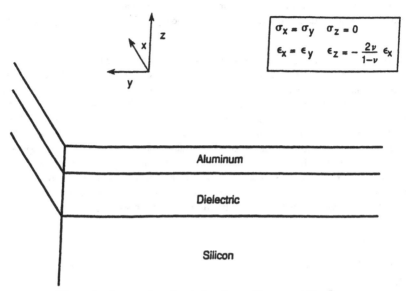

Figure 2: Stress and strain relations in a uniform metal film.[7]

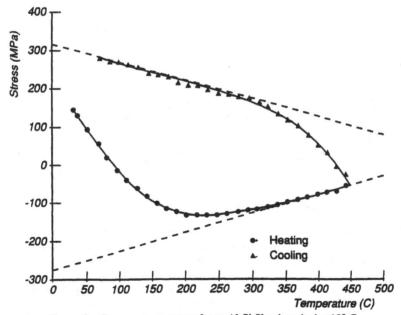

Figure 3: Stress vs. temperature for an Al-Si film deposited at 150 C.
Dashed lines indicate limits for plastic flow during heating and cooling.[8]

dielectric deposition, 300 - 400 C, aluminum is quite soft and prior to, and during, dielectric deposition, the stresses are quite low. During cooling after deposition, the constraint of the surrounding silicon and dielectric suppress the contraction of the aluminum and large triaxial tensile stresses result.

The stresses in metal films can also be determined by X-ray diffraction[10, 11, 12]. The method is applicable to metal lines under passivation as well as to uniform films. Typical results for uniform films, with and without passivation, are shown in Figure 4. The interplanar spacing of the {422} reflection is plotted as a function of $sin^2\psi$, where $\psi$ is the angle of observation relative to the sample surface. The dashed line indicates the interplanar spacing of unstrained aluminum. For low values of $\psi$, corresponding to observation of planes nearly parallel to the surface, the interplanar spacing is lower than that of the unstrained metal. Conversely, as $\psi$ approaches 1, corresponding to the observation of planes nearly perpendicular to the surface, the interplanar spacing is larger than the unstrained value. This is the situation for biaxial tension. Passivation has no direct effect in the case of uniform films; the small difference in stress between the passivated and unpassivated metal is due to the different thermal history.

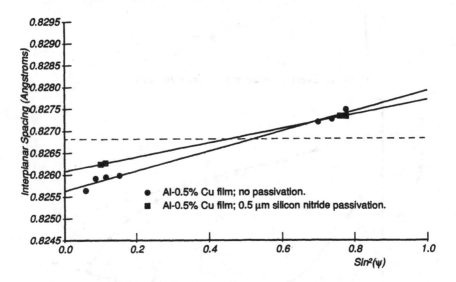

Figure 4: Interplanar spacings of unpatterned metal films.

Two sets of measurements are necessary for metal lines: one with the lines vertical in the X-ray apparatus, so that $sin^2\psi = 1$ corresponds to measurements across the width of the line (the $y$ direction, see Figure 5); and the other with the lines horizontal, so that $sin^2\psi = 1$ corresponds to measurements along the line (the $x$ direction). For both orientations, the result at $sin^2\psi = 0$ corresponds to measurement normal to the surface (the $z$ direction). Within experimental error, the values at $sin^2\psi = 0$ are the same, as they should be. We note that the strain is now positive in all three directions, since the presence of the passivation prevents contraction of the metal in any direction. It is smallest in the $z$ direction and largest in the $x$ direction. The results of recent elastic strain measurements[12] on Al-0.5% Cu are shown in Figure 6. Note that the room temperature stress is essentially unchanged by the high temperature anneal.

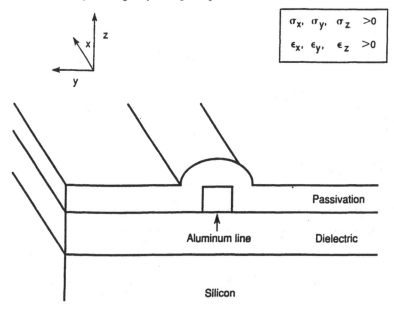

**Figure 5:** Average stress and strain relations in lines under passivation.[7]

The stresses can be calculated by the use of the Finite Element Method; some results obtained by Sauter[13] are plotted in Figures 7 and 8. If the aluminum is assumed to behave elastically, the calculation shows large stress gradients. The $z$ component of stress is maximum at the outer edge of the line, and the $y$ component is maximum at the center of the line. If, however, plastic flow during cooling is properly taken into account, on the basis of the data from uniform film measurements, the stress gradients are almost completely eliminated, as shown by the solid curves in Figures 7 and 8. This result is consistent with X-ray diffraction measurements, which show no significant line broadening, and thus indicate relatively uniform stresses throughout the lines. Since the dielectric does not undergo plastic flow, the large stress gradients in it remain; but they are irrelevant to the problem of void growth in the metal. The plastic flow does not remove the hydrostatic component of the stress. The stress tensor determined from the X-ray results was in excellent agreement with the results of Finite Element Calculations[12]. This is in marked contrast with the results of earlier X-ray measurements[11] on poorer quality material, where significant voiding occurred and the stresses were relaxed far below the calculated values. The stress increase and the potential volume of voiding which can occur by this mechanism is limited by the total differential volume decrease of the metal on cooling, typically of the order of 1%.

The role of the dielectric in this phenomenon has been a subject of much controversy. Contrary to widely held belief, the *intrinsic* stress in the dielectric is not a factor in the problem[11, 12]. The important role of the dielectric is simply that of constraining the metal; differences among dielectrics are primarily due to differences in elastic moduli[14]. The deleterious effect of highly

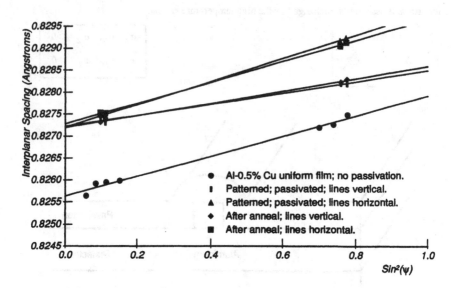

Figure 6: Interplanar spacings of Al-0.5% Cu lines under passivation; before and after anneal for 2 hours at 400 C.

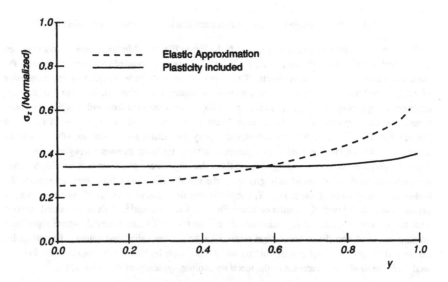

Figure 7: Normalized $z$ component of stress in an aluminum line under dielectric; section at 0.75 of line height. Line width/height ratio= 2. Replotted from Sauter[13], Figures 3.5 and 4.2.

compressive silicon nitride is probably due to the excess hydrogen in the nitride[15, 11, 16].

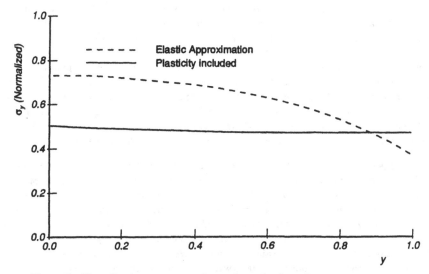

**Figure 8:** Normalized y component of stress in an aluminum line under dielectric; section at 0.75 of line height. Line width/height ratio = 2. Replotted from Sauter[13], Figures 3.5 and 4.2.

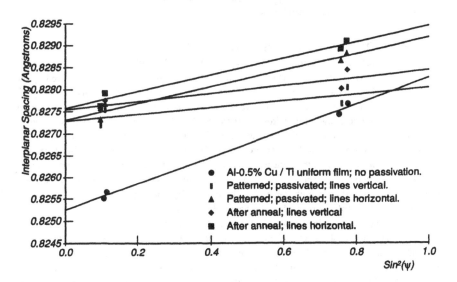

**Figure 9:** Interplanar spacings of Al-0.5% Cu / Ti lines under passivation; before and after anneal for 2 hours at 400 C.

Stress induced voiding has also been observed at high temperatures, even well above the dielectric deposition temperature, where the effect discussed above cannot be present. Another mechanism can operate when a composite metallization is used; for example, a layered structure of aluminum and titanium. At elevated temperatures, a reaction between aluminum and titanium occurs, with the formation of the intermetallic compound $Al_3Ti$, and a substantial decrease in volume[17]. The voiding reported by Murali et al[18], and Freiberger and Wu[19] probably was a consequence of this effect. Again, hydrostatic tension results and voiding can occur. The potential magnitude of the effect is limited in this case by the amount of material available for reaction. The effect of this reaction can be seen by X-ray diffraction, as shown in Figure 9. In contrast to the simple Al-Cu case (Figure 6), the stress at room temperature is considerably increased by the anneal, as a consequence of the volume decrease associated with $TiAl_3$ formation.

A third possibility is reaction between the aluminum and the dielectric at elevated temperatures. This can occur in multilevel interconnection systems, particularly where upper level metal is deposited at a relatively high temperature and the previously deposited layer of metal is thus subjected to a high temperature exposure[20, 21]. Two possibilities have been suggested for a reaction. One is a direct reaction between the aluminum and the silica of the dielectric[22]. The reaction is thermodynamically possible, and a volume decrease will occur, but one might expect the reaction to be relatively slow. Another possibility, since the reported observations have all involved a silica based dielectric that normally contains water, is a reaction between the aluminum and water from the dielectric, with the formation of hydrogen that can dissolve in the aluminum metal, and aluminum ions that are quite small and can dissolve easily in the silicon. Since reactions between metal and dielectric can continue, in principle, until the metal is exhausted, there is no limit to the amount of voiding which can occur by this mechanism.

## NUCLEATION OF STRESS INDUCED VOIDS

The presence of hydrostatic tensile stress increases the local equilibrium number of vacancies present at any given temperature. The free energy required to form a vacancy is reduced by the $p\Delta v$ work associated with vacancy formation under hydrostatic tension (negative hydrostatic pressure). This results in an increase in the vacancy concentration. If some of these vacancies coalesce to form a void, the local hydrostatic stress can be reduced. This is not necessarily energetically favorable, since the surface energy of the interior of the void must be included. The void must reach a certain critical size before it is stable. A void smaller than the critical size will evaporate; a void larger than the critical size will grow. For the case where the stress is simply due to thermal contraction and thus limited (typically to the order of 500 MPa, as discussed above) the critical radius is quite large, and the probability that it will be attained by thermal fluctuations is minuscule, as pointed out by Sauter[13]. In this situation, voids can form only as the result of heterogeneous nucleation; that is, there must be an internal interface of some sort with a relatively high interfacial energy. If internal free surface is formed at this interface, the energy required to form the surface is reduced by the amount of the energy of the existing interface. A modest reduction in the required surface energy will not help. The interface must have an energy quite close to that of free surface, so that the energy required for the void surface is only a small fraction of the normal surface energy. Clean grain boundaries have energies typically about one half of free surface energies, so that they cannot serve as nucleating sites. Similarly, the typical energy of an interface between a metallic precipitate and the aluminum is substantially less than the energy of a free surface, so such precipitates also are not suitable for nucleation sites.

We can now understand the observations of Klema et al[5] and Curry et al[6]. Their metal was deposited in the presence of significant amounts of nitrogen. Nitrogen adsorbed on grain boundaries evidently reduces sufficiently the energy required to form free surface to permit heterogeneous nucleation to occur. This can account for the slit-like voids reported in this early work. When the nitrogen content of the sputtering atmosphere is reduced sufficiently, this type of voiding is suppressed. More recent observations of voiding are typically of the sort shown in Figure 1(b), grain like, rather than slit or crack like, in appearance. They seem to be associated with the interface between the metal and the dielectric. Since aluminum bonds strongly to the commonly used dielectrics, the interfacial energy should be relatively low, and nucleation should not occur. Any etch residue or other foreign material, however, could provide a high energy surface and an opportunity for heterogeneous nucleation. When the processing conditions are sufficiently clean, no voids are observed and the stress in the metal, as measured by X-ray diffraction, is in excellent agreement with the results of Finite Element calculations[12]. Further evidence of the importance of a clean interface was recently reported by Abe et al[23]. They found that an argon sputter etch of the aluminum lines immediately prior to dielectric deposition essentially eliminated voiding.

## NUCLEATION OF ELECTROMIGRATION INDUCED VOIDS

The nucleation of voids under electromigration conditions is much more difficult to analyze, since the vacancy supersaturation is a consequence of dynamic processes and sensitive to both microstructural details and temperature gradients. Sensitive resistance change measurements[24, 25] show resistance increasing as soon as electromigration stress is applied; if the current is removed before too much damage is done, the effect is reversible. If the temperature is maintained, the resistance decays to close to its original value with a characteristic time scale of hours. As Lloyd pointed out[24], isolated vacancies alone cannot account for these effects; the decay is much too slow for that explanation. A possible interpretation is, that under dynamic conditions a distribution of microvoids of various sizes is formed; when the driving force is removed, they evaporate. Nix and Arzt[26] have shown that the homogeneous nucleation of voids cannot occur under normal electromigration conditions. We would expect then, that no stable voids will form until an adequate supersaturation of vacancies is reached in the neighborhood of a site suitable for heterogeneous nucleation.

This model is consistent with recent observations made with a STEM operated with a backscatter detector to serve as a high voltage SEM[4]. Some results from this work are shown in Figure 10. Even with high resolution no voids were observed in the line prior to electromigration stressing. A considerable period of stress was required to produce voids. When voids did appear, they clearly were nucleated at the interface between the dielectric and the metal, in a manner very similar to that observed for stress induced voids. The development of the lower void is particularly interesting. After nucleation as a long thin strip at the interface, it gradually became roughly equiaxed and eventually broke away from the interface and moved out into the line.

24

Figure 10: Nucleation and subsequent development of a void under electromigration conditions as observed with a 120 KV SEM[4].

## SUMMARY

As a consequence of either mechanical stress or electromigration, the presence of excess vacancies makes metal lines under passivation thermodynamically unstable with respect to void formation. Whether or not voids actually form depends on the magnitude of the driving force and the availability of sites for the heterogeneous nucleation of voids. For the case of stress due only to thermal contraction, the driving force is limited and voiding can be suppressed by limiting the availability of heterogeneous nucleation sites. When the mechanical stress is due to high temperature reaction, however, the driving force can be much larger, and voiding much more difficult to suppress. Vacancy concentrations produced by electromigration can be much higher than those due to thermal contraction stresses; consequently, material that is void free after processing can develop voids under electromigration conditions.

## ACKNOWLEDGEMENTS

This paper is based on a long term cooperative effort of Intel and Stanford University. Professors W. D. Nix and D. Barnett of Stanford University provided invaluable assistance and advice throughout the course of this work. Dr. G. A. Waychunas of Stanford and Dr. D. S. Gardner, and Dr. A. I. Sauter, formerly of Stanford, now Intel, have played a major role in the program. I also wish to thank P. Besser and T. Meribe of Stanford, and E. Abratowski, B. Greenebaum, C. Chiang, D. Fraser, and M. Madden of Intel for important contributions.

## References

1. I. A. Blech and E. S. Meiernn, "Electromigration in Thin Aluminum Films", *Applied Physics Letters*, Vol. 11, 1967, pp. 263-.

2. E. Castano, J. Maiz, P. Flinn, and M. Madden, "*In situ* Observations of DC and AC Electromigration in Passivated Al Lines", *Applied Physics Letters*, Vol. 59, 1991, pp. 129-131.

3. P. R. Besser, M. C. Madden and P. A. Flinn, "*In Situ* Observation of the Dynamic Behavior of Electromigration Voids in Passivated Aluminum Lines". Private Communication.

4. M. Madden, T. Meribe, E. Abratowski, and P. A. Flinn, "High Resolution Observation of Void Motion in Passivated Metal Lines under Electromgration Stress". Paper H1.4, this conference.

5. J. Klema, R. Pyle and E. Domangue, "Reliability Implications of Nitrogen Contamination During Deposition of Sputtered Aluminum/Silicon Metal Films", *Proceedings of the 22nd Annual International Reliability Symposium*, IEEE, New York, 1984, pp. 1-5.

6. J. Curry, G. Fitzgibbon, Y. Guan, R. Muollo, G. Nelson and A. Thomas, "New Failure Mechanism in Sputtered Aluminum-Silicon Films", *Proceedings of the 22nd Annual International Reliability Symposium*, IEEE, New York, 1984, pp. 6-8.

7. P. A. Flinn, "Stress in Passivated Films", *Thin Films: Stresses and Mechanical Properties II. MRS Symposium Proceedings Volume 188.*, Materials Research Society, Pittsburgh, PA, 1990, pp. 3-13.

8. P. A. Flinn, "Principles and Applications of Wafer Curvature Techniques for Stress Measurements in Thin Films", *Thin Films: Stresses and Mechanical Properties. MRS Symposium Proceedings Volume 130.*, Materials Research Society, Pittsburgh, PA, 1989, pp. 41-51.

9. P. A. Flinn, D. S. Gardner and W. D. Nix, "Measurement and Interpretation of Stress in Aluminum-Based Metallization as a Function of Thermal History", *IEEE Transactions on Electron Devices*, Vol. ED-34, 1987, pp. 689-699.

10. P. A. Flinn and G. A. Waychunas, "A New X-ray Diffractometer Design for Thin-film Texture, Strain, and Phase Characterization", *Journal of Vacuum Science and Technology*, Vol. B6, 1988, pp. 1749-1755.

11. P. A. Flinn and C. Chiang, "X-ray Diffraction Determination of the Effect of Various Passivations on Stress in Metal Films and Patterned Lines", *Journal of Applied Physics*, Vol. 67, 1990, pp. 2927-2931.

12. B. Greenebaum, A. I. Sauter, P. A. Flinn, and W. D. Nix, "Stress in Metal Lines under Passivation: Comparison of Experiment with Finite Element Calculations", *Appl. Phys. Lett.*, Vol. 58, 1991, pp. 1845-1847.

13. A. I. Sauter, *Modeling of Thermal Stresses and Void Growth Processes in Microelectronic Interconnect Structures*, PhD dissertation, Stanford University, 1991.

14. C. Chiang, G. Neubauer, K. Yoshioka, P. A. Flinn, and D. B. Fraser, "Hardness and Modulus Studies on Dielectric Thin Films". Paper H4.1, this conference.

15. H. L. Peek and R. A. M. Wolters, "Bubble and Cavity Formation in Aluminum-Plasma Silicon Nitride Structures", *Proceedings Third International IEEE VLSI Multilevel Interconnection Conference*, IEEE, 1986, pp. 165-172.

16. W. F. Filter and J. A. Van Den Ayle, "A Test Vehicle to Assess Stress Voiding Models and Acceleration Methods". Proceedings of the First International Workshop on Stress Induced Phenomena in Metallizations, American Physical Society, New York, 1992.

17. D. S. Gardner, T. L. Michalka, P. A. Flinn, T. W. Barbee, Jr. K. C. Saraswat and J. D. Meindl, "Homogeneous and Layered Films of Aluminum/Silicon with Titanium for Multilevel Interconnects", *Proceedings Second International IEEE VLSI Multilevel Interconnection Conference*, IEEE, 1985, pp. 102-113.

18. V. Murali, S. Sachdev, I. Banerjee, S. Casey and P. Gargini, "Metal-Voiding Phenomena in Aluminum and its Alloys", *Proceedings Seventh International IEEE VLSI Multilevel Interconnection Conference*, IEEE, 1990, pp. 127-132.

19. P. Freiberger and K. Wu, "A Novel Via Failure Mechanism in an Al-Cu/Ti Double Level Metal System", *Proceedings of the 30th Annual International Reliability Symposium*, IEEE, New York, 1992, pp. 356-360.

20. H. Shin, "A Sunken Phase in Aluminu-Copper Interconnects as a New Kind of Stress Void", *Proceedings Eighth International IEEE VLSI Multilevel Interconnection Conference*, IEEE, 1991, pp. 292-294.

21. H. Okabayashi, A. Tanikawa, H. Mori, and H. Fujita, "UHVEM Observations of Stress-Induced Voiding in Al Metallization". Proceedings of the First International Workshop on Stress Induced Phenomena in Metallizations, American Physical Society, New York, 1992.

22. W. T. Tseng and J. P. Stark, "Interface Reaction Model for Process Voiding in Aluminum Conductor Lines", *Applied Physics Letters*, Vol. 59, 1991, pp. 680-681.

23. H.Abe, S. Tanabe, Y. Kondo, and M. Ikubo, "The Influence of Adhesion between Passivation and Aluminum Films on Stress Induced Voiding". Japan Society of Appl. Physics, 39th Spring Meeting, Extended Abstracts, p. 658, April 1992.

24. J. R. Lloyd and R. H. Koch, "Study of Electromigration-Induced Resistance and Resistance Decay in Al Thin Film Conductors", *Proceedings of the 25th Annual International Reliability Symposium*, IEEE, 1987, pp. 161-168.

25. K. Hinode, T. Furusawa, and Y. Homma, "Relaxation Phenomena During Electromigration under Pulsed Current", *Proceedings of the 30th Annual International Reliability Symposium*, IEEE, New York, 1992, pp. 205-210.

26. W. D. Nix and E. Arzt, "On Void Nucleation and Growth in Metal Interconnect Lines under Electromigration Conditions". Private Communication.

PREDICTION AND PREVENTION OF STRESS MIGRATION
AND ELECTROMIGRATION DAMAGE IN PASSIVATED LINES

P. Børgesen, M. A. Korhonen, and C.-Y. Li
Department of Materials Science & Engineering
Cornell University, Ithaca, NY 14853

ABSTRACT

Based on recent theoretical progress we discuss current and potential new remedies for thermal stress and electromigration induced damage and failure in microelectronic interconnects. We present a new idea involving the patterning of alternately adhering and non-adhering interfaces between metal and surroundings, and briefly discuss the potential for predicting failure and/or extrapolating accelerated test results to service conditions and into the so-called 'six-sigma' range.

INTRODUCTION

The reliability of the narrow, passivated metal lines connecting devices in microelectronic circuits may well define the limits on continued increases in device density. Over the years a variety of techniques have been developed to improve the resistance of such lines to thermal stress (stress migration) and electromigration induced damage and failure. However, as line widths are reduced to fractions of a micron, and line lengths add up to tens of meters, reliability remains an issue of concern.

After decades of research, and some recent advances, a comprehensive picture of both stress migration and electromigration in passivated lines is beginning to emerge, according to which the two phenomena are directly correlated and depend strongly on microstructure and precipitate distribution [1]. All of this now begins to allow the systematic evaluation of the various remedies, as well as the development of new ones, and of the necessary models for extrapolating accelerated test results to service conditions and very early (so-called 'six-sigma') failure.

The present paper reviews our current understanding and discusses some of the technologically relevant implications. This is followed by the outline and discussion of a new idea, based on eliminating the triaxiality of the residual tensile stresses and thus inhibiting voiding. For details, derivations and discussions of the underlying theory and models the reader is referred to a number of recent publications [1 - 5], and references therein.

THEORY AND DISCUSSION

Consider first the case of narrow, aluminum based interconnects, annealed to stabilize the microstructure and then passivated with a thick ceramic layer at about 400°C. Because of the thermal expansion mismatch and the rigid surroundings, the subsequent cool-down leads to large tensile stresses in the metal. These stresses may redistribute by plastic flow, but volume conservation prevents a triaxial stress relaxation, except through the nucleation and growth of voids.

Stress induced void nucleation is favored by a combination of large stress concentrations, atomic mobility and "global" stresses. For typical cooling rates, grain boundary sliding at temperatures near 200°C provide the necessary stress concentrations at hard precipitates and at the interfaces between line edges and passivation. The resulting voids then continue to grow during further cool-down, as well as during subsequent storage.

At a given temperature, significant grain boundary sliding occurs only within a finite strain rate window. Since the position of this varies rapidly with temperature, stress induced void nucleation is often very sensitive to the cooling-rate [6, 7] and may, in principle, be suppressed by an appropriate choice of metallurgy, microstructure and thermal treatment [8, 9]. However, this would most likely only delay void nucleation until the application of an electrical current (see below).

Initially, stress induced void growth is determined by the transfer of atoms from the stress-free void surface to the connecting grain boundary network, and thus depends on the grain boundary diffusivity. However, the resulting accumulation of atoms leads to local grain boundary thickening and build-up of a "back-stress", eventually reducing void growth substantially [4]. The final relaxation of the back stress then occurs by dislocation creep (lattice diffusion), allowing only very limited void growth [10]. The greater benefit from adding, say, a few per cent Cu to aluminum metallizations is thus the reduction in grain boundary diffusivity. Also, a reduction in the total area of the connected grain boundary network will lead to a faster build-up of the back stress, which explains the reduced void growth observed for 'near-bamboo' grain structure [2, 11, 12].

Both stress induced void nucleation and growth would be inhibited by sufficient reduction in adhesion or rigidity of the passivation. For example, thermal stress induced voiding was substantially reduced in aluminum based lines by switching from silicon nitride to silicon oxynitride passivation [12], and in Cu lines by switching from an Al- to a Cr-adhesion layer [13], in both cases undoubtedly because of a weaker adhesion. Unfortunately, a finite adhesion is still required to ensure good thermal coupling to the surroundings, and a non-adhering interface would provide an obvious path for corrosion when, for example, etching holes for vias.

Even if individual thermal stress induced voids do not grow large enough to sever a line, subsequent current induced void migration may still lead to rapid failure during testing or service. Voids tend to migrate, by surface diffusion, with a size dependent velocity [14] in the opposite direction of the electron current. Whenever possible, they will move along grain boundaries, but they may get trapped at blocking boundaries, triple points or precipitates. For example, voids of diameter less than [15]

$$R_c = \{8\gamma\Omega/\pi Z^*eE\}^{1/2} \tag{1}$$

are trapped at a sufficiently large grain or interphase boundary of surface tension, $\gamma$. E is here the electric field, $\Omega$ the atomic volume, and $eZ^*$ the 'effective charge'. As soon as one or a few of the voids become large enough, however, migration and coalescence will rapidly lead to open failure somewhere along the line [16, 17]. Acceptable electromigration 'resistance' thus requires that static growth of any initial void to size $R_c$ takes much longer than the subsequent migration and coalescence step, something which is usually achieved by adding precipitates.

The application of an electrical current also causes atomic motion through the lattice and along grain boundaries, and variations in the grain boundary network may lead to significant flux divergencies and the build-up of local compressive and tensile stresses. Now, if thermal stress induced void nucleation was somehow suppressed in rigidly passivated lines with good adhesion, the residual stresses would already be very high [9, 18], and a further perturbation of the local stress distributions at typical electromigration test temperatures (150-250°C) would clearly favor void nucleation. Even if service conditions are not severe enough to cause void nucleation or passivation cracking, microelectronic circuits with steps, vias and W-plugs are not likely to be completely defect free, and we suggest that failure cannot be prevented by suppressing nucleation.

In principle, the lifetime may be optimized by maximizing the critical void size and minimizing the initial void sizes. Precipitates, in which the components are not particularly 'mobile', may here well have the larger $\gamma$ or even effectively block void motion. Unfortunately, 'hard' precipitates (e.g. Si in Al lines) are also obvious stress raisers and nucleation sites for thermal stress induced voids, which therefore tend to be larger in such lines. On the other hand 'soft' precipitates, such as $Al_2Cu$, clearly inhibit thermal stress induced voiding, but they may also shrink due to the same flux divergences causing the void growth. Optimization may require the simultaneous addition of several types of precipitates. Finally, we notice that no further improvement is achieved by $R_c$ increasing beyond half the line width, i.e. for very narrow lines and low currents the only important parameters are initial void sizes and static void growth rates.

Consider now the application of an electrical current to a passivated line with large grains and small static voids. Figure 1 illustrates the special, but technologically very relevant, case of alternate 'bamboo' sections (B-C) and so-called 'cluster' sections with connecting grain boundaries (A-B), but the same principles apply also to other structures [17]. Initially, the voids at 0, A and C all grow rapidly under the indicated current because of the local (positive) flux divergencies. The voids serve as sources and sinks, thus strongly limiting the local stress. However, in a rigidly passivated line accumulation of atoms will lead to a growing compressive stress at a negative flux divergency such as B, and the resulting stress gradient eventually counteracts the current driven atomic flux, thus reducing the void growth rate at A substantially. The simultaneous increase in the flux through the lattice from B to C eventually leads to the establishment of a local 'steady-state' with a constant flux, $J_{AC}$, everywhere between A and C. Of course, if a defect near B leads to passivation cracking and metal extrusion, the back stress is

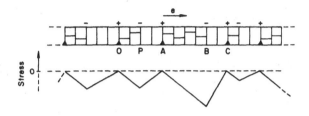

Figure 1: Sketch of line microstructure and 'steady-state' stresses.

'instantly' eliminated and the void at A may grow rapidly to sever the line.

During steady-state the void growth at A depends on the difference between the fluxes $J_{AC}$ and $J_{OA}$, both of which again depend on the steady-state stress gradients (Fig. 1), and thus on the relative lengths of bamboo and cluster segments. A completely repeatable structure, $(J_{AC} - J_{OA}) = 0$, would then eliminate void growth except at the line ends. At the interface with a W-plug, the flux divergency would of course equal $J_{AC}$. More importantly, $J_{AC}$ depends on the lattice diffusivity for the 'near-bamboo' structure in Figure 1, while it would be determined by the much larger grain boundary diffusivity in the absence of the bamboo-segment. Also, a long cluster segment next to a void allows substantial growth while establishing the back-stress. Since coalescence and failure may be initiated by a single void starting to migrate, it is therefore of essence to limit the length of the longest cluster segment. Unfortunately, particularly small grains are often found at steps, vias and W-plugs, many of which may also be preferred nucleation sites for thermal stress voids.

The presence of a rigid passivation layer thus affects the electromigration lifetime in two ways. It allows the establishment of large tensile stresses and stress concentrations, thus facilitating void nucleation, but also contributes to the rapid establishment of compressive 'back-stresses', thus reducing static growth. If the current is kept low enough, so that local heating remains negligible in spite of the poorer thermal conductivity, organic (polymer) passivation layers may benefit the reliability. For one thing, having larger coefficients of thermal expansion than aluminum such layers obviously prevent thermal stress induced voiding. In spite of the weaker constraint, sufficient current induced mass transport will eventually lead to the establishment of moderate local tensile stresses, and thus possibly void nucleation, but not until compressive stresses of a similar magnitude are established elsewhere. Even the initial void growth rate is therefore counteracted by a moderate stress gradient in this case. If currents are low and extrusion is not a problem, a superior electromigration resistance is indeed observed [19].

Based on studies of wide, unpassivated lines Ohmi and Tsubouchi [20] concluded that pure copper metallizations would be orders of magnitude more electromigration resistant than aluminum-based alloys. Certainly, the atomic mobilities are considerably lower in copper, but severe thermal stress induced voiding is still observed in passivated, narrow Cu-lines [13], if they adhere well to passivation and substrate. More work is needed to determine reliability under technologically relevant conditions.

Finally, in developing remedies it may be important to realize that the failure site observed in 'post-mortem' analysis of lines may often be quite far removed from the origin of the fatal coalescence.

## PREDICTIONS AND NEW REMEDY

The above discussion is by no means comprehensive. Notably, the use of 'sandwich' structures [21], in which thin layers of refractory metals maintain continuity, clearly warrants a detailed discussion. Other methods of improvement will be the subject of forthcoming publications, but we shall here briefly outline one of our ideas.

Figure 2 shows a sketch of a narrow metal line, which is assumed <u>not</u> to adhere well to the substrate itself. This is easily achieved in the case of Cu-lines [13]. Adhesion is thus only maintained in short regions by a thin, appropriately patterned adhering interlayer, say Al. In the regions of non-adhesion another interlayer is then deposited <u>over</u> the metal line, and the whole structure finally passivated with a dielectric which does not adhere to the metal. Obviously, the same effect could be achieved with non-adhering interlayers and a metal which <u>does</u> adhere to substrate and passivation. In both cases, the patterning strongly reduces the volume of metal under significant triaxial stress, and thus the thermal stress induced voiding.

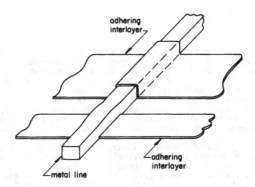

Figure 2: Metal line with patterned interlayers. Metal does not adhere to substrate and passivation. Interlayers adhere to these and to metal.

As mentioned above, some degree of metal-passivation adhesion is required to ensure thermal coupling to the surroundings, and to prevent corrosion at the interface when etching holes in the passivation. However, the above idea allows optimization in terms of a trade-off between voiding and thermal coupling, as well as the establishment of very good adhesion around the locations of later holes. More seriously, if the passivation does not adhere near B in Figure 1, some of the current induced compressive stress may be relieved along the free interface, delaying the build-up of a 'back-stress' limiting electromigration. Still, the improved reliability observed with polymer passivation layers [19] suggests, that a reduced back-stress may still be acceptable.

We are not yet in a position to predict absolute lifetimes solely on the basis of a known microstructure. Notably, we would still need a quantitative prediction of thermal stress induced void nucleation. However, the potential for extrapolating a limited set of accelerated test results to service conditions and very early failure is of great significance. For obvious reasons, extrapolations to the $6\sigma$-point of the cumulative failure distribution cannot be directly validated experimentally, so that a solid theoretical justification is of essence.

Currently, extrapolations of electromigration test data to early failure are often based on the assumption of a log-normal failure distribution, but according to our model the lifetime is determined by the growth of a 'worst' void, either to sever the line or to initiate migration and coalescence. A 'weakest-link' type of concept must therefore apply, which is irreconcilable with a log-normal distribution [22].

In a large number of practical cases, the slow 'steady-state' static void growth is the rate limiting stage for electromigration induced open failure of a narrow line. Now, for a given void the growth rates during both this and the initial 'transient' growth stage are proportional to the current density, $j$, as is the total atomic accumulation required to establish local 'steady-state'. If the initial (thermal stress induced) void volume is negligible, the volume, $\Delta V_o$, at the onset of steady-state is then also proportional to $j$. From Equation 1 follows that the critical void volume for onset of migration, $V_c$, varies roughly as $E^{-1}$, and thus as $j^{-1}$ [15]. The duration of the steady-state growth stage then varies as $(V_c - \Delta V_o)/j$, or approximately as $(j^{-2}-K)$, where K is proportional to $\Delta V_o/j$, and thus constant for given lengths of cluster and bamboo segments. Indeed, the median time to failure, $t_{50}$, observed during accelerated testing usually depends on current density in this way.

Of course, such a current density dependence can only apply for currents large enough that migration begins before the void severs the line. We thus need to know the binding energy for a small void at a given trap site, or $\gamma$ if Equation 1 applies. A program is under way to address this experimentally, as well as any role of precipitate coarsening. However, for the purpose of extrapolating to earlier failure it should also be realized that the shape of the failure distribution will vary with both temperature and current density.

For a microstructure like that sketched in Figure 1, long cluster segments (A-B) lead to large $\Delta V_o$, and thus large K. This again corresponds to the shorter lifetimes and the stronger current density dependencies, i.e. earlier failure times increase faster than $t_{50}$ with decreasing current. On the other hand, the severity of the current density dependence goes through a maximum. A cumulative failure at the ppm level or below may very well correspond to the occurrence of a cluster segment so long that a void starts migrating before steady-state is established. In this regime the earliest failure times then correspond to the weakest current density dependencies.

As far as the temperature dependence is concerned, the initial void growth rate clearly depends on the grain boundary diffusivity. However, for a near-bamboo structure (Fig. 1) the 'steady-state' rate varies as the lattice diffusivity [15]. For very early failures, initiated by a void starting to migrate before, or soon after, steady-state is established, the temperature dependence is thus much weaker than for $t_{50}$, i.e. the failure distribution may well get broader at lower temperatures.

In general, the electromigration induced failure distribution under service conditions may be broader or narrower than the one established during accelerated testing, depending on microstructure and precipitate distribution. Fortunately, it appears that this can all be addressed within the framework of our model.

## SUMMARY

Thermal stress induced voids nucleate and start to grow during cool-down from high temperature process steps. Electromigration leads to extrusions and/or further void growth, and eventually to migration and coalescence. Both phenomena depend strongly on microstructure and precipitate distribution. Good reliability requires reduced tensile stresses, slow grain boundary diffusion and strong trapping of voids. A basis is being established for safely extrapolating accelerated test results. Electromigration induced failure distributions cannot be log-normal. Observed dependencies of median failure times on temperature and current are not likely to apply to 'six-sigma' failure as well. The theoretical framework exists for addressing this quantitatively.

## ACKNOWLEDGEMENTS

This work was supported by the IBM Corporation. The authors wish to thank P. Totta, T. D. Sullivan, and P. J. Loos for helpful discussions.

## REFERENCES

1. Proc. 1st Int. Workshop on Stress Induced Phenomena in Metallizations (eds. C.-Y. Li, P. Totta, and P. Ho, AIP Conf. Proc., in press)
2  T. D. Sullivan, Appl. Phys. Lett. 55 (1989) 2399
3. M. A. Korhonen, C. A. Paszkiet, and C.-Y. Li, J. Appl. Phys. 69 (1991) 8083
4. M. A. Korhonen, W. R. LaFontaine, P. Børgesen, and C.-Y. Li, J. Appl. Phys. 70 (1991) 6774
5  P. Børgesen, M. A. Korhonen, and C.-Y. Li, accepted for publication in Thin

Solid Films
6.  J. T. Yue, W. P. Funsten, and R. V. Taylor, 23rd Annual Proceedings of
    Reliability Physics (IEEE, New York, 1985) 126
7.  P. Børgesen, M. A. Korhonen, C. Basa, W. R. LaFontaine, B. Land, and C.-Y.
    Li, Mat. Res. Soc. Symp. Proc. 225 (1991)143
8.  J. G. Ryan, J. B. Riendeau, S. E. Shore, G. J. Slusser, D. C. Beyar, D. P.
    Boulding, and T. D. Sullivan, J. Vac. Sci. Technol. A8 (1990) 1474
9.  B. Greenebaum, A. I. Sauter, P. A. Flinn, and W. D. Nix, Appl. Phys. Lett.
    58 (1991) 1845
10. M. A. Korhonen, P. Børgesen, C. A. Paszkiet, J. K. Lee, and C.-Y. Li, Mat.
    Res. Soc. Symp. Proc. 225 (1991) 155
11. T. D. Sullivan, L. Miller, and G. Endicott, in ref. [1]
12. P. A. Totta, in ref. [1]
13. D. D. Brown, P. Børgesen, D. A. Lilienfeld, M. A. Korhonen, and C.-Y. Li,
    Mat. Res. Soc. Symp. Proc. 239 (1992)
14. P. S. Ho, J. Appl. Phys. 41 (1970) 64
15. P. Børgesen, M. A. Korhonen, D. D. Brown, and C.-Y. Li, ref. [1]
16. C.-Y. Li, P. Børgesen, and T. D. Sullivan, Appl. Phys. Lett. 59 (1991) 1464
17. P. Børgesen, M. A. Korhonen, T. D. Sullivan, D. D. Brown, and C.-Y. Li,
    Mat. Res. Soc. Symp. Proc. 239 (1992)
18. P. A. Flinn, ref. [1]
19. J. R. Lloyd, Thin Solid Films 91 (1982) 175
20. T. Ohmi and K. Tsubouchi, Solid State Technol. 35, No.4 (1992) 47
21. E. Levine and B. Henry, in ref. [1]
22. J. R. Lloyd and J. Kitchin, J. Appl. Phys. 69 (1991) 2117

# HIGH RESOLUTION OBSERVATION OF VOID MOTION IN PASSIVATED METAL LINES UNDER ELECTROMIGRATION STRESS

MICHAEL C. MADDEN*, EDWARD V. ABRATOWSKI*, THOMAS MARIEB**
AND PAUL A. FLINN*
*Intel Corporation, 2200 Mission College Blvd., Santa Clara, CA 95052
**Department of Materials Science and Engineering, Stanford University, Stanford, CA 94305

## ABSTRACT

Using a 120 kV STEM equipped with a backscattered electron detector and operated as a conventional SEM, voids in metal lines can be detected through 1 $\mu$m of passivation. By applying current to passivated thin metal lines while in the microscope, voids can be observed while electromigration is in progress. Voids move significant distances during electromigration. On at least some occasions, failure of the line is not the result of a void growing until the width of the line is reached. On these occasions, when the size of the void approaches the width of the line, the void breaks up into smaller voids.

## INTRODUCTION

Susceptibility to electromigration failure is a major factor determining the dimensions of metal lines in integrated circuits. Incorporation of electromigration effects in the design of integrated circuits has been hampered by the absence of a accurate theoretical model which, in turn, is due in part to a lack of experimental data gathered under the conditions which metal lines experience in actual devices. The condition which is most significant and most difficult to deal with experimentally is the presence of passivation. Passivation is a 1 to 2 $\mu$m thick amorphous layer of silicon oxide or silicon nitride deposited over the exposed surfaces of the line for electrical isolation and protection from mechanical and chemical damage. The presence of passivation constrains the metal and has a profound effect on electromigration [1]. It also makes void imaging in a scanning electron microscope (SEM) impossible under normal operating conditions.

In earlier work in our facility, passivated electromigration structures were observed *in situ* using a 35 kV SEM equipped with field emission gun and backscattered electron detector [2]. Electromigration voids were clearly visible, and void sizes and velocities were measured. Later, this technique was expanded to allow sample heating (independent of Joule heating) using a hot stage [3]. Resolution using this technique is about 0.5 $\mu$m.

Since passivation is optically transparent, electromigration voids in passivated lines have also been observed using light microscopy [4]. However, only voids intersecting the top interface between the metal and passivation are visible, and the width of the metal lines currently being used in integrated circuits is below the resolution limit of light microscopy.

Mat. Res. Soc. Symp. Proc. Vol. 265. ©1992 Materials Research Society

## METHODS

Increasing the energy of the electron beam in a SEM significantly improves the resolution of voids imaged under passivation. In this study, we used an unmodified 120 kV JEOL 1200 EX scanning transmission electron microscope (STEM) equipped with a backscattered electron detector. Although this is a transmission electron microscope, it was operated as a conventional SEM, i.e., only bulk samples were examined, and lenses below the specimen were not used. For voids in a 1 $\mu$m thick Al-1% Si film with 1 $\mu$m passivation the resolution is about 0.05 $\mu$m.

A standard JEOL TEM sample holder was modified to include a thin film platinum heating element and four electrical feedthroughs: two for the heater and two to provide current to the electromigration test structure. Temperature of the stage was calculated from measuring the resistance of the heating element. Temperature of the test structure was calculated from the stage temperature and Joule heating caused by the electromigration stressing current.

The experimental apparatus also included an Intel System 120 computer equipped IRMX real time software to allow digital image storage and recording of heater current, electromirgation current, and other parameters. A diagram of the assembled components is shown in Figure 1. Digitally stored images were averaged over eight frames. Some images were recorded photographically using Polaroid Type 55 film.

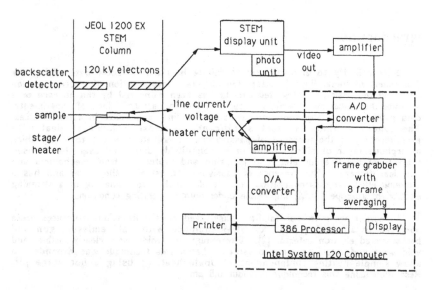

**Figure 1. Block diagram of electromigration test apparatus.**

The test structure used for this experiment was a straight 3 $\mu$m wide, 1 $\mu$m thick, Al-1% Si metal with 1 $\mu$m passivation. Bond pads were located at either end. Current density was approximately $4 \times 10^{6}$ A/cm$^{2}$. Sample temperature varied between 280°C and 405°C. The initial experiment was designed to investigate the effect on voids when electromigration stressing current is removed

## RESULTS AND DISCUSSION

The experiment was conducted over a five day period, although the temperature was not always above ambient and the electromigration current not always applied during that time. Temperature and stressing current data is summarized in Table I.

No voids were observed prior to the application of electromigration stressing current. A test current of 115 mA. and a line temperature of 300° C were initially selected. These conditions should have generated voids in a period of hours based on previous data [2,3]. When no voids had been observed after 6 hours, the test current was increased to 125 mA. This increased the temperature of the test structure to 315° C due to additional Joule heating. When voids still had not formed after an additional 2 hours, the stage temperature was increased 60° C to give a line temperature of 370° C and 1.5 hours later by an additional 30° C to give a line temperature 405° C. Voids then formed at a total elapsed time of 9.5 hours, as shown in Figure 2. At this point the stage temperature was reduced to produce a line temperature of 370° C and testing continued for an additional 4.5 hours to a total time of 14 hours. During this time the voids moved and changed shape somewhat as shown in Figure 3.

### TABLE I

Electromigration Stressing Current and Sample Temperature

| Date | Clock Time | Elapsed Time (hrs) | Current (mA) | Stage Temp (C) | Line Temp (C) | Comments |
|------|-----------|--------------------|--------------|----------------|---------------|----------|
| 11 Apr | 16:00 | 0 | 115 | 220 | 300 | Start |
| 11 Apr | 22:00 | 6 | 125 | 220 | 315 | incr. current |
| 11 Apr | 24:00 | 8 | 125 | 280 | 370 | incr. temp. |
| 12 Apr | 1:30 | 9.5 | 125 | 310 | 405 | voids form |
| 12 Apr | 1:30 | 9.5 | 125 | 280 | 370 | red. temp. |
| 12 Apr | 6:00 | 14 | 0 | 280 | 280 | curr. off |
| 13 Apr | 7:00 | 39 | 0 | 20 | 20 | heater off |
| 14 Apr | 9:30 | 65.5 | 125 | 280 | 370 | restart |
| 14 Apr | 11:30 | 67.5 | -125 | 280 | 370 | rev. current |
| 14 Apr | 15:30 | 71.5 | 0 | 280 | 280 | curr. off |
| 15 Apr | 8:30 | 88.5 | 0 | 20 | 20 | heater off |

At this point the electromigration stressing current was reduced to zero. The specimen was left in the microscope with the line temperature maintained at 280° C for 25 hours, to an elapsed time of 39 hours. A micrograph of the voids is sown in Figure 4. No significant changes in shape or position had occurred.

The heater current was reduced to zero and the specimen allowed to return to room temperature. Testing was resumed at an elapsed time of 69.5 hours, with a current of 125 mA. and a test temperature of 370° C, which were the conditions before testing was suspended. The position and shape of the voids had not changed during the time at room temperature. After testing resumed, the voids moved substantially. In Figure 5, failure of the line appears imminent, as one of the voids has nearly bridged the width of the line. At this point a crack appeared in the passivation at the opposite end of the line, as shown in Figure 6. This indicates that mass transport is occurring along the length of the line, a phenomena which does not occur with unpassivated metal. Failure did not occur, however. The void broke apart, as shown in Figure 7. This segment of the test was concluded after an additional 2 hours, at an elapsed time of 67.5 hours. The voids are shown at this point in Figure 8.

36

 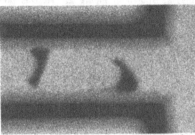

Figure 2. ET=11 hrs. Voids on the passivation sidewalls.

Figure 3. ET = 13 hrs. Voids shortly before electromigration stressing current was removed.

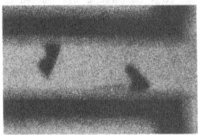 

Figure 4. ET=40 hrs. Voids after 25 hrs. at 280° C with no electromigration stressing current applied. Voids have not changed shape from Figure 3.

Figure 5. ET=66.75 hrs. Voids after electromigration current restored. The void on the left has nearly bridged the width of the line.

 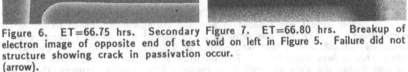

Figure 6. ET=66.75 hrs. Secondary electron image of opposite end of test structure showing crack in passivation (arrow).

Figure 7. ET=66.80 hrs. Breakup of void on left in Figure 5. Failure did not occur.

NOTE: All figures are backscattered electron images obtained at 120 kV except Figure 6. The metal line being tested appears and the bond pad are labelled in Figure 2. The intermediate grey area above and below the line and to the left of the bond pad is the passivation sidewall. The dark areas above and below the passivation sidewalls are spaces between adjacent structures. In all figures, electrons are moving from right to left, voids from left to right. ET = elapsed time. All Figures on this page are at the same magnification ▬ = 1.0 μm.

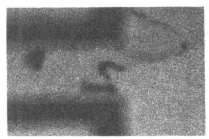 

Figure 8. ET=67.4 hrs. Just before electromigration current was reversed. A string of voids is in the bond pad.

Figure 9. ET=67.75 hrs. Shortly after current reversal. String of voids in bond pad have begun to heal.

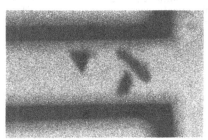 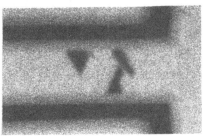

Figure 10. ET=68.5 hrs. Note parallel edges on two adjacent voids.

Figure 11. ET=69 hrs. Two voids on right have nearly bridged the width of the line.

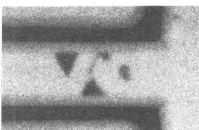 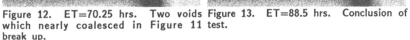

Figure 12. ET=70.25 hrs. Two voids which nearly coalesced in Figure 11 break up.

Figure 13. ET=88.5 hrs. Conclusion of test.

NOTE: Figures are backscattered electron images obtained at 120 kV. The metal line being tested appears as the bright horizontal band in the center. Description of features is given in the captions on the previous page and in Figure 2. In Figure 8, electrons are moving from right to left, voids from left to right, consistent with Figures 2-7. In Figures 9 - 12, electrons are moving from left to right, voids from right to left. ET = elapsed time. All Figures on this page are at the same magnification ▬ 1.0 $\mu$m.

The current to the specimen was then maintained at 125 mA but reversed in direction. Figures 9 and 10 were made shortly after the current was reversed. In Figure 11, failure once again appears imminent at 70.5 hours elapsed time but, as before, the void breaks apart, as shown in Figure 12.

The test was concluded after an additional 17 hours, at an elapsed time of 88.5 hours. The voids are shown at the conclusion of the test are shown in Figure 13.

## CONCLUSIONS

Based on these results, we can say that voids appear to be nucleated at the interface with the passivation sidewall and move towards the center of the line. They move large distances during electromigration stressing. In at least some instances, voids do not simply increase in size until failure occurs. When a void occupies most of a cross section, high local current density can cause the void to break up, and failure does not occur. After formation, extended holding of voids at temperature does not necessarily cause voids to disappear.

## REFERENCES

1. N. G. Ainslie, F. M. d'Heurle, and O. C. Wells. Appl. Phys. Lett. 20, 173 (1972).

2. E. Castaño, J. Maiz, P. Flinn, and M. Madden, Appl. Phys. Lett. 59, 129 (1991).

3. P. Besser, M. Madden, P. Flinn, submitted to J. Appl. Phys.

4. E. Levine and J. Kitcher, Proceedings of the 22nd Annual IEEE International Reliability Physics Symposium, USA (1984), p. 242.

ESTIMATION OF THERMAL STRESSES AND STRESS CONCENTRATIONS IN CONFINED INTERCONNECT
LINES OF RECTANGULAR CROSS SECTION

M.A. Korhonen, P. Børgesen, and Che-Yu Li
Department of Materials Science and Engineering, Bard Hall
Cornell University, Ithaca, NY 14853

ABSTRACT

Stress-induced voiding is an important reliability concern in narrow
aluminum based metallizations used as interconnects in very large scale
integrated circuits. The thermal stresses that arise in the interconnects after
excursions to elevated temperatures are tensile and extremely high. Void
nucleation is commonly found to take place at the line edges where the stress is
highest. In lines of rectangular cross section, in particular, there arise shear
stresses at line edges and corners which can be higher than the dilational stress
components. In this paper we consider analytically the stress states arising in
interconnect lines of rectangular cross section. We argue that void nucleation
is likely to be connected to the high shear stress at the edges and corners,
facilitating grain boundary sliding.

INTRODUCTION

Stress induced voiding has been identified as an important reliability
concern in aluminum based interconnects in integrated microcircuits [1-3].
Significant tensile stresses are generated upon cooling to room temperature after
excursions to higher temperatures.

The passivated interconnect lines in question can advantageously be
modelled as inclusions embedded in an infinite, elastic silicon matrix. The
thermally induced stresses can then be calculated from the Eshelby theory of
inclusions [4-6], which provides upperbound estimates for the stresses expected
[6]. Particularly, uniform thermal strains result in uniform stresses in
ellipsoidal inclusions, including elliptic cylinders as a particular case [4-6].
Modelling aluminum interconnects as long cylinders of elliptic cross section
embedded in the silicon matrix, the thermal stresses can readily be estimated
from the Eshelby theory of inclusions. Fig. 1 shows an example of the calculated
thermal stresses in aluminum interconnects at RT after cooldown from 400°C [6].
It is remarkable that the stresses are very large and tensile, and have a large
hydrostatic component which increases with the aspect ratio h/w, or the line
thickness (2h) to the width (2w). Because of the silicon confinement, plastic
deformation (or diffusional flow) can only redistribute the stresses so that the
volumetric strain is conserved [5,6]. The solid line in Fig. 1 gives the limiting
hydrostatic stress state in the case that the stress redistribution has proceeded
to completion, i.e. all shear stresses have become zero. Obviously, quite
substantial stresses remain in the interconnects. In general, void growth remains
the sole possibility for complete stress relaxation in confined interconnects
with h/w > 0. The limiting hydrostatic stress, Fig. 1, can be seen as the driving
force for the void growth. Accordingly, the tendency to void formation increases
with the aspect ratio, and reaches the maximum at h/w = 1.

It is generally agreed that voids nucleate during creep at elevated
temperatures at the sites of stress concentrations, often as a result of grain
boundary sliding [7]. Similar mechanisms are likely to operate in the passivated
interconnects during cooldown from a heat treatment [8]. Increasing passivation
thickness means higher stress state, and results in enhanced void formation [8].
It also appears that practically no voids nucleate after cooldown, which confirms
the importance of a high stress state and of grain boundary sliding for void
formation [8].

Most usually, thermal stress induced voids nucleate at line edges; however,
in the presence of precipitates, voids are occasionally seen also in the middle
of the lines [8]. Line edges provide heterogeneous nucleation sites for the
voids. Apart from this, it appears that the stress state near the line edges must

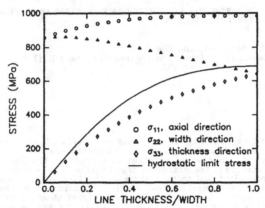

**Figure 1**  Thermally induced stresses in aluminum lines embedded in an infinite elastic silicon enclosure, after heat treatment at 400°C [6].

have been such as to enhance void nucleation. Real interconnects are rather rectangular than elliptical in cross section. It is the purpose of this paper to estimate the stresses analytically, particularly those of shear character, that arise in aluminum based interconnects of rectangular cross section during cooldown from a thermal treatment.

Although we address thermal stresses, the general concepts to be presented in this paper are applicable to the electromigration induced stresses [9] as well. As an example, consider electromigration of atoms to a sink region, where the atoms will be deposited in climbing dislocations. For the sake of argument, assume an isotropic distribution of edge dislocations over the entire cross section. Now, in analogy with the thermal stress case, homogeneous and isotropic deposition of atoms at dislocations in the sink regime creates different stresses in the direction of the line than transverse to it, see Fig. 1. Further, in lines of rectangular cross section, shear stress concentrations arise at line edges, analogously to the case of thermally induced stresses to be dealt with next.

## ESTIMATION OF THERMAL STRESSES IN A CUBOIDAL INCLUSION

Consider a cuboidal inclusion (interconnect), the center of which we select to be the origin of the rectangular system of coordinates. The length of the cuboid is 2l, width 2w, and heigth 2h in the x-, y- and z-directions; the tensorial indices 1, 2 and 3 refer to these directions, respectively. The inclusion is subjected to the thermal strain $\Delta\alpha\Delta T$, where $\Delta\alpha$ is the differential thermal expansion coefficient between the inclusion and the surroundings, and $\Delta T$ is the change in temperature. For simplicity we assume that the elastic properties of the inclusion and the surrounding matrix are the same. In the case of aluminum lines embedded in silicon, using the elastic properties of aluminum for both the inclusion and surroundings, introduces an inaccuracy of about 10 - 30% at h/w = 1, depending on the stress component, while there is no error at the limit of h/w = 0. For the purposes of the present paper, a qualitative picture of stress distributions is quite satisfactory, and the above errors are well tolerable. According to Eshelby theory [4] the constrained strain in the inclusion due to the thermal strain can readily be written as

$$\epsilon^C_{ij} = - (1+\nu) \, \Delta\alpha\Delta T \, \phi_{,ij} \, /4\pi(1-\nu),  \qquad (1)$$

where $\nu$ is Poisson's ratio and $\phi_{,ij}$ denotes the second derivative of the Newtonian potential function $\phi$. For a cuboid, with corners at $(\pm l, \pm w, \pm h)$, the Newtonian potential is

$$\phi = \int_{-h}^{h} \int_{-w}^{w} \int_{-1}^{1} dx'dy'dz'/r, \tag{2}$$

where $r = [x-x', y-y', z-z']$ specifies the vector from the point of observation $(x,y,z)$ to the position $(x',y',z')$ to be integrated over the volume of the cuboid. The potential $\phi$, although involving only simple analytic functions, is rather complicated [10]. However, the second derivatives needed for the stress calculation turn out to be relatively simple:

$$\phi,_{11} = -\int_{-h}^{h} \int_{-w}^{w} \int_{-1}^{1} atn\ [(y-y')(z-z')/r(x-x')], \tag{3a}$$

$$\phi,_{12} = \int_{-h}^{h} \int_{-w}^{w} \int_{-1}^{1} ln\ (r+z-z'), \tag{3b}$$

where the primed coordinates take the values $x' = -1$ to $1$, $y' = -w$ to $w$, and $z' = -h$ to $h$. The rest of $\phi,_{ij}$ are found from the cyclic permutation of the indices and of the coordinates.

Combining eqn. (1) with Hooke's law we find for the thermal stresses <u>inside and outside</u> a cuboidal inclusion

$$\sigma_{11} = E\Delta\alpha\Delta T\ (\phi,_{22}+\phi,_{33})/4\pi(1-\nu) \tag{4a}$$

$$\sigma_{12} = -E\Delta\alpha\Delta T\ \phi,_{12}/4\pi(1-\nu); \tag{4b}$$

where E is Young's modulus. The rest of the stresses are readily found by cyclic permutation of the indices. Eqs. (4), with the definitions (3), are very similar to the formulas given previously by Hu for the <u>outside</u> of a cuboidal inclusion subject to thermal strains [11]. Hu also provides formulas for the outside of the cuboidal inclusion in a semi-space. Combination of these with eqns. (4) yields useful stress estimates for the cases when the passivation layer can no longer be assumed to be infinitely thick [12]. A particular benefit of the Eshelby-method is that it can be readily extended to the case of general shape and size change of an inclusion [4]. The general method for the calculation of stresses in long, rectangular interconnect lines is outlined elsewhere [12].

RESULTS AND DISCUSSION

Thermal stresses in ellipsoidal inclusions embedded in an infinite matrix are uniform [4]. The most notable features of eqs. (4) are that the stresses in the rectangular lines depend on the position $(x,y,z)$. It is also remarkable that purely dilatational strains give rise to shear stresses in the principal axes system of the inclusion. Preliminary calculations showed that the stresses over the end faces of the line, at $x = \pm1$, depended only weakly on the line length when $1 > 5$ w. Conversely, the stresses in the middle portions of the line remained practically independent of the line length for $1 > 5$ w. We assume a line length $1 = 100$ w in our calculations, and concentrate in the center cross section at $x = 0$, and the end face at $x = 1$. Further, we follow the stress components along the y- and z-axes and along the diagonal $z = hy/w$, starting from the center of the cross section.

Fig. 2 displays the stresses over the cross section at $x = 0$ of a relatively long line with $h/w = 1$, along the width direction y (solid line) and the diagonal $z = y$ (dashed line). Along the diagonal, stresses $\sigma_{22}$ and $\sigma_{33}$ are equal, while in the y-direction $\sigma_{22}$ decreases towards the line edge while $\sigma_{33}$ increases. The stress $\sigma_{22}$ is continuos across the interface, while there is a discontinuity in $\sigma_{33}$ which is negative on the silicon side.

No shear stresses arise along the y- and z-directions, while there is a shear stress $\sigma_{23}$ along the diagonal, which approaches infinity at the line edge.

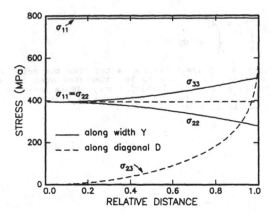

**Figure 2**    Stresses in the cross section at x = 0, in the width direction (y) direction and along the diagonal z-y.

The average stresses over the whole cross section are the same as the stress values at the origin. It is remarkable that the average stresses for the square cross section are the same as the uniform stresses for the circular cross section [12]. In fact, it appears that quite generally, the average stresses for a rectangular cross section of a given aspect ratio agree closely with the stresses for an elliptic cross section of the same aspect ratio [12]; hence also the limiting hydrostatic stress, see Fig. 1, is the same. This suggests that, as far as void growth is concerned, elliptic and rectangular interconnects behave similarly.

Fig. 3 shows the situation in the interface at the line end, x = 1. As above, the solid line corresponds to stresses along the width direction y, while the dashed line gives stresses along the diagonal. The stresses $\sigma_{22}$ and $\sigma_{33}$ are about equal over the end face, and generally larger than the stress $\sigma_{11}$ normal to the interface. Along the diagonal the shear stresses are relatively small, but approach infinity at the line corner. Along the y-direction the shear stress $\sigma_{12}$ is larger, and reaches magnitudes comparable to the dilational stresses well before the line edge. It should be noted that a stress singularity involving smaller volumes than the critical void size [7,8] is likely to be ineffective in promoting nucleation of thermodynamically stable voids. From this viewpoint, the

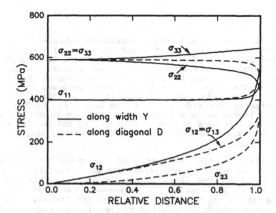

**Figure 3**    The stresses in the interface at x = 1, in the width direction (y) and along the diagonal z-y.

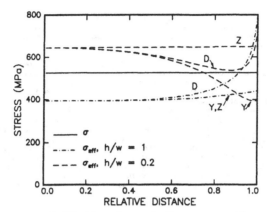

**Figure 4**     The average and effective stress in the cross section at x = 0, for the aspect ratios of h/w = 1 and 0.2

shear stresses encountered at the line edges are more dangerous than those found at the line corners, Figs. 2 and 3.

For comparison of stresses in interconnects of different aspect ratios h/w, it is advantageous to reduce the triaxial stress state to two characteristics, only. The average stress $\sigma = (\sigma_{11}+\sigma_{22}+\sigma_{33})/3$ represents the hydrostatic component of the stress state, while the von Mises effective stress $\sigma_{\text{eff}}$ is a measure of the shear stresses driving plastic flow. Please note, that stress redistribution by plastic flow can reduce the average stress $\sigma$ to zero when h/w = 0, while at h/w ≈ 1, the values of the average stress and the limiting hydrostatic stress (Fig. 1) are practically the same. Fig. 4 shows the average stress $\sigma$ and the effective stress $\sigma_{\text{eff}}$ over the cross section x = 0, for lines of aspect ratios of 1 and .2, along the width and thickness direction (Y and Z) as well as along the diagonal (D). For the aspect ratio 1, $\sigma_{\text{eff}}$ is generally smaller, although still of the same order as $\sigma$. However, near the line edge $\sigma_{\text{eff}}$ increases strongly. On the other hand, for the aspect ratio of .2, the effective stress is generally larger than $\sigma$; it increases, but not significantly, at the line corners; at the line edges it may even decrease.

Fig. 5 shows the average stress $\sigma$ and the effective stress $\sigma_{\text{eff}}$ at the end of the line, x = 1, for the aspect ratios of 1 and 0.2, along the width and thickness direction (Y and Z) as well as along the diagonal (D). The effective

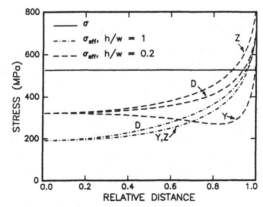

**Figure 5**     The average and effective stress in the interface at x = 1, for the aspect ratios of h/w = 1 and 0.2.

stresses are small in the center of the cross section, and increase strongly towards the edges and corners for h/w - 1. Basically, a similar type of situation is encountered also in lines with h/w - .2, particularly in the diagonal and the thickness directions.

Based on the above observations, it appears that overall plastic yielding is possible during cooldown in lines of relatively low aspect ratio. Plastic flow smoothes out the stress concentrations, and the average stresses relax to the limiting hydrostatic stress level, Fig. 1, where there is little danger of void formation. On the other hand, it seems that the lines most susceptible to voiding are the ones where the effective stress is small in the central portions of the line, so that no general yielding occurs, while, at the same time, the effective stress has large value at the line edges and corners. In such lines the limiting hydrostatic stress is high, Fig. 1, and provides the driving force for void growth, while the high shear stresses at line edges and corners provide the driving force for void nucleation. Grain boundary sliding is likely to be favored relative to slip at high stresses when the slip is limited due to interfacial constraint [13]. Grain boundary sliding, when blocked at the interface, results in large local stress concentrations able to nucleate voids.

## CONCLUSIONS

Large thermally induced stresses arise in passivated interconnects after excursions to elevated temperatures. In interconnects of aspect ratio in the neighborhood of 0.3 - 1, there is a large hydrostatic stress component which can not be relieved by plastic deformation; void growth appears as an alternative relaxation mechanism in these conditions. Thermal stresses are nonuniform in the interconnects of rectangular cross section. Particularly, large shear stresses arise at line edges and corners. The average stresses are still about the same as the uniform stresses in the interconnects of elliptic cross section; hence the driving force for void growth is the same in both cases. Particularly, lines with the aspect ratio in the neighborhood of 0.3 - 1, appear to be most susceptible to void growth. However, as to void nucleation, lines of elliptic and rectangular cross section behave differently. In interconnects of rectangular cross section, large shear stresses at line edges and corners enhance local, inhomogeneous deformation; voids are likely to be nucleated by the grain boundary sliding mechanism during cooldown.

## ACKNOWLEDGEMENTS

Portions of this work were supported by an IBM Shared University Project, by the National Science Foundation through the Materials Science Center, and by the National Nanofabrication Facility at Cornell.

## REFERENCES

1. T. Turner and K. Wendel, Proc. Int. Reliability Phys. Symp. 25 (IEEE, New York, 1985) p. 142
2. K. Hinode N. Owada, T. Nishida, and K. Mukai, J. Vac. Sci. Technol. B5, 518 (1987)
3. Che-Yu Li, R.D. Black and W.R. LaFontaine, Appl. Phys. Lett. 53, 31 (1988).
4. J.D. Eshelby, Proc. Roy. Soc. A241, 376 (1957)
5. H. Niwa, H. Yagi, H. Tsuchikawa, and M. Kato, J. Appl. Phys. 68, 328 (1990)
6. M.A. Korhonen, R.D. Black and Che-Yu Li, J. Appl. Phys. 69, 1748 (1991)
7. H. Riedel, "Fracture at High Temperatures" (Springer-Verlag, Berlin, Heidelberg, 1987)
8. M.A. Korhonen, W.R. LaFontaine, P. Børgesen, and Che-Yu Li, J. Appl. Phys. 70, 6774 (1991)
9. M.A. Korhonen, P. Børgesen, and Che-Yu Li, in "Thin Films: Stresses and Mechanical Properties III" (Mat. Res. Soc. Symp. Proc. 239, Pittsburgh, PA, 1992)
10. J.K. Lee and W.C. Johnson, Scripta Metall. 11, 477 (1977)
11. S.M. Hu, J. Appl. Phys. 66, 2741 (1989)
12. M.A. Korhonen, P. Børgesen, and Che-Yu Li, in "Thermal Stress and Strain in Microelectronics Packaging", edited by J.H. Lau (Van Nostrand Reinhold, New York, 1992)
13. M.A. Korhonen, P. Børgesen, and Che-Yu Li, in "Materials Reliability Issues in Microelectronics" (Mat. Res. Soc. Symp. Proc. 225, Pittsburgh, PA, 1991), p. 133

# A SIMPLE MODEL FOR STRESS VOIDING IN PASSIVATED THIN FILM CONDUCTORS

J.R. Lloyd* and E. Arzt, Max Planck Institut für Metallforschung, Institut für Werkstoffwissenschaft, Seestraße 71, D-7000, Stuttgart 1, Germany * also with Digital Equipment Corporation, 77 Reed Road, Hudson MA 01749-2895 USA

## ABSTRACT

A model is proposed for stress voiding in passivated thin film conductors. The rate limiting step is argued to be the formation of vacancies at dislocation jogs which then diffuse to void sites.

## INTRODUCTION

Considerable effort has been expended to understand an insidious failure mode in passivated thin metal film conductors called "stress voiding". [1-7] It has been pretty well accepted that the failure mechanism is a form of diffusive creep relieving stress caused by the difference in the coefficients of thermal expansion between the metal conductor and the surrounding media and the temperature excursions experienced during integrated circuit processing. It has been demonstrated, both theoretically and experimentally that these thermally induced stresses are quite large, nearly hydrostatic and independent of the retained stress in the overlying passivation material. They are dependent primarily on the temperature history and to a lesser degree on the geometry. [8]

In this paper, a simple model of the process of stress voiding is proposed where void formation and growth are the products of a stress induced vacancy supersaturation generated by dislocation climb, and the subsequent diffusion of these vacancies to voids. The following model also has the feature of being consistent with hydrostatic stress relief.

## MODEL

Consider a metal line deposited at an elevated temperature onto a Silicon substrate (or one which had been held at high temperature for a sufficiently long time so as to permit stress relief). While at high temperature, a strong passivation material possessing a much smaller thermal coefficient of expansion (TCE) than the metal, in this case, identical to the substrate, is applied. The sample is then cooled relatively rapidly. Since the enclosed metal film has a larger coefficient of expansion than the passivation, it will "want" to assume a smaller volume than that enclosed by the passivation. If the adhesion is good and the passivation is so thick as to not be appreciably distorted, the amount of stress, $\sigma$, retained in the metal conductor is;

$$\sigma = \beta \Delta \alpha \Delta T \qquad (1)$$

where $\Delta \alpha$ is the difference in the TCE'S of the metal and the substrate/passivation system and $\Delta T$ is the difference in temperature between the application of the passivation and the temperature of observation. $\beta$ is the bulk modulus. In this idealized picture, the stress will be in hydrostatic tension. Under more realistic conditions, the picture would become somewhat more complicated, but the complexity will not contribute to understanding the general mechanism of void formation and growth.

Note for a temperature difference of a few hundred degrees, typical of IC processing, the expected stress is extremely high. Stresses close to 0.5% of the Bulk Modulus can be expected, or about five times the tensile yield stress. [8] The nearly hydrostatic nature of the stress keeps the stress from being relieved by dislocation glide.

Although dislocation glide cannot relieve hydrostatic stresses, dislocation climb can until the vacancy concentration in thermal equilibrium with the stress is attained Dislocation climb is accompanied by the formation or the annihilation of vacancies depending on the sign of the motion. A dislocation climbing under the influes the vibrational frequency of the lattice ($10^{13}$/sec) and $\Delta H^f_v$ is the activation free energy of vacancy formation. The vacancy formation will continue until;

$$C_\sigma = C_v \exp(-\sigma\Omega / kT) \qquad (2)$$

where $C_v$ is the thermal equilibrium vacancy concentration in the absence of stress and $\Omega_v$ is the vacancy volume. For Al, this can be shown to be a very efficient process even at low temperatures. Each dislocation leaves behind the equivalent of a sheet of vacancies which disperse throughout the crystal. It can be shown that only a single dislocation need climb through a 1 um cube crystal to produce a vacancy concentration in equilibrium with the thermal stress.

It is interesting to note that despite their near ubiquity, it can be shown that the nucleation of voids is theoretically impossible without the unsatisfying necessity of "deus ex machina" stress concentrations. [9] These stress concentrations, which undoubtedly exist, determine the location of void nucleation, characteristically at film edges or sites of known high stress such as at steps in the underlying substrate. It has also been suggested, however, that delaminations between the surface metal and the passivation are sufficient to nucleate voids without the classical nucleation problem. [7] The shape of the voids is probably determined by random fluctuations in grain orientation rather than surface energy considerations. [10]

Once formed, voids grow by the diffusion of vacancies provided by dislocation climb to the stress free void surface. Consider a thin narrow conductor with a "bamboo" structure containing voids at some of the grain boundaries with an average spacing of $2l_v$. (Fig.1) Vacancy diffusion along grain boundaries to the voids is assumed to be very rapid (in fact, instantaneous) compared to lattice diffusion. The flux of vacancies into a void will, therefore, be equal to the flux of lattice vacancies into a grain boundary with a void. This approximation simplifies the geometry for the calculation significantly, leaving us with a simple 1 dimensional boundary value problem. The case of a polycrystalline film will require the use of a more complicated geometry, which will be the subject of future work, but will not differ in concept from what is treated here.

As vacancies diffuse to the grain boundaries, they are replaced via dislocation climb at a rate given by

$$K = \frac{N_d}{a^2 v_d}$$

where Arrhenius is the dislocation density expressed in length of dislocation per unit volume, a is the lattice constant and $v_d$ is the drift velocity of the climbing dislocation given by

$$v_d = \frac{D_d F}{kT} \qquad (3)$$

where $D_d$ is the dislocation diffusivity and F is the driving force on the dislocation given by;

$$Fb = kT\ln\left(\frac{C_\sigma}{C}\right) = \sigma\Omega_v - kT\ln\left(\frac{C_o}{C}\right) \qquad (3a)$$

where b is the burgers vector of the dislocation. F will be the chemical potential gradient from the dislocation jog generating site to the surrounding medium. If vacancy formation at jogs is the rate limiting step, the activation energy for dislocation climb will be the same as that for vacancy formation. Since vacancy diffusion is a faster process than vacancy formation, the second term in eqn. (3a) can be neglected and a steady state concentration profile will be achieved expressed by the solution to;

$$D_v \left( \frac{\partial^2 C}{\partial x^2} \right) + K = 0$$

(4)

where $D_v$ is the vacancy diffusivity according to the boundary condition that the vacancy concentration is at the thermal equilibrium value, $C_o$, at the grain boundaries ($x = +- l_v$). Grain boundaries not containing voids can be ignored since they will be in thermal equilibrium with the lattice. Grain boundary grooving is not considered. The solution to (4) according to the stated boundary conditions is simply;

$$C - C_o = N_d \left( \frac{D_d}{D_v} \right) \frac{\sigma}{2kT} \left( l_v^2 - x^2 \right)$$

(5)

recognizing that $a^2 b \sim \Omega$.

The vacancy flux contributing to void growth will be twice the flux calculated from the spatial derivative of eqn. (5), since each grain boundary is fed by both of the adjacent diffusion fields on either side. Void growth rate will be;

$$\frac{\partial V(t)}{\partial t} = 2D_d \Omega_v w\tau \frac{\partial C(l_v)}{\partial x} = \frac{2D_d \sigma N_d \Omega_v (l_v w\tau)}{kT}$$

(6)

where w is the film width, and $\tau$ is the film thickness.

Since the reliability of an integrated circuit can be determined by a stress void which grows to a size on the order of the line width, stress induced void growth kinetics are an important engineering topic. Curiously, reliability engineers appear to have a passionate love affair with the Arrhenius relation, to the exclusion of entertaining all others. Unfortunately, as any metallurgy student can tell you, M. Nature does not always favor such simple kinetics. Stress voiding is one such process which does not cooperate. In order to relate to previous work, we will define kinetics in terms of the derivative of the logarithm of the void growth rate with the Metallizations temperature as one would for a purely thermally activated process.

During void growth, thermal stresses are continuously being relieved by the dislocation climb/vacancy diffusion process outlined above. Since stress appears as the driving force for dislocation climb, and therefore void growth in eqn. (3) stress relief must be considered when discussing void growth kinetics. Realizing that the stress will be reduced by vacancy diffusion to the void we need to rewrite eqn. (6) to incorporate this;

$$\frac{\partial V(t)}{\partial t} = \left( \frac{2D_d \Omega_v \beta}{kT} \right) \left[ \Delta \alpha (T_o - T) - \left( \frac{1}{l_v w\tau} \right) \int_0^t \frac{dV(t)}{dt} dt \right]$$

(7)

Realizing that the integral in eqn. (7) is simply the void volume, V(t);

$$\frac{dV(t)}{dt} + \Phi V(t) - \Phi \Psi = 0$$

(8)

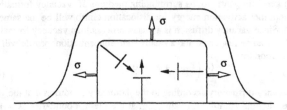

**Figure 1.** Schematic of cross section of typical passivated integrated circuit conductor. The differential thermal expansion provides for a nearly hydrostatic stress. Dislocations, possibly generated as a response to non-hydrostatic components of the thermal stress, interior to the grain are sources of vacancies contributing to void growth.

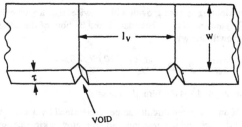

**figure 2.** Schematic of the geometry represented in the model. Voids are assumed to be nucleated at grain boundariesseperated by a distance, $l_v$, in a bamboo structure thin film w wide and $\tau$ thick.

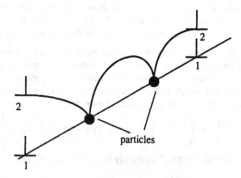

**Figure 3.** Schematic of possible mechanism to inhibit voids. small particles can inhibit the climb of dislocations and thereby slow the generation of vacancies required for void growth.

where

$$\Psi = l_v w \tau \Delta \alpha \Delta T$$

and

$$\Phi = \frac{2D_d \Omega_v \beta}{kT}$$

The void volume as a function of time is;

$$V(t) = V_\infty \left[1 - \exp(-\Phi t)\right] \tag{9}$$

where it is recognized that $\Psi$ is the ultimate void volume after infinite time, $V_\infty$. The change in $D_d$ with reduced stress has been ignored for computational simplicity and should not appreciably affect the solution. If it is assumed, incorrectly, that the void growth rate is purely thermally activated, an apparent activation energy can be obtained from the differentiation of the logarithm of the void growth rate with the reciprocal of the temperature.

$$\Delta H_{app} = \Delta H_f (1 - t\Phi) - \frac{kT^2}{(T_o - T)} \tag{10}$$

which hardly looks like an Arrhenius relation. It is dependent on the temperature of the observation, the temperature at which the passivation was applied and with time.

For finer grain sizes, the 1 dimensional problem above must be replaced by a solution to an appropriate 2 dimensional geometry, but will, in principle, be similar.

DISCUSSION

The preceding model correctly predicts the distinctly non-Arrhenius behavior observed in stress voiding. This is not unique to this model, of course. A distinct feature, however, is that it offers a mechanism for relaxation of the principally hydrostatic stresses observed in passivated thin films. The behavior predicted in this model suggests the possibility of producing a series of TTT type diagrams for films of various alloys passivated at different temperatures. TTT diagrams may be useful in deciding whether and to what extent a problem will exist given various thermal histories.

The major difference between this model and the others is that in this model, we are concerned with the generation and migration of vacancies to relieve the stresses induced by thermal expansion. The major determining parameter for void growth will be in the intervoid distance. Grain size, although of some consequence will be less important.

An important component of this model is role of dislocations in the void growth process. It is interesting that a recent UHVTEM study has observed a high dislocation density in grains with well developed voids. [11] One important feature of this model is that the perceived panacea of producing "bamboo" structure films will not work. Grain boundaries are not needed to transport metal atoms away from the void. This process could, in principle, be operating in a single crystal, providing voids could be nucleated. In fact, a scenario could be created where a bamboo film with fewer voids on grain boundaries of nearly parallel orientation would be a much greater reliability risk than a fine grain film with more voids. It is conceivable that void growth, with equal stress, could be faster in a bamboo film than in one with many grain boundary paths. A recent study suggests such an effect. [12]

The model also suggests methods for reducing the propensity for voiding. Dislocation climb is a structure sensitive process, in that it can be impeded by obstacles. Metallizations containing finely disbursed particles or precipitates should be less vulnerable to stress voiding. Sigma in eqn. (3) could be replaced by $\sigma - \sigma_p$, where $\sigma_p$ is a threshold stress required to free the climbing dislocation from the particle given by $\sigma_p = K_p$ (G b/$l_p$) where $K_p$ is a constant, G is the shear modulus and $l_p$ is the particle spacing. Opportunities for new alloy development considering this may be there, providing high conductivity is maintained.

The design of accelerated tests is an important part of any reliability engineer's job. These tests must be capable of providing information for conditions typical of use, yet must be obtained in a short time. It is, therefore, very important to fully understand the mechanism being investigated. With the plethora of stress void models proposed over the past few years, this is an uncomfortable task. Clever experimentalists must attack this problem and perform the "critical test" of the theories. Only then can we feel comfortable using any model for predicting reliability in integrated circuits.

## ACKNOWLEDGMENTS

One of the authors (JL) would like to thank the members of "Gruppe Arzt" at MPI in Stuttgart for stimulating discussions leading to the "Verfeinerung des Modells" and to the enlightened management at Digital Equipment Corporation (Dr. Maria Menendez and Paul Stoltze in particular) for encouragement and for supporting the collaboration that made this work possible.

## REFERENCES

1) J. Klema, R. Pyle and E. Domangue, Proc. 22nd Ann. Int'l. Reliab. Phys. Symp., IEEE, 1 (1984)

2) J. Curry, G. Fitzgibbon, Y. Guan, R. Muollo, G. Nelson and A. Thomas, 22nd Int'l Reliab. Phys. Symp. IEEE, 6 (1984)

3) J.T. Yue, W.P. Funsten and R.V. Taylor, Proc. 23rd Ann. Int'l. Reliab. Phys. Symp. IEEE 4 (1985)

4) T. Turner and K. Wendel, Proc. 23rd Ann. Int'l Reliab.Phys. Symp. IEEE, 142 (1985)

5) K. Hinode, N. Owada, T. Nishida and K. Mukai, J. Vac. Sci. Technol. B5, 518 (1987)

6) J.W. McPherson and C.F. Dunn, J. Vac. Sci. Technol. B5, 1321 (1987)

7) F.G. Yost, A.D. Romig, Jr. and R.J. Bourcier, Sandia Report SAND88-0946, Sandia National Laboratories (1988) and F.G. Yost, D.E. Amos and A.D. Romig, Proc. 27th Ann. Int'l. Reliab. Phys. Smp. IEEE, 193 (1989)

8) P.A. Flinn and C. Chiang, J. Appl. Phys, 67, 2927 (1990)

9) E. Arzt and W.D Nix, to be published in Met. Trans. (1992)

10 H. Kaneko, M. Hasanuma, A. Sawabe, T. Kawanoue, Y. Kohanawa, S. Komatsu and M. Miyachi, Proc. 28th Ann. Int'l Reliab. Phys. Symp. IEEE, 194 (1990)

11) A. Tanikawa, H. Okabayashi, H. Morii and H. Fujita, Proc. 28th Ann. Int'l. Reliab. Phys. Symp., 209 (1990)

12) D. Pramanik and V. Jain, Proceedings SPIE Vol. 1596, 132 (1991)

# ELECTROMIGRATION RELIABILITY SIMULATOR

J. Niehof, D.C.L. van Geest and J.F. Verwey
Faculty of Electrical Engineering, University of Twente
P.O. Box 217, 7500 AE Enschede
The Netherlands

## Abstract

In our development of a reliability circuit simulator, capable of indicating the most vulnerable spots for failure caused by electromigration, we use two levels: circuit and physical level. On circuit level a new approach towards reliability called the stressor/susceptibility method is used. On physical level 2D electromigration simulations can be performed on metallization structures. The atomic redistribution and the consequent resistance change due to electromigration can be calculated.

## Introduction

Quality analysis and quality optimisation are getting increasingly important, as demands for quality are increasing, while products are getting more complicated. There is a trend from adding reliability to a product towards building-in reliability by design. This means that the design must be robust against the conditions under which it will be produced and used.

Test-products will not be available yet in an early stage of the design, so CAD tools have to be used to describe the failure behaviour. This paper focuses on the failure mechanism electromigration and presents a method to translate physical failure behaviour to designable parameters on circuit level in order to make a design robust.

This method, called "stressor/susceptibility optimisation", is based on the susceptibility of physical failure mechanisms for external influence factors, called stressors [4]:

- A stressor $\Psi$ is a physical entity influencing the lifetime of a component or circuit. In case of electromigration the most important stressors are the current(-density) and the temperature.
- The susceptibility $S_\Psi$ of a component to a certain failure mechanism is defined as the probability function indicating the probability that a failure occurs under a given combination of stressors.

Due to variations in material (tolerances) and user conditions stressor values in a batch of circuits will not be identical. Figure 1 shows the probability density function of the stressor-values of a batch of circuits. The susceptibility represents for each stressor value the probability that the component fails. If the stressor probability function and the susceptibility probability function are overlapping, there is a chance that a failure occurs. To make a design robust the overlap of stressor and susceptibility must be minimised.

Figure 1 is based on a short-term failure mechanism where a failure occurs as soon as the failure mechanism is activated. In case of electromigration activation of the failure mechanism causes degradation and a failure does not occur until a certain degradation level is exceeded. The degradation model that has been used is explained in detail in the circuit level modeling part.

The research on incorporating electromigration into this methodology is divided into two parts, viz. circuit level and physical level (see figure 2).

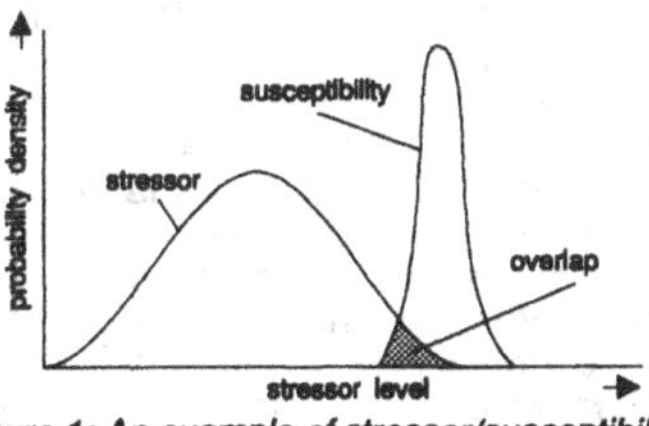

On circuit level the distributions of the stressors (current and temperature) are determined by performing circuit simulations. The circuit simulator that has been used, is Pstar (version 1.10) [6]. To get a proper result the simulation models must be detailed enough to give a good representation not only in normal operating modes but also in extreme operating modes as practical failures are most likely to occur in extreme operating

figure 1: An example of stressor/susceptibility

Mat. Res. Soc. Symp. Proc. Vol. 265. ©1992 Materials Research Society

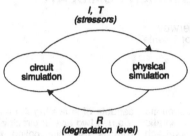

**I, T**
*(stressors)*

circuit simulation

physical simulation

**R**
*(degradation level)*

*figure 2: Circuit simulation and physical simulation*

modes. The same is true for the external conditions under which the circuit is used. The distributions of stressors from a batch of circuits are determined by Monte-Carlo analysis in which all variation is included. The model for the susceptibility and degradation behaviour is a simplified model based on the results of physical simulations. After determining the sensitivity of the amount of overlap of stressor and susceptibility for all designable parameters in the circuit, these parameters are modified towards minimum overlap and minimum sensitivity. This may be an iterative process. The software package used for optimisation is Minnie [7]. For more information about the optimisation algorithms the reader is referred to [8].

On physical level the degradation behaviour is simulated. The simulator used for this purpose is TRENDY [11]. The current density distribution in a metal line is simulated based on stressor values found in the circuit simulation. The atomic flux under these stressor conditions is then calculated giving rise to changes in the physical structure of the metal line. These changes can then be expressed in changes in the metallization resistance.

In this paper the models on circuit level and on physical level are discussed. Beside, examples are given on both levels. On circuit level the reliability of a clock-generator IC in a specific application is optimised, taking into account functional demands as well. On physical level results of simulations performed on a via structure are given.

## Circuit level electromigration modeling

The most important difference between short-term and long-term failure mechanisms, like electromigration, is that the susceptibility not only depends on the momentary stressor-value but on the degradation-level (determined by stressor-values in the past) as well.

In case of a stressor distribution that is constant in time and a degradation behaviour that is constant and equal for each individual circuit from the batch, this can be seen as a susceptibility function that moves to the left (see figure 3).

In practice however, stressor-distributions are not necessarily constant in time. Due to resistance increase the functional behaviour may depend on the degradation level and thus vary during the lifetime of the conductor. At the end of the lifetime the degradation behaviour may be different from the beginning as voids locally cause very high current densities that cause faster degradation. Due to variations in material properties the degradation behaviour will not be constant for each circuit from the batch.

To incorporate these effects in the stressor/susceptibility method it is necessary to introduce an extra dimension that represents the degradation level (see figure 4). At t=0 the degradation-level of all circuits from the batch equals zero. The degradation-level at a later time of each individual circuit depends on the stressor-value and the degradation behaviour of this individual circuit between t=0 and that later time.

In this example the susceptibility only depends on the degradation level. When the degradation exceeds the threshold $\beta$ the conductor fails with probability 1. In this example failures occur for some of circuits in the batch at t=$t_3$.

A more detailed degradation and susceptibility model where degradation and susceptibility depend on all known stressors and degradation level will be developed in future based on the results of the physical simulations.

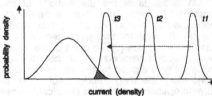

*figure 3: Electromigration stressor and susceptibility in case of constant stressor-distribution*

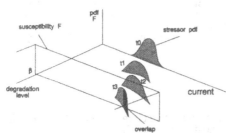

*figure 4: Electromigration stressor and susceptibility with degradation as an extra dimension*

To avoid simulations over very long time intervals, the degradation process can be calculated using a Markov approach. Based on the stressor and degradation distribution at $t=t_1$ the degradation at $t=t_2$ is calculated. Due to drift the electrical behaviour may change resulting in a different stressor distribution at $t=t_2$.

The possibility to take into account non-linear degradation occurring in a batch of circuits with variation in both electrical and degradation behaviour is an important advantage of this approach. Other electromigration simulators, like BERT [9], only calculate the TTF, based on an extrapolation at t=0.

The electromigration degradation model used for the simulations in this paper is a simplified model only depending on the current. As mentioned before other stressors and more detailed models are easy to implement and will be implemented in future. Figure 4 shows the susceptibility function of this model and possible stressor distributions on $t = t_0$, $t_1$, $t_2$ and $t_3$.

$$\Psi_{EM}(t) = \int_{t}^{t+\Delta} I^2 dt \qquad (1)$$

$$D_{\Psi_{EM}}(t_{n+1}) = D_{\Psi_{EM}}(t_n) + \alpha \Psi_{EM}(t_n)\frac{t_{n+1}-t_n}{\Delta} \qquad (2)$$

$$S_{\Psi_{EM}}(\Psi_{EM},D_{\Psi_{EM}}) = \begin{cases} 0 \ if \ D_{\Psi_{EM}} < \beta \\ 1 \ if \ D_{\Psi_{EM}} \geq \beta \end{cases} \qquad (3)$$

$$R_{conductor} = R_0 \cdot (1+D_{\Psi_{EM}}) \qquad (4)$$

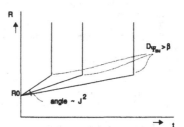

*figure 5: Resistance increase and failure under DC-conditions*

The stressor in this failure model $\Psi_{EM}$ is the integral of the square current in the time-interval [t, t+$\Delta$] (eq. 1). This interval must be large enough to contain all occurring current-values and small enough to have negligible degradation during this interval. The degradation level $D_{\Psi_{EM}}$ increases linearly with the stressor value (eq. 2). The conductor fails when the degradation level exceeds $\beta$, which appears from $S_{\Psi_{EM}}$ (eq. 3). The resistance increase due to degradation depends linearly on $D_{\Psi_{EM}}$ (eq. 4). In case of a conductor, stressed by a DC current, the resistance increases linearly with time (see figure 5). The slope of the curve is proportional with $J^2$. The conductor fails when the resistance has reached a certain threshold value, determined by $\beta$. This causes the time-to-failure to be proportional to $1/J^2$ which corresponds to literature when other influence factors like temperature are neglected [9,10].

## Physical level electromigration modeling

The tool into which the physical electromigration model is incorporated is the 2D process and device simulator TRENDY [11]. TRENDY is capable of simulating process-related steps such as ion-implantation, etching, deposition, oxide growth and diffusion. Moreover TRENDY is capable of performing two-dimensional device simulations. For this purpose a general partial differential equation solver has been developed which allows the implementation of any set of continuity equations of the following form:

$$\gamma H \frac{dC_k}{dt} - \nabla \cdot J_k + R_k = 0 \qquad k=1..N \qquad (5)$$

where k is the equation number, $C_k$ denotes the concentration of the $k^{th}$ species, $\gamma$ denotes the steady state (0) or transient (1) mode, H is the 'capacity' term, $J_k$ is the flux which is generally a function of $C_1..C_N$ and finally $R_k$ is a recombination/generation term.

Generally, device simulation, even the most complex ones, consist of solving a set of equations of this form. For our electromigration simulations, both the potential and the atomic density are solved simultaneously.

The atomic transport due to electromigration is modelled by using a Huntington-like formula for the atomic flux $J_a$ [12,13]:

$$J_a = -D\nabla N_{Al} + \frac{N_{Al}D}{kT} Z_b^* e \frac{1}{\sigma} \qquad (6)$$

where $Z_b^*$ effective charge number
  $N_{al}$ atomic density.

All other symbols have their usual meaning. No threshold current density term is used, since this is already accounted for by the gradient term in eq 6.

An atomic density distribution is assigned to metal strips. A density unequal to the normal material density implies material transport. A higher density means material surplus, a lower density the onset of the development of voids. The change in atomic density at a point due to a divergence in atomic flux must be in accordance with the mass continuity equation:

$$\frac{\partial N_{Al}}{\partial t} = -\nabla \cdot J_a \qquad (7)$$

This equation is used to calculate the redistribution of the atomic density after a certain period of time.

TRENDY allows us to perform electromigration simulations on 2D multi-metal (Al, W, Ti, TiW) metallization structures like vias and plugs. The stress distribution and the resistance increase due to the atomic redistribution can be calculated. From these distributions the time to void formation or the time for the resistance to increase a certain percentage can be calculated. Some first results were obtained on void formation simulations, this required changing the geometry of the metallization during simulation. Void formation and void growth can be simulated. This part of the simulator is still under development at the moment.

In the near future results from electromigration tests also performed on via and plug structures will be used to adjust the model parameters in order to match simulation results as close as possible to experimental results.

## Example circuit level: A clock-generator IC in its application

As an example the application of a clock-generator IC, susceptible to electromigration, was optimised. Figure 6 shows the circuit diagram. The output stage of the IC has been simulated on transistor-level including tolerances due to process variation. The IC is considered to be non-designable as this is fabricated by an external supplier. The IC has two outputs, one connected with one clock-in at a distance of 4cm, the other with three clock-ins at a distance of 20cm. Due to the high clock-frequency the lines have to be modelled as transmission-lines. To prevent reflections the lines are closed with the resistances $R_1 ... R_4$ These resistances have a tolerance of 20% and are designable. To obtain all occurring current-levels one total clock period was simulated.

Beside the demands for reliability (electromigration), the circuit has to fulfil its functional demands (clock-skew, signal levels, power and false pulses). The nominal values of resistances $R_1 ... R_4$ are optimised towards optimum reliability and optimum functional behaviour. In this example three iterations were done.

Figure 7 shows the nominal wave forms of the current through output 1 before and after optimisation. It is not necessarily true that the *nominal* circuit after optimisation gives the best reliability and functional behaviour. The effect of optimisation is that a *batch* of circuits, having tolerances on IC-parameters and resistance values, gives the best reliability and functional behaviour.

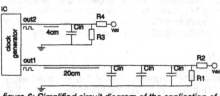

figure 6: Simplified circuit diagram of the application of the clock-generator IC

The optimisation algorithm divides the batch of circuits into two groups, one passing, and one failing functional and reliability specifications. For each designable parameter a so-called pass-fail diagram (see figure 8) gives a relation between the fraction of circuits passing all specifications for all values within the tolerance space. It is used to optimise the value of all designable parameters (e.g. increase R1 as higher values result in more passes). The advantage of this method above sensitivity analysis is that one monte-carlo simulation can be used for all designable parameters and correlations are incorporated.

| Parameter | Before | After |
|-----------|--------|-------|
| R1 | 100Ω | 180Ω |
| R2 | 390Ω | 390Ω |
| R3 | 100Ω | 100Ω |
| R4 | 390Ω | 560Ω |

| Yield | Before | After |
|-------|--------|-------|
| Electromigration | 69% | 100% |
| Clock skew | 35% | 75% |
| Total | 5% | 64% |

table 1: Results of optimisation

Figure 9 shows the distribution of the degradation-level over the batch. The fraction of the circuits in the batch that have not exceeded the susceptibility limit at the end of demanded lifetime has increased from 69% to 100% (see table 1:electromigration yield). One of the functional demands, the clock-skew, is given in figure 10. The fraction of the circuits meeting this demand has increased from 35% to 75% (see table 1: skew yield). The fraction of the circuits meeting all reliability and functional demands has increased from 5% to 64% (see table 1: total yield).

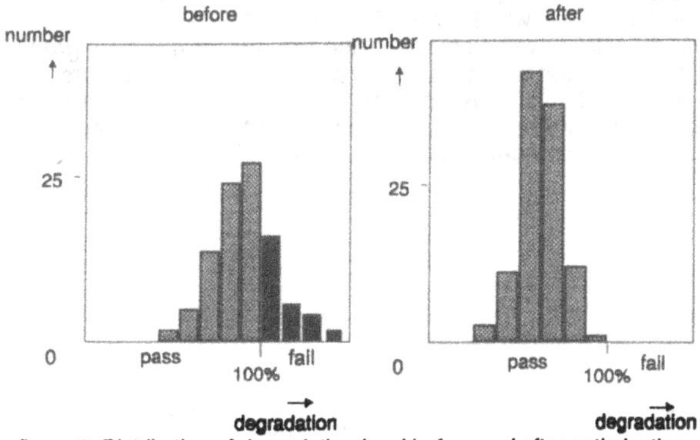

## Example physical level: Via structure

A simulation is performed on a via structure. The via structure together with the current density distribution is given in figure 11. This via structure is typically used for connecting different metal layer levels in multilayer interconnect structures. The simulation result is shown in figure 12, where the atomic density distribution after 2500 min. is given. It clearly shows that there is negligible atomic transport in W and, in agreement with [15], that the depletion of Al starts immediately at the W-Al interface. Another interesting result is the mass transport induced by the left current density peak in figure 11. The major part of this peak is situated in the W, therefore only little mass transport has occurred in that region (refer to figure 12).

figure 9: Distribution of degradation-level before and after optimisation

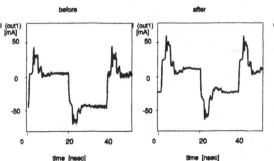

figure 7: Simulated nominal current before and after optimisation

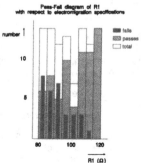

figure 8: A pass-fail diagram

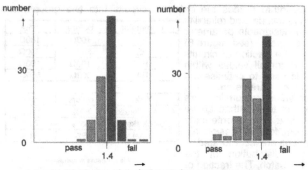

figure 10: Distribution of clock-skew before and after optimisation

## Conclusions

It is possible to optimise the reliability of a batch of IC's, susceptible to electromigration, in a specific application taking into account tolerances and functional demands

The stressor/ susceptibility method extended with an extra degradation-dimension is suitable for changing stress-distributions due to drift or non-identical degradation behaviour in a batch.

The electromigration module as implemented allows simulations of atomic transport in 2D metallization structures which can be made of Al, W, Ti or any other metal. The model gives much insight in the complex phenomena of electromigration by enabling us to study electromigration as a function of macroscopic parameters.

The approach as presented here is a first step in implementing an electromigration model. The incorporation of microstructural features of the materials (grain distribution, grain size) into the model is being developed at present.

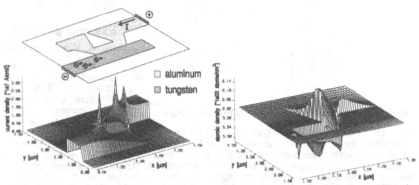

figure 11: Via simulation structure and current    figure 12: Atomic density distribution, t=2500 min

## Literature

[1] MIL-HDBK-217E, United States Department of Defense, 1987
[2] D.C.L. van Geest et al., Productronica 1991, München, Germany, November 1991
[3] P.D.T. O'Connor, Solid State Techn., August 1990
[4] A.C. Brombacher, J. Wiley & Sons, January 1992
[5] F. Jensen, IEEE Ann. Rel., Av. and Maint. Symp., Atlanta (GA), USA, 1989
[6] Philips Eindhoven, the Netherlands
[7] Interactive Solutions Ltd., Hammersmith, London W6 9LZ, UK (1990)
[8] R. Spence, R.S. Soin, Addison-Wesley, England, ISBN 0-201-18242-4 (1988)
[9] B.K. Liew, P. Fang, N.W. Cheung, C. Hu, IRPS 1990
[10] J.R. Lloyd, J. Kitchin, J. Appl. Phys. 69(4), 15 February 1991
[11] E. van Schie et al., Dublin: Boole, pp. 155-161, 1989.
[12] J.W. Harrison Jr, IEEE Tr. El. Dev., vol 35, no 12, pp. 2170-2179, 1988.
[13] P.J. Marcoux, et al., Hewlett-Packard Journal, pp. 79-84, June 1989.
[14] J. Tao, K.K. Young, N.W. Cheung, C. Hu, IRPS 1992, p338.

# VACANCY SUPERSATURATION MODEL FOR ELECTROMIGRATION FAILURE UNDER DC AND PULSED DC STRESS

J. J. Clement
Digital Equipment Corp., 77 Reed Road, Hudson, MA 01749

## ABSTRACT

Time-dependent vacancy concentration profiles are calculated numerically as solutions to the electromigration transport equation under dc and pulsed dc current stress conditions in finite-length conductors. An electromigration model based on the buildup of a critical vacancy supersaturation to initiate failure agrees well with several reported experimental observations.

## INTRODUCTION

Electromigration reliability testing is carried out at temperatures and currents which are much higher than those experienced at circuit operating conditions. In order to extrapolate the test results to operating conditions with confidence, a good physical model of the failure mechanism is essential. In early work, Black [1] proposed that the electromigration median-time-to-failure $MTF$ is related to the temperature $T$ and current density $j$ by

$$MTF = Aj^{-n} \exp\left(\Delta H/kT\right) \qquad (1)$$

where $A$ and $n$ are constants, $k$ is Boltzmann's constant, and $\Delta H$ is the activation energy. Black contended that $n = 2$ based on a simple electron-ion momentum exchange model, but critical re-examination of his argument left the current density dependence in question [2,3]. Nevertheless, the preponderance of experimental evidence has shown that, in the absence of Joule heating, $n = 2$, or nearly so.

An electromigration failure model based on the buildup of vacancies in the conductor at a blocking boundary under the opposing forces of electromigration and concentration gradient was shown by Shatzkes and Lloyd [4] to yield $n = 2$. In their model, the conductor is treated as semi-infinite continuum, and electromigration failure is assumed to depend on the buildup of vacancies at the blocking boundary to a critical level. While it permitted the derivation of an analytical solution for the vacancy buildup as a function of time, the semi-infinite boundary condition is unrealistic, and precludes a steady-state solution. In this study, the time-dependent vacancy distributions were determined using numerical calculation for a finite-length conductor with various boundary conditions.

It has been long known that a metal line will survive longer under repetitive pulsed current stress compared to constant current stress. That is, when the peak pulse current is the same as the dc current, the ratio of the $MTF$ for pulsed current to that for dc current stress will be larger than the inverse of the duty ratio $r$ $(0 < r \leq 1)$.

Several workers [5,6,7] have found a $r^{-2}$ dependence for this ratio in lifetimes. For $n = 2$, under pulsed current stress Equation (1) then becomes

$$MTF_p = Aj_{avg}^{-2} \exp\left(\Delta H/kT\right) \qquad (2)$$

where $j_{avg}$ is the average current density which is the product of the duty ratio $r$ and the peak pulse current density for rectangular pulses.

To explain this enhancement in lifetime under pulse current stressing, it is widely assumed that some of the electromigration damage introduced when the pulsed stress is applied anneals during the period that the pulsed current is off. For the model presented here, the vacancy buildup with time under pulsed current stress is calculated and found to be proportional to $\tau^{-2}$.

## MODEL

Assuming that the metal atoms migrate via vacancies, the flux of metal atoms is equal and opposite to the flux of vacancies. The vacancy flux is described by

$$\mathbf{J} = -D\nabla C + \frac{DC}{kT}Z^*e\mathbf{E} \tag{3}$$

where $C$ is the vacancy concentration, $D$ is the diffusivity, $Z^*e$ is the effective charge, and $E$ is the electric field. In one dimension this can be written

$$J = -D\frac{\partial C}{\partial x} - \frac{DZ^*e\rho j}{kT}C \tag{4}$$

where $\rho$ is the resistivity of the metal, and $E = \rho j$. The electric field was taken to be in the $-x$ direction.

The concentration as a function of space and time $C(x,t)$ can be obtained by solving the continuity equation

$$\frac{\partial C}{\partial t} + \frac{\partial J}{\partial x} = 0 \tag{5}$$

subject to the appropriate boundary conditions.

It is convenient to introduce the following dimensionless quantities for length, concentration, stress, and time:

$$\xi \equiv \frac{x}{l} \qquad \eta \equiv \frac{C}{C_0} \qquad \gamma \equiv \frac{Z^*e\rho jl}{kT} \qquad \tau \equiv \left(\frac{\gamma}{l}\right)^2 Dt \tag{6}$$

where $l$ is the conductor length, and $C_0$ is the equilibrium concentration in the absence of stress. Before the electromigration stress is applied, the vacancy concentration is assumed to be uniform at the equilibrium value, that is

$$C(x,0) = C_0 \qquad \text{or} \qquad \eta(\xi,0) = 1. \tag{7}$$

It is also useful to define a dimensionless variable for flux from Equation (4):

$$\phi \equiv \frac{Jl}{DC_0} = -\frac{\partial \eta}{\partial \xi} - \gamma\eta. \tag{8}$$

With these definitions, the continuity equation becomes

$$\gamma^2\frac{\partial \eta}{\partial \tau} - \frac{\partial^2 \eta}{\partial \xi^2} - \gamma\frac{\partial \eta}{\partial \xi} = 0. \tag{9}$$

The time-dependent vacancy concentration profiles were determined by numerically solving this equation for various boundary conditions using the simulation program PROMIS [8,9].

## DC STRESS

### Blocking boundary

Following the treatment of Shatzkes and Lloyd [4], the boundary at $x = 0$ was initially assumed to be completely blocking such that no vacancies are allowed to pass, that is

$$J(0,t) = 0 \qquad \text{or} \qquad \phi(0,\tau) = 0. \tag{10}$$

Two choices for the boundary condition at $x = l$ were examined initially [9]. In one case (Case A), this boundary is also taken to be completely blocking, that is

$$J(l,t) = 0 \qquad \text{or} \qquad \phi(1,\tau) = 0. \tag{11a}$$

This could represent the situation, for example, that both ends of the metal line are contacted to silicon or some other barrier material. Moreover, in this case the quantity of vacancies is conserved which would preclude changes in the volume of the conductor. This situation could be maintained in a system where the metal line is encased in a thick strong passivation.

In the other case (Case B), the vacancy concentration at the boundary is maintained at a constant level, that is

$$C(l,t) = C_0 \qquad \text{or} \qquad \eta(1,\tau) = 1. \tag{11b}$$

This could correspond to several possible situations. One example might be the case that one end of the conductor is attached to a unpassivated bonding pad, which could act as a reservoir of vacancies. This might also be a good representation of the situation when there is a void present, such as might be formed by stress relief in a passivated line, which could act as a vacancy source. These boundary conditions might also be a good approximation of an unpassivated aluminum line in which the thin native oxide blocks the surface from acting as a sink for vacancies, yet is not so strong as to prevent vacancy creation.

There is no tractable general solution for the concentration as a function of length and time that is known to us for Case B. However, the steady-state solution is given by

$$\frac{C(x,\infty)}{C_0} = \exp\left(\gamma(1 - \xi)\right). \tag{12}$$

A solution does exist for Case A in the form of an infinite series [10]:

$$\frac{C(x,t)}{C_0} = A_0 + \sum_{m=1}^{\infty} A_m \exp(-B_m \tau) \tag{13}$$

where

$$A_0 = \frac{C(x,\infty)}{C_0} = \frac{\gamma e^{-\gamma\xi}}{1 - e^{-\gamma}}$$

is the steady-state solution,

$$A_m = \frac{16m\pi\gamma^2[1 - (-1)^m e^{\gamma/2}]}{(\gamma^2 + 4m^2\pi^2)^2}\left(\sin m\pi\xi - \frac{2m\pi}{\gamma}\cos m\pi\xi\right)e^{-\gamma\xi/2}$$

and

$$B_m = (m\pi/\gamma)^2 + 1/4.$$

In Figure 1, the distribution of vacancies as a function of reduced length $\xi$ have been plotted at several values of the reduced time $\tau$ for $\gamma = 2$. Numerical solutions for Case A and Case B boundary conditions are represented by the symbols in Figure 1a and 1b, respectively. The solutions calculated from Equation (13) using the first 50 terms in the series are indicated by the solid lines in Figure 1a. The solid line in Figure 1b is the steady-state solution given by Equation (12).

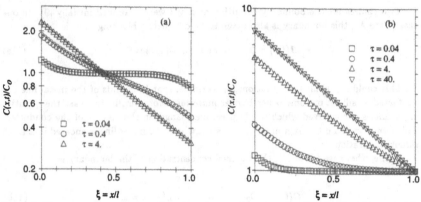

Figure 1. Vacancy concentration profile vs length at various times for $J(0,t) = 0$ with $\gamma = 2$. (a) Case A, $J(l,t) = 0$. (b) Case B, $C(l,t) = C_0$.

The buildup in the vacancy concentration at the blocking boundary at $x = 0$ versus the reduced time $\tau$ for different values of reduced stress $\gamma$ is shown in Figure 2. The open symbols are solutions for Case A and the filled symbols are for Case B. There are several features in this figure to note.

The first is that for a finite length $l$ the concentration buildup will saturate. The level at which this saturation occurs depends on $\gamma$, that is the stress conditions, and, therefore, on the stress current $j$ and the length $l$. The saturation level decreases with decreasing $j$ and $l$.

In addition, the choice of boundary conditions affects the level of the saturation; the saturation level is lower for Case A than for Case B. This can be interpreted as confirmation of the effectiveness of a strong passivation layer in increasing the electromigration lifetime, which has been well established [11,12].

If it is assumed that the vacancy concentration at the blocking boundary at $x = 0$ must reach a critical value to intiate void formation and failure, then it follows that a threshold value of $jl$ is required to initiate void formation [13]. This agrees with results of edge displacement measurements [12,14].

In the region below saturation, all of the curves in Figure 3 are coincident, which means that the critical vacancy buildup corresponds to a single value $\tau_{crit}$ in this region. Therefore, from Equation (6) the time to reach a critical vacancy concentration to initiate failure $t_f$ is proportional to $j^{-2}$, that is $t_f \propto j^{-2}\tau_{crit}$. In other words, if $jl$ is large enough to reach the critical vacancy concentration to cause failure we should see a $j^{-2}$ dependence to initiate failure.

Also shown in this figure is the solution for the semi-infinite case derived by Shatzkes and Lloyd [4,15]. This agrees very well with the solutions obtained for finite-length lines in the region below saturation. This is not too surprising, as one would expect that the vacancy buildup at this blocking boundary at $x = 0$ initially would not depend on the boundary condition at $x = l$.

### Permeable boundary

Until this point the metal line has been modeled as a continuum in one-dimension in

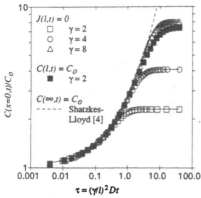

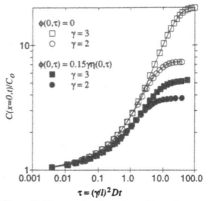

Figure 2. Vacancy buildup at $x = 0$ as a function of reduced time $\tau$ showing $j^{-2}$ dependence in the region below saturation where the curves are coincident, and saturation level as a function of $\gamma$ and boundary conditions.

Figure 3. Vacancy buildup at $x = 0$ as a function of reduced time $\tau$ with $C(l,t) = C_0$ comparing a perfectly blocking barrier to a permeable barrier.

which $l$ is the line length. In this simple view, electromigration failures would occur near the ends of the line. We know from experience, however, that the line often fails at a random site along its length. At typical test and use temperatures the electromigration mass transport in metallic thin films proceeds predominately along the grain boundaries [16]. Atomic mass flux divergences can occur at various points in the conductor due to the variation in film microstructure.

A metal line, therefore, might be envisioned as a serial combination of segments with permeable or leaky boundaries [17]. With our numerical modeling tool, we can examine the vacancy buildup as a function of time at a permeable boundary. At $x = 0$, we let

$$J(0,t) = f\left(\frac{Z^* e \rho j}{kT}\right) D C(0,t) \quad \cdot \quad \text{or} \quad \phi(0,\tau) = f\gamma\eta(0,\tau) \qquad (14)$$

where $0 \le f \le 1$, and, at $x = l$, we use the boundary condition given in Equation (11b).

In Figure 3 the filled symbols indicate the results obtained with these boundary conditions for $f = 0.15$, which can be compared to those from Case B with the completely blocking boundary shown by the open symbols. A couple of points should be noted.

The level at which the vacancy buildup saturates decreases as the blocking barrier becomes less effective. In addition, the vacancy buildup with time is slowed although there is still a $j^{-2}$ dependence below saturation. Not unexpectedly, this confirms that the more permeable barrier is less likely to be a failure site.

## PULSED CURRENT STRESS

A periodic pulsed dc stress can be described by

$$\gamma = \begin{cases} \gamma_{peak} & \text{for } mP < t \le mP + t_{on} \\ 0 & \text{for } mP + t_{on} < t \le (m+1)P \end{cases} \qquad m = 0, 1, 2, \ldots$$

where $P$ is the period, and $t_{on}$ is the pulse width. The duty ratio $r = t_{on}/P$. When $\gamma_{peak}$ is held constant as $r$ is varied, several experimenters have found a $r^{-2}$ dependence on lifetime, as noted earlier. It will be useful, therefore, to define a normalized dc-equivalent time $r^2\tau$.

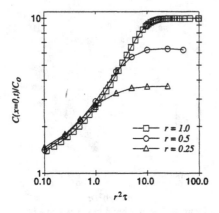

Figure 4. Vacancy buildup as a function of normalized time $r^2\tau$ for $\gamma = 10$ showing $r^{-2}$ dependence in the region below saturation where the curves are coincident, and saturation level as a function of $r$.

While the current stress is applied, the vacancy concentration will build up at the blocking boundary at $x = 0$. When the stress is removed, the vacancy concentration will start to relax toward equilibrium conditions. Numerical calculations were performed to compute the vacancy concentration distribution in the conductor alternating between having the current stress on, then off. This procedure was repeated many times to simulate pulsed current stressing. For each calculation the resulting distribution from the previous calculation was used as the initial condition.

Obtaining results over several orders of magnitude in time presents a problem. The pulse width must be small compared to the time of interest so that the results are "averaged" over several stress cycles, yet we want the width to be as large as possible in order to minimize computation time. To meet these objectives, the pulse width was increased in roughly logarithmic steps during the simulation. The pulse width was chosen such that the pulse "on" time $t_{on}$ was typically less than 5% of the normalized time $r^2\tau$.

The simulation results are presented in Figure 4 in which the vacancy concentration at the blocking boundary at $x = 0$ is plotted versus the normalized time $r^2\tau$ for three values of $r$. These results were calculated for the case that both boundaries $(x = 0, l)$ are completely blocking. There are a couple of features in this figure which deserve comment.

First, the level at which the curve saturates decreases with decreasing $r$. This is similar to the effect of decreasing $j$ and/or $l$. This implies that there is a threshold value of $r$ as well as $jl$ required to initiate failure.

However, the curves are nearly coincident below saturation. Therefore, the time to reach a critical vacancy concentration is proportional to $r^{-2}$ in this region. This agrees with the experimentally observed dependence of electromigration on duty ratio [5,6,7].

## CONCLUSION

Under dc stress conditions:

- The electromigration-induced vacancy concentration buildup at a blocking boundary saturates with time for finite length $l$. The level at which vacancy concentration buildup saturates decreases with decreasing $j$ and $l$.
- A threshold value of $jl$, therefore, is required to cause failure, assuming that the vacancy concentration at the blocking boundary must reach a critical value to initiate void formation. This agrees with results of edge displacement experiments.

- For a permeable barrier the vacancy buildup is slower than that for a perfectly blocking boundary, and the buildup saturates at a lower level.
- The vacancy buildup as a function of time has a $j^{-2}$ current density dependence below saturation.

Under pulsed dc stress conditions:

- The level at which vacancy concentration buildup saturates decreases with decreasing duty ratio $r$. This is similar to the effect of decreasing $j$ and/or $l$ and implies that there is a threshold value of $r$ as well as $jl$ required to initiate failure.
- The vacancy buildup with time is proportional to $r^{-2}$ below saturation in agreement with the empirically derived "average current" model.

### Acknowledgement

It is a pleasure to thank Prof. Carl Thompson of Massachusetts Institute of Technology and Jim Lloyd for some stimulating discussions.

## REFERENCES

1. J. R. Black, Proc. 6th Annual International Reliability Physics Symposium (IEEE, New York, 1967), p. 148.
2. G. L. Hoffman and H. M. Breitling, Proc. IEEE (Lett.), **58**, 833 (1970).
3. J. C. Blair, P. B. Ghate, and C. T. Haywood, Proc. IEEE (Lett.), **59**, 1023 (1971).
4. M. Shatzkes and J. R. Lloyd, J. Appl. Phys. **59**, 3890 (1986).
5. J. M. Towner and E. P. van de Ven, Proc. 21st Annual International Reliability Physics Symposium (IEEE, New York, 1983), p. 36.
6. L. Brooke, Proc. 25th Annual International Reliability Physics Symposium (IEEE, New York, 1987), p. 136.
7. J. A. Maiz, Proc. 27th Annual International Reliability Physics Symposium (IEEE, New York, 1989), p. 220.
8. P. Pichler, W. Jüngling, S. Selberherr, E. Guerrero, and H. Pötzl, IEEE Trans. Computer Aided Design **CAD-4**, 384 (1985).
9. J. J. Clement and J. R. Lloyd, J. Appl. Phys. **71**, 1729 (1992).
10. S. R. deGroot, Physica **9**, 699 (1942).
11. J. R. Lloyd and P. M. Smith, J. Vac. Sci. Technol. **A1**, 455 (1983)
12. C. A. Ross, J. S. Drewery, R. E. Somekh, and J. E. Evetts, J. Appl. Phys. **66**, 2349 (1989).
13. R. Kirchheim and U. Kaeber, J. Appl. Phys. **70**, 172 (1991).
14. I. A. Blech, J. Appl. Phys. **47**, 1203 (1976).
15. There is a typographical error in the solution given in Reference [4].

$$\frac{C(0,t)}{C_0} = 1 + 2\left(\beta^2(1 + \mathrm{erf}\beta) + \frac{1}{2}\mathrm{erf}\beta + \frac{\beta}{\sqrt{\pi}}\exp(-\beta^2)\right)$$

is the correct solution, where $\beta = \sqrt{\tau}/2$.
16. F. M. d'Heurle and P. S. Ho, in *Thin Films - Interdiffusion and Reactions*, edited by J. M. Poate, K. N. Tu, and J. W. Mayer (Wiley, New York, 1978), p. 243.
17. J. R. Lloyd and J. Kitchin, J. Appl. Phys. **69**, 2117 (1991).

# RELIABILITY OF INTERCONNECTS EXHIBITING BIMODAL ELECTROMIGRATION-INDUCED FAILURE DISTRIBUTIONS

H. Kahn, C.V. Thompson, and S.S. Cooperman*, Department of Materials Science and Engineering, Massachusetts Institute of Technology, Cambridge, MA 02139
*presently at Digital Equipment Corporation, Hudson, MA 01749

## ABSTRACT

Bimodal electromigration-induced open failure time distributions are reported for both single-level and double-level interconnect structures due to inherent microstructural variations. For flat interconnect lines, the bimodality arises when some lines in a test population have only bamboo boundaries, and other lines contain some triple junctions as well as bamboo boundaries. For via structures, two failure mechanisms result if some of the vias are not in contact with grain boundaries in the underlying metal, and others are. For both lines and vias, a large apparent deviation in failure times, $\sigma$, is a signal of a possible bimodal distribution, which can also be confirmed by microscopic examinations, resistance measurements, or an investigation of the metal microstructure. Insufficient testing can result in an incorrect asumption of a monomodal failure distribution, leading to either optimistic or pessimistic reliability predictions, depending on the size of the test sample population and the degree to which the test population is representative of the actual overall population.

## INTRODUCTION

Accurate reliability predictions can only be made if the distribution of failure times for the limiting failure mechanism is properly determined. An incorrect selection of the most critical failure mechanism, as well as the incorrect assumption of a monomodal distribution for bimodal failures will lead to imprecise projections. Bimodal failure distributions have been previously reported for flat interconnect lines which had been annealed such that some of the lines contained both large and small grains, while the rest of the lines had only large grains.[1] For typical industrial processes, as devices shrink, the average metal grain sizes become larger than the line and via dimensions. This results in distinct microstructural differences, which can lead to distinct mechanisms for electromigration-induced failures. For flat lines, this is manifested by some lines in a test population having a complete bamboo structure, and some lines containing a a few triple junctions. For vias, a bimodal failure distribution can be obtained if a fraction of the vias lies on top of grain boundaries in the underlying metal, and the others lie within a single grain.

## EXPERIMENTAL PROCEDURE

For accelerated lifetime tests on flat interconnect lines, multiple-line test structures[2] were fabricated from a 0.75 μm thick Al-2%Cu-0.3%Cr alloy film. This material has been shown to exhibit abnormal grain growth,[3,4] leading to very large grain sizes. The as-deposited films were annealed at 550°C for 20 minutes, to achieve a median grain size of 20 μm. The test structures were then patterned and etched, and a second anneal at 560°C for 45 minutes was performed to remove any triple junctions and to lead to rotation of the grain boundaries, to produce bamboo microstructures. The lines were stressed under constant voltage at 305°C, at a current density of $1.2 \times 10^6$ A/cm$^2$, while the stress current was monitored to detect electromigration-induced open circuit failures.

**Mat. Res. Soc. Symp. Proc. Vol. 265. ©1992 Materials Research Society**

For tests on electromigration-limited via reliability, the 2-level metal multiple-line skewed-Kelvin structure, described previously,[5] was used. The first-level metal consisted of the same 0.75 μm Al-2%Cu-0.3%Cr alloy, which was subjected to a 20 minute, 540°C anneal to produce a grain size distribution (before patterning) with a median of 5.8 μm and a lognormal standard deviation of 0.58. The interlevel dielectric was a 1.3 μm polyimide layer, and the second-level metal was a 2.2 μm tungsten layer on a 0.1 μm layer of $Ti_{0.1}W_{0.9}$. The vias were 2.8 μm in diameter. The first-level metal lines were 3.8 μm wide, and the second-level lines were 10 μm wide. The structures were stressed at 290°C, at a via current density of $4.8 \times 10^5$ A/cm$^2$ (corresponding to a curent density in the first-level metal lines of $1.0 \times 10^6$ A/cm$^2$), with the first-level positively biased, so that the electron flow was down through the via, and the voiding occured in the first-level metal.

## RESULTS AND DISCUSSION

### Flat interconnect lines

Open failure time results for an eleven line interconnect test structure with a line width of 1.85 μm are plotted on a lognormal graph in Fig. 1. The extracted median failure time, $t_{50}$, is 76 hours, and the lognormal deviation in failure times, $\sigma$, is 1.5. There appears to be a distinct separation between two sets of failures in Fig. 1, and the extracted $\sigma$ is anomolously high. Tests on similar samples typically exhibited $\sigma$'s around 0.6.[6] Microscopic examination of the failure sites revealed that four of the failures resulted from electromigration voiding at triple junctions which were not eliminated during the high temperature anneal. The other seven lines contained no triple junctions, and the morphology of their voids was characteristic of bamboo failures. Sketches of the two different microstuctures are included in Fig. 1. It is probable that the four earlier failures were those which resulted from triple junctions.

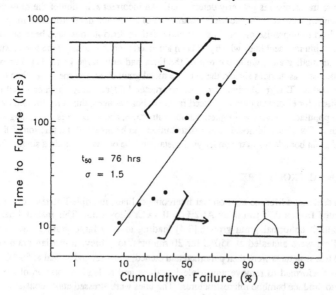

**Figure 1.** Open failure time results for flat interconnect lines, including sketches of the two different microstructures represented.

The failure times in Fig. 1 are properly replotted in Fig. 2 as two separate distributions. (The incorrect single distribution is included in the figure as circles.) The extracted σ's are now much more reasonable. The x-axis in Fig. 2 is extended to very low cumulative failures (which will be most important in terms of reliability), and the fitted lines are extrapolated to 0.01 percent failure. The effect of the bimodality in failure times is obvious. The extrapolation assuming a single distribution would predict an early failure time of about 0.3 hour, while the correct fits project early failure times of 4 and 15 hours for the two different mechanisms. In other words, if the bimodal distribution is incorrectly assumed to be monomodal, the reliablity prediction will be excessively pessimistic. Since the opens due to triple junctions occur faster than the bamboo failures, the appropriate prediction for the 0.01 percent failure time is 4 hours.

The results for an identically processed ten line structure with a line width of 1.15 μm are plotted in Fig. 3. Partially due to the narrower line width, and also due to chance, none of these lines contained triple junctions, and the prediction for the 0.01 percent failure time is around 20 hours. Reliability predictions for a metallization taken from this test would be overly optimistic. The second failure mechanism applicable to these structures - the one initiated by triple junctions, which will exhibit lower failure times - is not accounted for. (Even though the appearance of triple junctions depends on line width, and the probability of finding a triple junction is lower in the narrower lines, they will still exist if enough lines are sampled.) In addition, the two failure mechanisms demonstrate different temperature dependences. The true bamboo failures exhibit a higher activation energy than those failures resulting from grain boundary triple junctions in the near-bamboo lines.[6] As a result, at low operating temperatures, the near-bamboo failure lines will be much less reliable than the true bamboo lines. Therefore, predictions using only the data for lines without triple points, as demonstrated in Fig. 3, will be even more improperly optimistic.

## Vias

Electromigration-induced open failure times for the 2-level metal structures are shown in Fig. 4. As was seen for the flat line results, the extracted σ is very high, indicating a bimodal distribution. The skewed-Kelvin structures used in this test included two contact pads for each

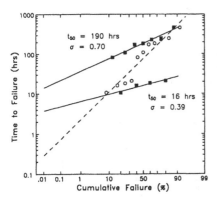

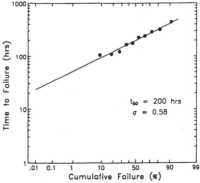

Figure 2. The open failure times for flat lines, plotted correctly as two separate distributions and incorrectly as a monomodal distribution (circles), and extended to low failure percent.

Failure 3. Open failure results for a flat line structure with a line width of 1.15 μm. Note that none of the lines exhibited failures due to triple junctions.

metal level. As a result, resistance measurements between various pairs of contact pads can be made to determine the open failure site. Of the nineteen failures plotted in Fig. 4, eleven occured at the via, while eight occured in the first-level metal line. These two separate mechanisms can be explained by microstructural differences between the lines.

The Al-Cu-Cr film from which the metal-1 lines were patterned exhibited a grain size distribution with a median grain diameter larger than the line width and also larger than the vias. As a result, the microstructure of the lines was near-bamboo - i.e., the lines contained both bamboo boundaries and triple junctions. Two different cases are possible for these structures, which are sketched in Fig. 4. If the vias lie directly over grain boundaries, the open failure will take place beneath the via. However, if the vias lie within a single bamboo grain, electromigration-induced voiding will most readily occur at a site along the interconnect where triple junctions exist. Since the via is the natural site of highest flux divergence, those lines with grain boundaries beneath the via will fail faster than those lines without, which must then fail along the interconnect. Like the flat line results, the 2-level structure failure times are properly re-plotted as separate distributions in Fig. 5, along with the incorrect monomodal distribution and its coresponding erroneous reliability prediction.

It should be noted that the fraction of vias lying within a single grain could have been predicted by comparing the continuous film grain size distribution and the via diameter. This fraction can be estimated from the ratio of the available area for grain boundary-free vias to the total film area. The available area, assuming circular grains, is given by

$$\text{Available Area} = \frac{\pi}{4} \int_{d_v}^{\infty} (d_g - d_v)^2 f(d_g) dd_g = \frac{\pi}{4} \int_{d_v}^{\infty} \frac{(d_g - d_v)^2}{\sigma_g d_g \sqrt{2\pi}} \exp\left[-\frac{1}{2}\left(\frac{\ln d_g - \ln d_g^*}{\sigma_g}\right)^2\right] dd_g , \quad (1)$$

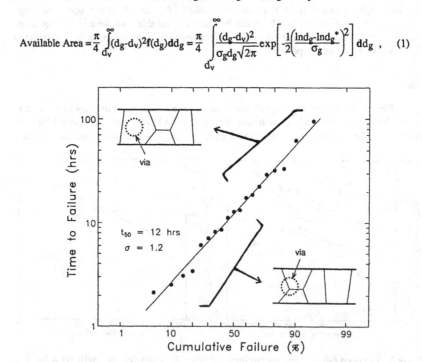

Figure 4. Open failure results for 2-level metal via structures, including sketches of the two different microstructures leading to two failure mechanisms - via and line failure.

where $d_g$ is the grain diameter, $d_v$ is the via diameter, $d_g^*$ is the median grain diameter, and $\sigma_g$ is the standard deviation of the grain diameters; and the total area is

$$\text{Total Area} = \frac{\pi}{4} \int_0^\infty d_g^2 f(d_g) dd_g = \frac{\pi}{4} \int_0^\infty \frac{d_g}{\sigma_g \sqrt{2\pi}} \exp\left[-\frac{1}{2}\left(\frac{\ln d_g - \ln d_g^*}{\sigma_g}\right)^2\right] dd_g \ . \tag{2}$$

This fraction can be calculated numerically, and for the 2.8 μm vias, this turns out to be 0.39. In other words, given measurements for $d_g^*$ and $\sigma_g$ for the film microstructure, it is predicted that 39 percent of the 2-level metal failures will occur in the lines, and 61 percent will occur at the vias, which is very similar to the actual results.

In addition, results for electromigration tests on the first-level metal lines only (This is possible with the skewed-Kelvin structure.[5]) are included in Fig. 6. These results indicate that the line failures - the more reliable of the two distributions represented in the 2-level metal tests - should have a $t_{50}$ close to 28 hours and a $\sigma$ around 0.67. If no other information regarding the fraction of each type of failure were available, this result would allow an appropriate separation between the two distributions. A portion of failure times at the higher end could be re-plotted as a separate distribution, varying the number of failure times taken, until the distribution exhibited a $t_{50}$ and $\sigma$ close to the correct values for line failures.

## CONCLUSIONS

Bimodal failure distributions are reported for both flat line and via structures due to inherent microstructural differences. For both cases, a large apparent $\sigma$ is a signal of a possible bimodal distribution. Microscopic examinations or resistance measurements can be used to confirm variations in failure mechanism. These variations can also be predicted by a characterization of the film microstructure or a comparison with the failure statistics of a known system, such as comparing flat line tests with multilevel structures.

Insufficient testing can lead to an incorrect assumption of a monomodal failure distribution. Two possible cases exist: i) if the limiting failure mechanism does not occur in the

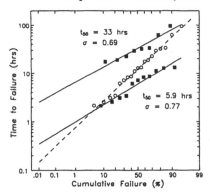

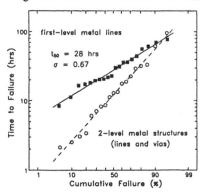

Figure 5. The open failure times for 2-level structures, plotted both as two distributions and as a single distribution, and extended to low failure percent.

Figure 6. Open failure times for the 2-level structures, compared to open failure time results obtained from stressing the first-level lines by themselves.

test sample population, optimistic reliability predictions will be made, as seen in Fig. 3; ii) if both failure mechanisms are present but are not recognized as being independent, pessimistic predictions will be made, as seen in Fig. 1 for lines and Fig. 4 for 2-level structures. As device dimensions shrink, the interconnect line widths and via diameters will be increasingly smaller than the metal grain sizes. As a result, not only will bimodal failure distributions be likely, but the less reliable failure mechanism (associated with grain boundary triple junctions) will become increasingly rare. This will necessitate more extensive testing in order to obtain a representative sample population. Sufficient testing and complete microstructural characterization are necessary to accurately predict the electromigration behavior and reliability of a metallization system.

This work is sponsored by the Semiconductor Research Corporation. Additional support is provided by Digital Equipment Corporation.

## REFERENCES

1. J. Cho and C.V. Thompson, J. Electronic Materials, **19**, 125 (1990).
2. C.V. Thompson and J. Cho, IEEE Elec. Dev. Lett., **EDL-7**, 667 (1986).
3. F. d'Heurle and I. Ames, Appl. Phys. Lett., **16**, 80 (1970).
4. H.P. Longworth and C.V. Thompson, J. Appl. Phys., **69**, 1 (1991).
5. H. Kahn and C.V. Thompson in Materials Reliability Issues in Microelectronics, edited by J.R. Lloyd, F.G. Yost , and P.S. Ho (Mater. Res. Soc., Pittsburgh, PA 1991) pp. 15-20.
6. S.S. Cooperman, MS thesis, Massachusetts Institute of Technology, 1992.

## PART II

# Microstructure and Electromigration

# MICROSTRUCTURE EVOLUTION
# IN METAL INTERCONNECTS

D. A. SMITH, T. KWOK, M. B. SMALL and C. STANIS, IBM Research Division,
Thomas J. Watson Research Center, P.O. Box 218 Yorktown Heights, New York 10598

## ABSTRACT

Grain growth is the key process in the development of the microstructure of deposited metal films. Grain boundaries in turn exert a dominant influence on the properties of interconnects in integrated circuits. In addition to the practical importance of the phenomena, grain growth in films is of considerable fundamental interest because of the possibility of comparing the predictions of models with observations. The thermodynamic basis for the limiting grain size resulting from grain growth in sheets (or films) was laid by Mullins and subsequently extended by Walton et al. to the case of interconnects. In their treatments grain growth stagnates when the forces resulting form the grain boundary excess energy are balanced by the pinning force resulting from grooving. Surface energy anisotropy can drastically modify this equilibrium and even lead to the result that there is no limiting grain size. The basic phenomenology of grain growth in films has been modelled and investigated experimentally. In situ experiments provide a direct comparison with models and confirm many of the predictions based on models. Specifically in the context of lines the effects of grooving on the top and side surfaces are predicted to result in a critical ratio of width to thickness above which a fully bamboo structure cannot be achieved. This is verified by observations.

## INTRODUCTION

The performance of metal interconnects in the context of high current densities and mechanical stress is a classical structure property issue. The demands of ULSI in the sub-micron regime require the understanding of properties and control of microstructure in a situation where one or more of the characteristic dimensions of the sample are of the same order as the grain size of the material. Consequently the effects of finite size and of surfaces become as important as the intrinsic bulk properties. However the grain structure remains the first order factor governing the properties of a material. The constitutive laws for grain boundary dominated properties all involve some power of the grain diameter, ( reviewed in [1] ), so that control of the grain structure offers direct leverage on the properties. From a practical point of view there are considerable complications because the grain structure evolves during deposition and processing [2]. The practical issue is to define a processing sequence, consistent with other manufacturing constraints and which will lead to the reproducible formation of a desirable grain structure. "Desirable" means stable under service conditions, homogeneous and, for reliability purposes, free of divergences. It has been found that divergences in the grain structure can correlate with a decrease in electromigration resistance characterized as either or both of a decrease in mean time to failure [3] or an increase in its standard deviation [4]. In the language of physical metallurgy a "desirable" grain structure means a narrow grain size distribution with the grains having similar morphologies and separated by boundaries having similar properties such that the microstructure is in local equilibrium i.e. stagnant.

## DRIVING FORCES FOR GRAIN GROWTH

Grain boundaries in a metal are never thermodynamically stable defects i.e. there is no equilibrium concentration of grain boundaries, in contrast to the case of point defects. Grain growth of a material proceeds so as to decrease the interfacial excess energy associated with the grain boundaries ( reviewed in [5] ). The grain boundary excess energy results in a surface tension which in turn can lead to grain boundary curvature; a consequence of this curvature is that a chemical potential gradient is set up biassing atomic motion from the concave side of a boundary to the convex side. In the simplified context of isotropic grain boundary energy in two dimensions, on average, grains with five or fewer sides shrink and those with seven or more sides grow. A microstructure consisting entirely of hexagonal grains in such an idealized material would be in local equilibrium i.e. stagnant. However there are various combinations of kinetic and thermodynamic factors which can result in the stagnation of grain growth. Essentially a grain boundary ceases to move when the driving forces are balanced by the dragging forces. An example which has been analyzed relatively rigorously and dynamically is the case of pinning by surface grooving. Mullins [6] treated the geometries and kinetics of formation of the grooves produced at the intersection between a free surface and (a) a static grain boundary and (b) a moving grain boundary. For a moving boundary the groove profile is asymmetrical; a steady state configuration is possible for which the groove and boundary migrate together. Deviation from steady state results in either pinning or breakaway from the groove. Mullins derived a stagnation condition for grain growth by balancing the curvature driving force against the grooving pinning force. His result was the prediction of a limiting grain size in thin films; $D \approx h$: (D is the grain diameter and h is the film thickness).

Walton et al. [7,8] extended the work of Mullins to include pinning by the grooves formed where grain boundaries cut the sides of lines. Their main result is the prediction that there is a critical ratio of the line width w to the line height h above which stagnation occurs before a fully bamboo structure is developed. The critical ratio w/h is estimated to be between 2 and 3 for a pure metal. In the context of a films and interconnects there are other major influences on the grain boundary movement. These are grain to grain variations in surface energy, alloy additions in the form of partitioned solute atom concentrations or precipitates and constraints arising from passivation of the substrate. It is useful to make a simple quantitative assessment of the orders of magnitude of the predominant driving forces which may act on a grain boundary. A change in free energy accompanies grain boundary migration; this energy may be thought of as the product of a force / area acting on the grain boundary and the area swept by the boundary ( Stuwe in [5] ). This is a mean field approach based on a static analysis. Consequently the results are limited to self-similar structures and specifically do not apply to contexts such as the transition from three dimensional growth to two dimensional growth. Table I gives expression for the the major forces acting on a columnar grain boundary in a film. .

In the table $y_b, y_{sv}$ and $y_{su}$ are the surface energies of grain boundaries, the solid vapor interface and the solid substrate interface respectively; h is the film thickness; $\Delta$ is the change in the quantity which follows; D is the grain diameter; d is the precipitate diameter and f the volume fraction of precipitate and a is the atom diameter.

**Table I**

FORCES ON A COLUMNAR GRAIN BOUNDARY IN A FILM

| Grain boundary energy | $\dfrac{2\gamma_b}{D}$ |
|---|---|
| Surface energy variation | $\dfrac{\Delta\gamma_{rv} + \Delta\gamma_{su}}{h}$ |
| Precipitates | $\dfrac{4\gamma_b f}{d}$ |
| Segregation | $\dfrac{4\gamma_b C_b}{a}$ |
| Grooving | $\dfrac{\gamma_b^2}{h\gamma_{rv}}$ |

In a one component thin film, with columnar grains, stagnation of grain growth occurs when the curvature and surface energy driving forces balance the retarding force arising from grooving. The Mullins stagnation condition is now modified by the addition of a term $\dfrac{\Delta\gamma_{rv}}{h}$ and a stagnation condition may be written as follows:

$$\frac{2\gamma_b}{D} + \frac{\Delta\gamma_{sv}}{h} = \frac{\gamma_b^2}{h\gamma_{sv}} \tag{1}$$

With the order of magnitude approximations that $3\gamma_b = \gamma_{rv}$ and $\Delta\gamma_{rv} = 1/3$ there is no limiting grain size. The surface energy term is equal in magnitude to the groove pinning term; in this example both grooving and surface energy anisotropy at the substrate-deposit interface have been neglected. Table II gives expressions for the major forces acting on a grain boundary in a line where they differ from the case of the columnar grain.

**Table 2**

FORCES ON A GRAIN BOUNDARY IN A BAMBOO STRUCTURE

| Grain boundary energy | $\dfrac{\gamma_b}{D}$ |
|---|---|
| Surface energy variation | $\dfrac{\Delta\gamma_r}{h} + \dfrac{2\Delta\gamma_z}{w} + \dfrac{\Delta\gamma_{su}}{h}$ |
| Grooves | $\dfrac{\gamma_b^2}{\gamma_{rv}}\left[\dfrac{1}{h} + \dfrac{2}{w}\right]$ |

Note the reduction in driving force by a factor two; the sides of the line introduce additional surface energy and grooving terms. The simulations of Walton et al. [8] show that the number of surfaces which are free to form grooves can have a significant effect on pinning and therefore the final grain size.

The velocity, V of a boundary is given by the product of the grain boundary mobility and the net force acting on the boundary.

$$V = MP \tag{2}$$

where M is the mobility and P the driving force, $\Delta\mu$ On a one dimensional argument the velocity of the interface may be described more explicitly by the expression [9]:

$$V = rAv\left(\frac{\Delta\mu}{kT}\right)\exp\left(-\frac{\Delta g}{kT}\right) \tag{3}$$

where r is the distance moved per activated event, $v$ is the Debye frequency ,A is a geometric factor, $\Delta\mu$ the chemical potential change driving the process, $\Delta g$ the free energy of activation and kT has its usual meaning; r, A, $v$ and $\Delta g$ are all in principle structure dependent. Walton et al. express the relationship between velocity and curvature as

$$V = MK \tag{4}$$

where M is a so called mobility constant and K is the grain boundary curvature. Thus

$$2M\gamma\Omega = M \tag{5}$$

here $\Omega$ is the atomic volume.

## OBSERVATIONS OF GRAIN STRUCTURE

We now report a number of observations of grain structures in films and lines made from nominally pure aluminum, copper and some dilute aluminum alloys. The grain structures were observed by transmission electron microscopy in a Philips EM 430 operating at 300 kV or a JEOL 4000FX operating at 400kV and by scanning ion microscopy. Ion microscopy was performed on a JEOL JIBL 106D focused ion beam (FIB) system using a 20 pA, 100 keV Ga$^+$ beam with an estimated diameter of 70 nm. In FIB images, the orientation of individual crystallites relative to the incident beam direction causes large variations in ion-induced secondary electron emission, enabling pronounced crystallographic contrast to be observed.

A transmission electron micrograph of an aluminum film is shown in Figure 1; the film is 90nm in thickness and was e-beam evaporated onto a silicon nitride window substrate at room temperature in a diffusion pumped vacuum system with a base pressure of $5.10^{-8}$ Torr.

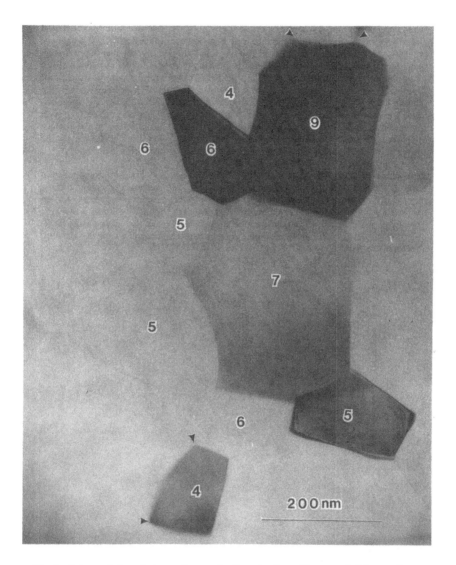

Fig. 1. A transmission electron micrograph of an aluminum film annealed for one hour at 400°C. The numbers indicate the number of sides possessed by particular grains and the arrows designate boundaries which appear to be doubly curved.

The film was annealed in situ at 400° C for one hour after which the grain structure appeared stable. Neither the as-deposited nor annealed grain structures revealed any evidence for abnormal grain growth nor was any preferred orientation detected by selected area diffraction. A survey of individual grain orientations was quite consistent and also offered no evidence for a preference for any particular surface normal. The stagnant grain structure is shown in Fig. 1 is not all like a net of regular hexagons. Indeed in this quite typical field of view there are grains with as many as nine sides and as few as four. Using the intercept method the grains in the field of view range in diameter from about 73nm to 290nm. The largest grains are thus substantially greater in diameter than the film thickness. $D_{max} \approx 3h$. Many grain boundaries are curved but there is no direct evidence for grooving. Mullins [6] pointed out that a doubly curved surface such as a catenoid has zero overall curvature. Such a surface viewed in projection, in the absence of thickness fringes, would appear as curve with a finite width and extended on the convex side. The arrowed boundaries seem to have this characteristic.

Addition of 0.5wt% copper results in a reduction in mean grain size (expressed in terms of the film thickness) and secondary recrystallization as shown in Figure 2. $D \approx$ h but the diameter of the secondary grain exceeds 4h. The film was sputtered onto an oxidized silicon wafer and annealed for one hour at 400°; the film thickness was 50nm. The largest grains are much larger than the those in unalloyed films of the same thickness and subjected to similar annealing. Again characteristic doubly curved stagnant structures are visible.

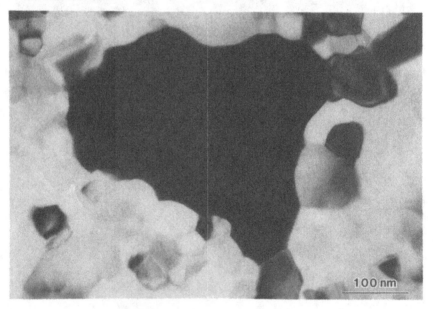

Figure 2 is a transmission electron micrograph of a sputtered film of Al-0.5wt%Cu after annealing for one hour at 400°C.

Figure 3 shows the grain structure of Al-4wt%Cu lines. In figures 3a-d the aspect ratios are 1,1,2 and 4 respectively. Figure 3a shows the microstructure in the as deposited condition and figures 3b-d show the microstructures resulting from an anneal for 60 minutes at 400°C. The lines were deposited by e-beam evaporation from a source with a composition of Al-8wt%Cu and patterned by liftoff. All the lines are 0.5µm in thickness. The oxidized silicon substrate was at ambient temperature during deposition. For annealing, samples were placed in a quartz tube which was purged with forming gas after which the samples were moved slowly into the center of the furnace. After annealing, samples were moved out slowly from the center of the furnace but remained in the quartz tube which was cooled slowly to room temperature. It took around 30 minutes to reach room temperature from 400°C. It is clear that the lines with w/h = 1 approach, but do not achieve a perfect bamboo structure after this annealing treatment. This treatment overages the precipitates so that the theta phase precipitates are coarse and offer rather weak grain boundary pinning. Segregated copper atoms, as distinct from second phase copper containing precipitates are likely to have a larger effect which is discussed later.

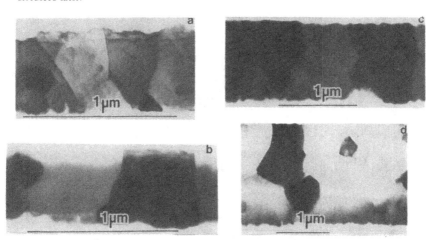

Figure 3 shows the grain structure of Al-4wt%Cu lines. Figure 3a shows the microstructure in the as deposited condition and figures 3b-d show the microstructures resulting from an anneal for 60 minutes at 400°C.

When it is desired to examine the grain structure of a line without removal of the substrate it is convenient to use FIB imaging. Figure 4 compares transmission electron microscope, figures 4a and 4b, and FIB images, figures 4c and 4d of aluminum 0.3% palladium 0.3% niobium alloys in the form of lines one and two microns wide and one micron thick. The fine grained, dark material in figures 4c and 4d is tungsten. The substrate is oxidized silicon and the heat treatment was for 30 minutes at 450°C. The microstructural data from the two instruments consistently reveal a closer approach to a bamboo structure for the narrower line. Isolated particles of an as yet unidentified second phase (most likely the intermetallic compound $Al_4Pd$) can be seen as light contrast in the FIB images and as dark contrast in the electron micrographs. The bright edge in the FIB images is a result of the combination of enhanced sputtering at the edges of lines and enhanced secondary electron emission. The more advanced stage of evolution towards a bamboo structure for the thinner line w/h ≈ 1 is very similar to the behavior observed for Al-4wt%Cu lines with the same aspect ratio.

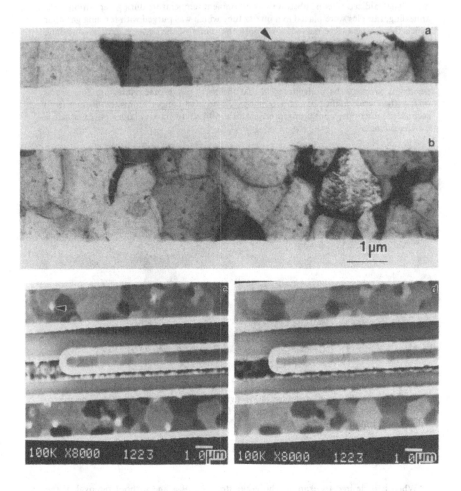

Fig. 4. Transmission electron microscope, 4a and 4b and FIB images, 4c and 4d of aluminum 0.3wt% palladium 0.3wt% niobium alloys in the form of lines one and two microns wide and one micron thick. The arrow in fig. 4a indicates a non-bamboo region. The arrow in fig. 4c indicates a precipitate which is not visible in fig. 4d because of the sputtering action of the ions used to stimulate the emission of secondary electrons in the FIB technique.

A fully bamboo structure was achieved by heating in situ to 600°C as shown in Figure 5.

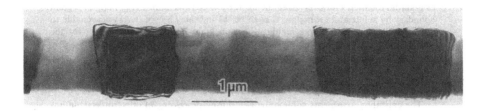

Fig. 5. Transmission electron micrograph showing a fully bamboo structure for a one micron line of Al-0.3wt%Nb-0.3wt%Pd heated to 600°C.

## DISCUSSION

The largest grains in the aluminum and aluminum 0.5wt%Cu films heated to 400°C for one hour approach stagnation and exceed the diameter predicted by the widely used Mullins' criterion. It is likely that this behavior is a result of the extra driving force available from surface energy anisotropy as discussed in [10] for thin films. It was shown earlier that an anisotropy of surface energy such that $\Delta y_t > \Delta y_b$ is sufficient to prevent stagnation of grain growth. On a broken bond model this corresponds to the difference in surface energy of a {111} and a {100} surface in a face centered cubic metal. Lesser differences in $y_t$ lead to stagnation at grain diameters $>>$ than the film thickness. In principle passivation of an interconnect on a substrate may decrease or eliminate grooving and thus eliminate one cause for stagnation. In view of the reactivity of aluminum it is possible that there is intrinsic resistance to grooving because of an oxide film. However by the same token evaporated aluminum films are notorious for oxide contamination so that there is likely to be some pinning by impurities which would serve to mask any decrease in pinning by grooves. Consequently in the absence of knowledge of the surface topography and the oxygen distribution in the grain boundaries further quantitative analysis of the stagnant structure is not possible. The stagnant grain size in Al-4%Cu lines is less than that in the blanket films reported by Iyer and Wong [11]; a line approaching a bamboo structure suffers a decrease in driving force and an increase in pinning forces from surface grooving ( Tables 1 and 11 ). For the narrowest line D is $\approx$ h. However this line with w/h = 1 is not entirely bamboo after a one hour anneal at 400°C. This is consistent with the analysis of Walton et al. [7,8] who defined a dimensionless time parameter $\tau$ in their simulations; $\tau = \dfrac{Mt}{w^2}$. According to their model, stagnation occurs after a normalized time $3\tau$. Using plausible values for the material constants in eq. 3 and T = 400°C gives a real time to stagnation calculated for a pure aluminum line of about 2 hours. On this basis the lines shown in figure 3 would not be stagnant even for pure aluminum; with this in mind the theory is a remarkably consistent with the experiments. The microstructures of Al-0.3wt%Nb-0.3wt%Pd lines are very similar to those of Al-4wt%Cu for comparable geometries and annealing treatments. Very few data are available concerning the concentration and binding energy of solute to grain boundaries. However using the expression in Table I for the pinning of a grain boundary by solute and estimating C b. to be 0.1 indicates how it is very reasonable to expect solute to exert a large pinning force on grain boundaries and except for the smallest grains to be a dominant factor.

## CONCLUSION

The annealing treatments commonly used during processing of interconnects do not necessarily result in stagnant grain structures. The microstructures observed in lines are qualitatively consistent with the trends predicted by Walton et al. [7,8]. Quantitative correlation between experimental data and models requires more detailed microstructural analysis especially of texture and grain boundary composition.

## ACKNOWLEDGEMENT

We are grateful to P. Blauner for assitance with FIB.

## REFERENCES

1. D.A.Smith, Ultramicroscopy, **40**, 321 (1992).

2. C.R.M. Grovenor, H.T.G. Hentzell and D.A. Smith, Acta Metall. **32**, 773 (1984).

3. F. M. d'Heurle and R. Rosenberg in Physics of Thin Films, Ed. E. Hass, M. H. Francomb and R. W. Hoffman, Academic Press, New York, Vol. 7 p. 257 (1973).

4. E. Kinsbron, Appl. Phys. Lett., **36**, 968 (1980).

5. Recrystallization of Metallic Materials, Edited by F.Haessner, Dr. Reiderer Verlag, Stuttgart, Germany, (1971).

6. W.W. Mullins, Acta Metall., **6**, 414 (1958).

7. D.T. Walton, MS thesis, Dartmouth College, (1991).

8. D.T. Walton, H.J. Frost and C.V. Thompson, Appl. Phys. Lett., in press.

9. D. Turnbull, Trans. AIME, **191**, 661 (1951).

10. C.V. Thompson and H.I. Smith, appl. Phys. Lett. **44**, 603 (1984).

11. S.S. Iyer and C.Y. Wong, J. Appl. Phys., **57**, 4594 (1985).

# MICROSTRUCTURE AND THE DEVELOPMENT OF ELECTROMIGRATION DAMAGE IN NARROW INTERCONNECTS

A.L. GREER AND W.C. SHIH
University of Cambridge, Department of Materials Science and Metallurgy, Pembroke Street, Cambridge CB2 3QZ, United Kingdom

## ABSTRACT

The microstructure in narrow (1.1, 1.5 and 2.1 μm) unpassivated lines of Al-4wt.% Cu is found to be 'near-bamboo', with $Al_2Cu$ grains being a significant feature correlated with thermal hillocking and with the development of damage on electromigration. The development of damage is shown to be closely related to the median time to failure, with its initiation being at a rate proportional to the square of the current density. The mechanisms of damage development are discussed, with particular reference to near-bamboo, two-phase microstructures.

## INTRODUCTION

Largely because of decreasing linewidths, electromigration (EM) in Al-based metallizations continues to be subject to intensive study to understand the mechanisms of failure, and to enable the prediction of failure behaviour for a particular microstructure or the design of superior microstructures. Lines are now sufficiently narrow that their widths can be comparable with the grain size of the metallization, giving "near-bamboo" or "bamboo" grain configurations. In such lines the predominant atomic transport paths must be different from the grain boundaries with which the EM flux is clearly associated in wider lines. Associated differences in the mechanisms of EM damage development are related to the finding (e.g., in [1]) that as the ratio of linewidth to grain size decreases, the median time to failure (MTF) decreases until near-bamboo structures are attained, when a further decrease leads to a rapid rise in MTF. Unfortunately, this rise in MTF is accompanied by a rise in the deviation of times to failure (DTF), the overall effect being a rather small change in device failure rate [2].

Further interest in narrow lines comes from the observation of failure modes not necessarily related to EM. Especially when passivated, stress-induced voiding (e.g., [3]) is found to be a major source of open-circuit failure. While passivation affects EM failure in narrow lines, it has been shown that the development of damage (void growth and coalescence) and failure modes can be very similar with and without passivation [4,5].

It is well established that addition of Cu to Al improves lifetimes under EM stressing [6]. Recently, there has been further study of the rôle of the Cu, concentrating on the $Al_2Cu$ particles which are found when the solubility limit is exceeded. It has been suggested that these particles can act as nucleation sites for voids [4,5] and (rather than Cu segregated to the Al grain boundaries) as barriers for diffusion [7].

In this work, and in a companion paper [8], we study EM in narrow, unpassivated interconnects of Al-4wt.%Cu. While some of the detailed results may be specific to this Cu-content, and to unpassivated lines, it is hoped to show some general aspects of the development of EM damage and failure in near-bamboo lines with $Al_2Cu$ particles. In such lines, it is possible to examine the development of damage at particular microstructural configurations. As will be clear, the mechanisms are quite different from those normally assumed (i.e., grain boundary transport and atomic flux divergences at grain boundary junctions) in computer models of EM failure (e.g., [9,10]) and a need for different models is demonstrated.

Mat. Res. Soc. Symp. Proc. Vol. 265. ©1992 Materials Research Society

## EXPERIMENTAL

The metallization is Al-4wt.%Cu, 1 μm thick, with a nominal sheet resistance of 32.5 mΩ/sq. Linewidths are 1.1, 1.5 and 2.1 μm, and the line length is 1.4 mm. Each line is bounded by two 4.5 μm wide guard rails (to detect short-circuiting due to hillocking, but this test was not performed in the present study). After patterning the samples are subjected to stress-relief annealing at 435 °C for 30 mins. To facilitate later examination, there is no passivation layer. Life-tests, described in [8], are at temperatures of 200, 230, 260 and 300 °C and current densities of 1, 2 and $3 \times 10^{10}$ A m$^{-2}$.

Failed lines from the life-tests are examined using scanning electron microscopy (SEM). Data for different times of electromigration (EM) stressing are obtained by observing lines with different lifetimes. The microscope (Camscan S2) has a storage frame to integrate 8 or 16 images to produce a high signal-to-noise ratio. Images are obtained both using secondary electrons and backscattered electrons. The former mode gives primarily topographical contrast, the latter atomic number contrast. Figure 1a shows a secondary electron image and Fig. 1b its backscattered electron counterpart obtained at an accelerating voltage of 10 kV. The bright regions in the latter are grains of Al$_2$Cu and are detected only near the surface. Figure 1c is a backscattered image obtained at 30 kV, showing how the greater electron penetration can detect Al$_2$Cu grains deeper in the line, but with impaired resolution due to beam-spreading. Apart from Fig. 1c, all SEM micrographs were obtained at 10 kV.

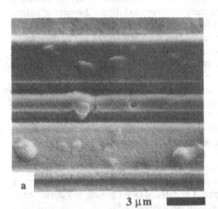

3 μm

*Fig. 1 — Scanning electron micrographs using (a) secondary electrons at 10 kV, (b) backscattered electrons at 10 kV, and (c) backscattered electrons at 30 kV. The EM stressed line is central, between two 4.5 μm unstressed guard rails. [1.5 μm effective linewidth, 260 °C oven temperature, 1 × 10$^{10}$ A m$^{-2}$ nominal current density, 640 hr test time, electron flow left to right]*

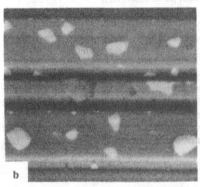

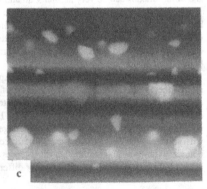

## ALLOY MICROSTRUCTURE

Standard EDS analysis of the brighter regions in backscattered images (e.g. in Fig. 1b), shows them to be Cu-rich. The measurement (average 20 at.% Cu) underestimates the Cu-content because of excitation in the surrounding Al grains. According to the phase diagram [11], Al-4wt.%Cu should in equilibrium be a mixture of $\alpha$-Al and $Al_2Cu$, the equilibrium volume fraction of $Al_2Cu$ varying from 5.1% at 200 °C (the lowest test temperature in this work) to 2.4% at 435 °C (the stress relief temperature); the microstructural observations in this work are in reasonable agreement with these values.

Figure 2 shows thermal grooving revealing the grain structure in a pad at the end of a test-line. Comparison of the secondary electron and backscattered images shows that most of the grains are Al, with some $Al_2Cu$. Thermal grooving is evident, suggesting the Al grain size is 2 to 3 $\mu m$. The grooving pattern does not correspond to the pattern of hillocks on the surface; since hillocks are often associated with grain structure, it seems that there can be significant recrystallization or grain growth after the initial pattern of hillocks has developed. That the grooves indicate the current grain boundaries is shown by the shape of the $Al_2Cu$ grains. As seen in Fig. 2 and elsewhere, the Al grains can vary widely in size, tending to be largest away from the edges of the line where boundary pinning can occur. Unfortunately, thermal grooving is not so clear on the narrow lines for EM stressing, or even on the guard rails. Thus while it is probable that the Al grain structure in the EM test-lines is near-bamboo, further study (by TEM) is necessary for confirmation.

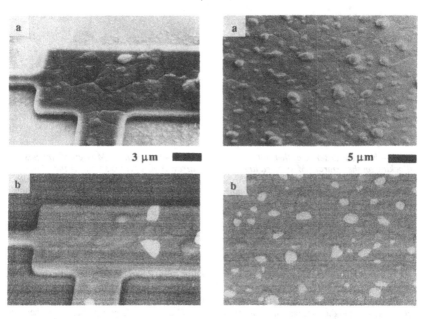

Fig. 2 — Thermal grooving clearly shows the grain structure in this pad. (b) shows that some of the grains are $Al_2Cu$. [SEM using (a) secondary, (b) backscattered electrons]

Fig. 3 — Metallization showing inferior lifetimes on EM stressing. There is pronounced thermal hillocking, mainly associated with $Al_2Cu$. [SEM using (a) secondary, (b) backscattered electrons]

The main function of SEM studies in this project is to characterize the types and pattern of development of EM damage, to assist in the elucidation of damage and failure mechanisms. Even without any understanding of mechanisms, however, it is possible that SEM of unstressed metallization can detect different microstructures correlated with different failure behaviour. An example of this is shown in Fig. 3; samples with this unusual microstructure (compared to the typical microstructure in Fig. 2), had a failure time of ~1/3 of the MTF for normal samples under the same testing conditions. The reasons for the different microstructure have not been analysed.

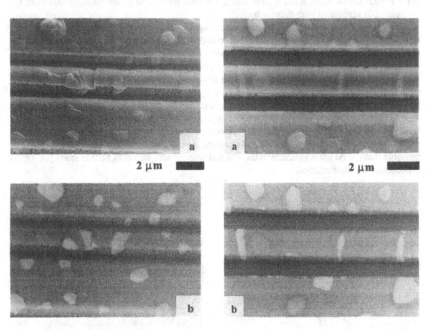

Fig. 4 — Note the absence of thermal hillocking on the central EM stressed line. On that line there are blocking $Al_2Cu$ grains which lead to damage and smaller non-blocking grains (on the right) which do not. [1.1 μm line, 260 °C, $1 \times 10^{10}$ A $m^{-2}$, 251 h] [SEM using (a) secondary, (b) backscattered electrons]

Fig. 5 — The central EM stressed line has narrow $Al_2Cu$ grains spanning the cross-section; these do not lead to damage. Note also the concentration of small $Al_2Cu$ grains at the top edges of the lines. [1.5 μm line, 260 °C, $1 \times 10^{10}$ A $m^{-2}$, 779 h] [SEM using (a) secondary, (b) backscattered electrons]

The $Al_2Cu$ grains in the typical microstructure (Fig. 2) are smaller than the Al grains. This may be expected, since coarsening of the Al grains is limited only by grain boundary mobility, whereas coarsening of $Al_2Cu$ grains is limited by the long-range transport of Cu. The detection, using higher accelerating voltage (Fig. 1c) of buried $Al_2Cu$ grains shows that there are grains that do not span the metallization thickness. Even in the narrowest lines, there are $Al_2Cu$ grains that do not span the linewidth (e.g., Fig. 4). Thus while the microstructure may be bamboo or near-bamboo as regards the Al grains, the same is not true for $Al_2Cu$. In two respects the distribution of $Al_2Cu$ grains is not random, being affected by the configuration of the line. There is a tendency for $Al_2Cu$ grains to span the width of the line as shown in Fig. 5

(and in Figs. 4 and 6, also seen in refs. [4,5]). Also shown in Fig. 5 is the tendency for $Al_2Cu$ to be concentrated along the top edges of the lines. These microstructural features must arise from the evolution of the alloy microstructure after patterning. The $Al_2Cu$ grains spanning the linewidth would be expected in a bamboo microstructure, but often the grains may not occupy the full line thickness, and in any case they are narrower (in the direction along the line) than would be found for a true bamboo structure. When a grain does span the full cross-section (in width and thickness), there is a particularly strong association with the development of EM damage, as discussed later. Some large areas of $Al_2Cu$ are found on the top surface of lines, suggesting a tendency for the phase to form preferentially at the surface.

THERMAL HILLOCKS

Thermal hillocks (i.e., arising without EM stressing) are common in pure metal films as a relief mechanism for compressive in-plane stresses from deposition or from thermal expansion mismatch with the substrate. In alloy films they can be associated with composition changes, as shown by Chang and Vook [12] for Al-15wt.%Cu films. They suggested that the Cu-rich regions all corresponded to hillocks, and that prolonged annealing (in the range 200 to 300 °C, as in the present work) would give fewer, larger hillocks, the kinetics of this coarsening fitting classical Ostwald ripening behaviour for surface transport.

In the present work, thermal effects in the absence of EM stressing are readily studied in the guard rails and bond pads, and the use of backscattered electrons in SEM enables a straightforward confirmation that hillocks are indeed strongly correlated with $Al_2Cu$ grains (Figs. 3 and 4). In contrast to the work of Chang and Vook [12], we find that, despite the correlation, there are hillocks not apparently associated with $Al_2Cu$, and $Al_2Cu$ grains not giving rise to hillocks (Figs. 2 and 3). This may be because the stress-relief anneal used for the metallization in this work would permit coarsening to proceed to an advanced stage, in contrast to the early stages studied by Chang and Vook. (In agreement with this, there is no evidence for coarsening of the $Al_2Cu$ grains during the EM stressing in this work.) A striking aspect of the hillocking, seen for example in Fig. 4, is the absence of thermal hillocks on the narrow central line (which has been EM stressed). This may be because relief of compressive stress is easier in a narrow line, or because of the distribution of $Al_2Cu$ grains (spanning the line or along edges); hillocking appears to be found for $Al_2Cu$ grains surrounded by Al grains. When a Cu-rich region is formed surrounded by Al, there is a net transport of atoms into the region causing a build-up of compressive stress. The net transport arises because at typical annealing temperatures, the diffusivity of Cu in Al is somewhat greater than the self-diffusivity of Al, and further evidence for it is provided by observations of line thinning around hillocks [12,13]. The absence of hillocks on narrow lines is in agreement with recent work on metallization of lower Cu content (0.5 wt.%, below the solubility limit), showing a complete absence of hillocks for widths between 0.6 and 3 μm [14].

MICROSTRUCTURE AND TYPES OF ELECTROMIGRATION DAMAGE

We now consider the development of EM damage and its relationship to the metallization microstructure. The observations on thermal hillocking (above) suggest that a correlation with the $Al_2Cu$ grains should be expected, and this is found. Sanchez et al. [4,5] have found voiding associated with $Al_2Cu$ grains; here we consider other types of damage as well. An example is given in Fig. 6 where a grain of $Al_2Cu$ blocking the line has led to a build-up of material (the flow of electrons and the EM flux of atoms is from left to right); there has been a thickening of the line, with the Al$_2$Cu grain being pushed upwards. With similar Al$_2$Cu configurations, other responses to the compressive stress have been observed in the form of hillocks on the top or the side of the line. The converse effect is seen in Fig. 7, where the line on the "downstream" side of the Al$_2$Cu grain shows thinning and grooving. Voids have also been observed at the interface with the substrate in a similar situation. In general, if an Al$_2$Cu grain does act as a region of the line in which transport is comparatively slow, both accumulation (Fig. 6) and depletion (Fig. 7) effects would be expected. Which is dominant may depend on the average stress level in that region of the line. Further examples of damage resulting from blocking grains are in Fig. 4. It is noteworthy that in Figs. 5 and 6 there are blocking grains of Al$_2$Cu which are narrow and which have not led to damage. As analysed by Blech and Herring [15] and others [e.g., 16], the stresses arising from divergences in EM flux give rise to mechanodiffusion fluxes. If the divergence is over a short lengthscale, the stress gradients can be sufficiently steep for the mechanodiffusion fluxes to balance the divergences in EM flux and for damage to be avoided. The lack of damage for narrow Al$_2$Cu grains may be an example of such a critical length effect.

It should be emphasized that while there is a strong correlation between Al$_2$Cu grains and damage, there are examples of damage not associated with Al$_2$Cu, for example the voiding in Figs. 8 and 9. Also in Fig. 8 is a hillock with a cap of Al$_2$Cu. Other hillocks have shown Al on top of Al$_2$Cu, or simultaneous growth with one half of the hillock being Al the other Al$_2$Cu.

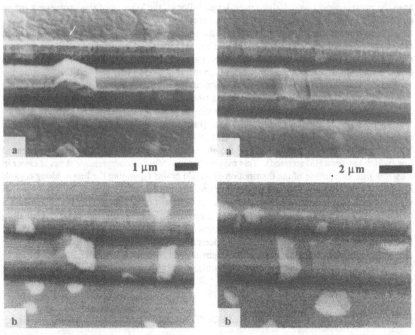

*Fig. 6 — The central EM stressed line has a blocking Al$_2$Cu grain, leading to a build-up of material. The smaller grain on the right does not lead to similar damage. [1.1 μm line, 300 ℃, 1 × 10$^{10}$ A m$^{-2}$, 180 h, electrons from left] [SEM using (a) secondary, (b) backscattered electrons]*

*Fig. 7 — The central EM stressed line has a blocking grain of Al$_2$Cu, leading to depletion of material. [1.5 μm line, 260 ℃, 1 × 10$^{10}$ A m$^{-2}$, 251 h, electrons from left] [SEM using (a) secondary, (b) backscattered electrons]*

## FAILURE SITES

A key issue in attempting to link the SEM observations with failure statistics is whether or not failure does indeed arise from the types of EM damage shown in Figs. 1 to 9. It is at least possible that failure occurs at special sites having little to do with the general development of damage, for example the slit-like voids found in [5] to appear late in testing and to lead rapidly to failure. In the life-testing accompanying this study, only open-circuit and not short-circuit failures are detected; failure is consequently associated with voiding. Examination of sites at which failure has occurred is not normally useful. The configurations leading to failure are destroyed by the catastrophic melting resulting from the extreme localized Joule heating as the line cross-section is reduced by voiding. On the other hand, there are failed lines with no region showing the effects of melting. An example of a probable failure site where the surrounding line is not destroyed is shown in Fig. 9, which also illustrates the high density of damage sites which can be found in these lines, making impossible the identification of the actual failure site from several possible sites. (There are up to 400 damage sites on the 5.6 mm of line [4 × 1.4 mm lengths] examined for each datum in this study.) There are some probable failure sites where the original topography of the line is less disrupted than in Fig. 9. Together with the correlations between the development of damage populations and MTF (described later), the observation that some "ordinary" damage sites can lead to failure suggests that the observations of damage can be useful in studying the mechanisms of failure.

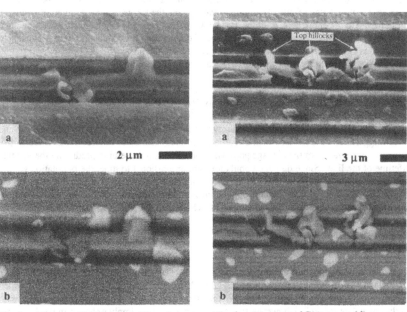

Fig. 8 — The central EM stressed line shows voiding not obviously associated with Al₂Cu and a hillock with an Al₂Cu tip. [1.1 μm linewidth, 260 °C, 1 × 10¹⁰ A m⁻², 685 h] [SEM using (a) secondary, (b) backscattered electrons]

Fig. 9 — The central EM stressed line shows a region of high damage density. [2.1 μm line, 260 °C, 1 × 10¹⁰ A m⁻², 640 h, electron flow from left] [SEM using (a) secondary, (b) backscattered electrons]

## RELATIONSHIP BETWEEN ELECTROMIGRATION DAMAGE AND TEST CONDITIONS

A feature distinguishing the present study from others of local damage configurations and microstructure is the extensive statistical correlation of the types and populations of damage with testing conditions and with MTF. As detailed in [8], it is found that the populations of damage of all types (hillocks and line thickening / voids and line thinning / grooves) rise linearly with testing time (from zero up to ~2 × MTF), with no evidence for incubation times. The balance of different types of damage therefore does not change with test time. The total damage rate can be compared at different temperatures to derive an activation energy. For 2.1 μm wide lines tested in the range 230 °C to 300 °C this is 0.85 eV atom$^{-1}$, matching the activation energy for the MTF of the same lines (0.86 eV atom$^{-1}$). The balance of different damage types is not strongly dependent on temperature, but there is a tendency for a larger fraction of hillocks at lower temperature.

As shown in [8], total damage populations are not significantly dependent on linewidth. This suggests that the number of damage sites is proportional to the line length, not to the area of its top surface, at least for the limited range of linewidths studied here. While the balance of different damage types is not strongly dependent on linewidth, there is a consistent trend for a higher relative proportion of hillocks in wider lines. This could arise from a greater compressive stress in wider lines (also correlating with the thermal hillocking behaviour) or from a greater population of grain boundary triple junctions. It would be expected that, for lines with a random polycrystalline structure of grain size much less than the linewidth, there would be a scaling between the population of hillocks on top of the line and the top area. The population would be proportional to the entire area, or to the area between dead zones arising from stress relaxation at the edges of the line. The weaker dependence of population on width found here is consistent with, and is evidence for, near-bamboo grain structures in the lines.

After correction for temperature rises due to Joule heating, the development of damage can be compared at different current densities [8]. The balance of different damage types is not significantly dependent on current density. The rate of appearance of total damage sites is proportional to the square of the current density (exponent ~1.9), again matching the behaviour of the MTF, which has a current density exponent for the same lines of ~2.2.

The matching of the activation energy and current density exponent for damage rate and MTF suggests that the development of damage is closely related to failure. As detailed in [8], however, observation of the damage population on a line is not useful in predicting the residual lifetime of the line. Since damage populations are proportional to time, the population simply indicates the time or fraction of MTF expended. Of course, prediction of residual lifetime on the basis of SEM observation of damage population would not be very useful in any case. Of more importance here is elucidation of fundamental mechanisms.

## DISCUSSION

Within the range of current densities (1 to 3 × 10$^{10}$ A m$^{-2}$) used, the lack of change in the balance of different damage types suggests that there is no change of mechanism which would complicate the interpretation. Within the temperature range (230 to 300 °C, with preliminary data at 200 °C), however, the change in balance may imply a change in mechanism.

The observed linear rise of damage populations with time is of particular interest. Rather than this behaviour, it might have been expected that there would be a fixed population of sites at which damage would develop; with time, the *size* of individual damage events would then increase and not their *number*. Since the detectability of damage sites must depend on their size, an increase in the apparent population would be expected in the early stages of a test, even if the real population were constant. However, the extended linear increases observed do strongly suggest that there are real increases in damage populations. Linear increases, however, could not continue without limit. Based on the assumptions that damage sites are associated with grain boundaries and that grain size is similar to linewidth, a reasonable upper limit to the total population of damage sites would be $1/_w$ sites per unit length of line, where $w$ is the linewidth. For the 1.1 μm lines tested here, this upper limit would be $\sim 7 \times 10^5 \, m^{-1}$. The highest observed damage population density in the lines is $\sim 7 \times 10^4 \, m^{-1}$, and therefore still far from the expected saturation value. However, locally the suggested saturation value can be reached or even exceeded (Fig. 9).

A continuing increase in the number of damage sites can arise if there is a wide spread of magnitude in the flux divergences in the lines. (The origin of these divergences is considered below.) Each divergence leads to a build-up of stress, or a deviation of the local vacancy concentration from the equilibrium value, and when the stress or concentration reaches a critical value damage is initiated. (The stress and vacancy concentration are related [17].) Damage will first be seen corresponding to the largest divergences, but will continue to appear at smaller divergences. It seems reasonable to suppose that any damage site could ultimately lead to failure, but that the larger the scale of the damage the higher the probability of failure. Early damage sites should then be particularly important because they have longer to grow, and because (arising from larger divergences) they should grow faster.

Atomic flux divergences can arise from variations in alloy resistivity changing the effective charge of the ions, but are most likely to be dominated by variations in atomic mobility. In wide lines, the atomic motion is conventionally associated with grain boundary diffusion, and divergences consequently associated with changes in grain size or with individual grain boundary junctions. In bamboo lines different sources of divergence must be sought, as grain boundaries cannot be the only significant transport paths. (They can still contribute to local transport when not perpendicular to the line edges [18].) Transport may also be by bulk diffusion through the grains, along the interface between the alloy and the substrate, or along the interface between the alloy and its oxide. The first of these cannot lead to variations in mobility between Al grains as diffusivity in cubic α-Al is isotropic. However, divergences could arise at the junctions between Al and $Al_2Cu$ grains, as the damage patterns suggest. The interface diffusivities could vary from grain to grain as a function of both grain orientation and phase identity.

In near-bamboo structures, it might be expected that damage would be especially associated with the local grain boundary configurations deviating from the ideal bamboo structure. While the SEM observations (the examples here are all for a test temperature of 260 °C) do show damage at such sites (Figs. 1 and 4), there is no evidence that the scale of damage is greater than at sites consistent with a bamboo structure (e.g., Fig. 7). Apart from microstructure, a further source of variation in atomic mobility is temperature variation. This can arise from variation in Joule heating in the patterned metallization, or from heat produced in other active regions of the device. However, the damage observed in the present work is clearly associated with microstructural features, and does not show the variations in hillock/void ratio associated with longer scale temperature variations [19]. On the other hand, damage itself could lead to localized Joule heating which may have a role in the development of the damage.

The empirical Black equation [20] is commonly used to describe the dependence of MTF on test conditions. Two parameters are important for damage and failure mechanisms: the current density exponent and the activation energy. The current density exponent is commonly found be be $\sim 2$ (as in this work), and this has attracted considerable interest as all simple models would predict an exponent of unity (arising from the linear relationship between atomic flux and current density). The exponent has been discussed in detail by Shatzkes and Lloyd [21], who show that an exponent of 2 is found if it is assumed that failure occurs when a critical vacancy concentration is reached, and that the flow of vacancies is influenced both by EM and by vacancy concentration gradients opposing the EM flux. Of particular significance in the present work is that an exponent of 2 has been found not only for the MTF, but also for the rate of development of damage (i.e., the rate of population increase). If the build-up of stress or vacancy concentration deviation were linear with the magnitude of flux divergences, such behaviour would not arise. Thus the observations in the present work provide strong evidence for the influence of diffusional back-fluxes as proposed by Shatzkes and Lloyd, though it is suggested that the critical stress or vacancy concentration value is associated with the initiation of damage and not with failure. This link with the initiation of damage is similar to later developments of the Shatzkes and Lloyd model focussing on void nucleation [22]. However, in the present work in which there are many sites of considerable damage which could lead to failure, there is no evidence for failure arising from rapid void growth following nucleation, as has been suggested [22]. The Shatzkes and Lloyd model [21] is normally considered in its application to atomic transport along grain boundaries. We note, however, that the basic concepts of the model, a characteristic diffusion length and a set of barriers to transport, are applicable also to other transport paths, such as may be more appropriate for bamboo grain configurations.

As shown in [8], the activation energy for MTF varies from 0.86 eV atom$^{-1}$ for 2.1 $\mu$m lines to 1.07 eV atom$^{-1}$ for 1.1 $\mu$m lines. These values can be compared with the activation energy for MTF in wider lines (i.e., with width >> grain size) which is found to have a maximum value of ~0.8 eV atom$^{-1}$ at 4wt.%Cu in Al [6]. In the wider lines this value is clearly associated with grain boundary diffusion. The higher activation energy in the narrow lines in this study would be expected as grain boundaries, acting as short-circuit paths, become less important, especially in the narrowest lines. The activation energies are, however, still significantly less than that for lattice self-diffusion of Al (1.4 to 1.5 eV atom$^{-1}$).

In general it is found that the MTF decreases as linewidth is decreased, reflecting the greater susceptibility of narrow lines to damage. However, as lines become so narrow that the grain structure becomes predominantly bamboo, the MTF then rises with decreasing width. In [8] both types of behaviour are found for the lines in the present study, the MTF (though changing only slightly) decreasing with decreasing width at higher temperatures (260 and 300 °C) and increasing with decreasing width at lower temperatures (200 and 230 °C). Since the microstructure does not change with test temperature some further explanation is needed. The balance of different damage types is well characterized only for the widest lines (2.1 $\mu$m), for which it is found [8] that there are significantly more hillocks and fewer grooves at lower temperature. Thus the pattern of damage at lower temperature is more characteristic of a non-bamboo structure. At lower temperatures, then, both the variation of MTF and the balance of damage types are consistent with the widest line deviating significantly from a bamboo structure. However, the same deviation from bamboo structure seems to have a less detrimental effect on the MTF at higher temperatures. This may be because the atomic mobilities in different transport paths become more similar (converging because of the different activation energies) at higher temperature. If effective mobilities are similar in bamboo and non-bamboo regions of the line, this would explain the lack of microstructural effect on MTF and would be consistent with the observed scale of damage at higher temperatures being roughly equal at bamboo and non-bamboo sites (discussed above).

Although the present study shows that EM damage is associated with the presence of Al$_2$Cu grains, those grains act as a source of Cu to inhibit the development of damage in the $\alpha$-Al portions of the lines. Thus the overall rôle of the particles in determining MTF and failure rate is not clear, and work is needed to determine the optimum Cu content, which is likely to be a function of linewidth.

CONCLUSIONS

SEM with backscattered electron imaging is useful for detecting the distribution of $Al_2Cu$ grains in Al-Cu metallization. SEM can identify some inferior (lower MTF) metallization before EM stressing. SEM also shows clearly the development of damage on EM stressing.

In the narrow Al-4wt. %Cu lines, the grain structure is near-bamboo for the $\alpha$-Al phase, with $Al_2Cu$ grains spanning the line or concentrated at its edges. The $Al_2Cu$ grains are an important feature of the microstructure, and their distribution shows that coarsening is at an advanced stage following the stress-relief anneal. The microstructure is affected by the fact that this anneal followed the patterning. Microstructure-based modelling of failure for these lines should include the effects of the $Al_2Cu$ grains.

Thermal hillocking is not seen on the narrow lines, but is found on wider areas of metallization, where it has a strong (but not absolute) correlation with $Al_2Cu$ grains.

Damage development on EM stressing has a strong (but not absolute) correlation with $Al_2Cu$ grains. Both accumulation of atoms (line thickening and hillocks) and depletion of atoms (line thinning and voids) are found near $Al_2Cu$ grains, and the association is particularly strong when the grains span the cross-section of the line. However, such blocking grains appear to be effective only if their dimension along the line exceeds a critical value.

The populations of damage sites measured in a companion study [8] are correlated with test conditions and with failure. During EM stressing the populations of all types of damage rise linearly with time. The activation energy of damage rate is intermediate between those for grain boundary and lattice diffusion in Al, and is larger in narrower lines. The activation energy (e.g., 0.85 eV atom$^{-1}$ for 2.1 $\mu$m lines) and current density dependence of damage rate (exponent of ~2) match those of the MTF. The balance of different damage types is independent of current density, suggesting no change of mechanism within the range of test conditions (1 to $3 \times 10^{10}$ A m$^{-2}$). The balance is dependent on temperature, with more hillocking at lower temperature.

The total damage populations are approximately independent of linewidth, but show a higher fraction of hillocks in wider lines, consistent with near-bamboo grain configurations in the range of width studied. The hillocks may be due to the presence of $Al_2Cu$ particles.

The inverse scaling of damage rate and MTF leads to the population of damage sites at MTF being constant, for a given linewidth. For 2.1 $\mu$m width, for example, the population at MTF is ~$10^5$ sites m$^{-1}$, about 20% of the expected saturation value of the reciprocal linewidth.

The linear rise of damage populations is consistent with a wide spread in the atomic flux divergences in the lines, and suggests that failure is likely to be associated with sites which appear early in testing.

There is no evidence for damage being especially prominent at sites where the grains deviate from a bamboo configuration, consistent with the MTF being not strongly dependent on linewidth (i.e., not strongly dependent on how large the deviation is from bamboo structure).

The current density dependence of damage rate suggests that the model of Shatzkes and Lloyd [21] may apply to the initiation of damage.

Since the $Al_2Cu$ particles are associated with damage, but also stabilize the Al against EM damage, the optimum Cu content needs further investigation for near-bamboo lines and may be dependent on linewidth (for a given grain size).

ACKNOWLEDGEMENTS

This work forms part of a larger project "IC Interconnect Performance" supported by the Science and Engineering Research Council (UK) and the Department of Trade and Industry (UK), and involving also GEC Plessey Semiconductors Ltd, GEC Marconi Materials Research

94

Ltd, BNR Europe Ltd, and the Universities of Kent and Lancaster. Samples for this study were supplied by GEC Plessey Semiconductors, and life-testing was by BNR Europe. Interactions with other partners in the project and with C.V. Thompson are gratefully acknowledged, as is the provision of laboratory facilities by Prof. C.J. Humphreys.

REFERENCES

1.  J. Cho and C.V. Thompson, *Appl. Phys. Lett.* **54**, 2577 (1989).

2.  J. Cho and C.V. Thompson, *J. Electron. Mater.* **19**, 1207 (1990).

3.  Q. Guo, C.S. Whitman, L.M. Keer and Y.-W. Chung, *J. Appl. Phys.* **69**, 7572 (1991).

4.  J.E. Sanchez, Jr. and J.W. Morris, Jr., in *Materials Reliability Issues in Microelectronics*, edited by J.R. Lloyd, F.G. Yost and P.S. Ho (*Mater. Res. Soc. Proc.* **225**, Pittsburgh, PA 1991) pp. 53-58.

5.  J.E. Sanchez, Jr., L.T. McKnelly and J.W. Morris, Jr., *J. Electron. Mater.* **19**, 1213 (1990).

6.  F.M. d'Heurle and P.S. Ho, in *Thin Films — Interdiffusion and Reactions*, edited by J.M. Poate, K.N. Tu and J.W. Mayer (Wiley, New York 1978), pp. 243-303.

7.  J.R. Lloyd, *Appl. Phys. Lett.* **57**, 1167 (1990).

8.  W.C. Shih, T.C. Denton and A.L. Greer, "A Microscopical and Statistical Study of Electromigration Damage and Failure in Al-4wt.%Cu Tracks", in this volume.

9.  P.J. Marcoux, P.P. Merchant, V. Naroditsky and W.D. Rehder, *Hewlett-Packard J.* June 1989, 79.

10. R. Kirchheim and U. Kaeber, *J. Appl. Phys.* **70**, 172 (1991).

11. L.A. Willey, in *Aluminum, Vol. 1*, edited by K.R. van Horn (Amer. Soc. Metals, 1967), pp. 359-381.

12. C.Y. Chang and R.W. Vook, *J. Mater. Res.* **4**, 1172 (1989).

13. F. d'Heurle, L. Berenbaum and R. Rosenberg, *Trans. Metall. Soc. AIME* **242**, 502 (1968).

14. C.A. Pico and T.D. Bonifield, *J. Mater. Res.* **6**, 1817 (1991).

15. I.A. Blech and C. Herring, *Appl. Phys. Lett.* **29**, 131 (1976).

16. C.A. Ross, in *Materials Reliability Issues in Microelectronics*, edited by J.R. Lloyd, F.G. Yost and P.S. Ho (*Mater. Res. Soc. Proc.* **225**, Pittsburgh, PA 1991) pp. 35-46.

17. R. Kirchheim, *Acta Metall. Mater.* **40**, 309 (1992).

18. H.P. Longworth and C.V. Thompson, "Electromigration in Bicrystal Al Lines", in this volume.

19. A.P. Schwarzenberger, C.A. Ross, J.E. Evetts and A.L. Greer, *J. Electron. Mater.* **17**, 473 (1988).

20. J.R. Black, *IEEE Trans. Electron Devices* **ED-16**, 338 (1969).

21. M. Shatzkes and J.R. Lloyd, *J. Appl. Phys.* **59**, 3890 (1986).

22. J.R. Lloyd, *J. Appl. Phys.* **69**, 7601 (1991).

ELECTROMIGRATION IN BICRYSTAL AL LINES

HAI P. LONGWORTH* AND C. V. THOMPSON**
*IBM Corp., Technology Products Division, E. Fishkill Facility, Dpt. 295, Z/35A, Hopewell Jct., NY 12533
**Department of Materials Science and Engineering, Massachusetts Institute of Technology, Cambridge, MA 02139

ABSTRACT

We have developed a new experimental technique to study electromigration in bicrytal Al lines as a function of the type and location of the grain boundary as well as the testing temperature. The failure times of these lines are found to be lognormally distributed. The median time to failure (MTTF) depends more strongly on the boundary orientation than the type of grain boundary. The dependence of lifetimes on the type and orientation of grain boundaries, the location and appearance of the failure sites, and the measured activation energy ($E_a$) of 0.94eV suggest that both interfacial and grain boundary diffusion contribute to failure in bicrystal lines, and likely in bamboo and near-bamboo lines as well.

INTRODUCTION

It has been established that the reliability of interconnects depends strongly on their microstructures and dimensions [1]. In general, the electromigration resistance of a line increases with the average grain size and the degree of preferred grain orientation, and decreases with the spread in grain size distribution [2]. Furthermore, both the MTTF and the lognormal standard deviation in the time to failure (DTTF) are found to depend strongly on the ratios of the line width and the average grain size [3, 4]. Cho and Thompson [4] reported that as the ratio of the line width to the grain size decreases, the MTTF first decreases to a minimum and then increases exponentially. However, they also found that the DTTF continuously increases with decreasing line width. These trends were qualitatively explained using a simple "failure unit" model in which the number of failure sites or units is assumed to scale with the number of grain boundaries, and an interconnect is assumed to be composed of a series of parallel units. The probability of failure of a line, G(t), is related to the probability of failure of individual "failure units", F(t), by

$$G(t) = 1 - [1 - F(t)^{w/d}]^{1/d}. \qquad (1)$$

where w, l, and d are the line width, line length, and grain size, and G(t) and F(t) are the cumulative distribution functions (cdf's) for lines and units respectively. In this model, the cdf for the failure times of the unit, F(t), is assumed to be that of a lognormal (LN) distribution function. This results in the cdf for a multiple lognormal (MLN) distribution for the lines, G(t). While the LN assumption may not affect the qualitative results of Cho and Thompson, a quantitative understanding of the distributions of failure times for a unit is necessary for accurate failure predictions.

As device densities are increased, line dimensions have decreased to micron and sub-micron levels, making individual failure units, i.e. grain boundaries, more significant in contributing to line failures. Our goal is to provide an improved understanding of the statistics and mechanisms of interconnect failure by concentrating on the failure of individual units. In order to do so, we need to control the type, as well as the number and locations of grain boundaries. Furthermore, most of the fundamental knowledge of electromigration behavior has been obtained through studies of polycrystalline lines, but near-bamboo and bamboo structures have become more common in sub-micron lines. Thus there is a need for new studies focusing on these types of microstructures to provide more accurate reliability predictions and to develop electromigration-resistant materials which are appropriate for these fine lines. Our study of failures in bicrystal lines is a start in this direction.

## EXPERIMENTAL PROCEDURE

Figure 1 schematically illustrates the technique used to create bicrystal test lines (see [5,6] for further details). First, NaCl single crystals with the desired orientations and geometry were cut, polished and welded together to produced NaCl bicrystals. Next, Al films were epitaxially deposited on the NaCl bicrystal substrates. The films were then transferred to an oxidized Si wafer and patterned to provide populations of Al lines, each with a single, identical grain boundary, and with controlled orientations and locations. Prior to testing, the wafers were annealed to bond the Al films to the substrates through the reduction of $SiO_2$ to form $Al_2O_3$ at the interface. The Al film thickness was about 5000Å, and the lines were 1mm long and 2 μm wide. Lines were tested at a constant voltage and at a current density of $2.5 \times 10^6$ A/cm$^2$.

Two types of grain boundaries: Σ13 [100] symmetric tilt boundaries and more general, (115)/(100) [(115)//(100) and [230]//[100] with the boundary on (321) for the (115) crystal and (010) for the (100) crystal] boundaries were selected [5,6] to study the effect of grain boundary type on electromigration. In all other parts of our study, lines were patterned from bicrystal films with Σ13 [100] symmetric tilt boundaries.

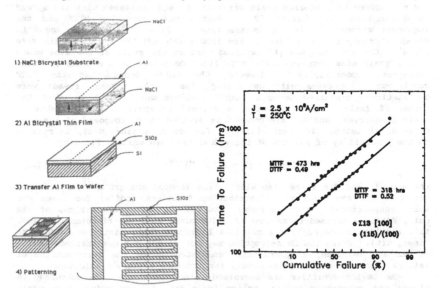

1) NaCl Bicrystal Substrate

2) Al Bicrystal Thin Film

3) Transfer Al Film to Wafer

4) Patterning

$J = 2.5 \times 10^6 A/cm^2$
$T = 250°C$

MTTF = 473 hrs
DTTF = 0.49

MTTF = 318 hrs
DTTF = 0.52

○ Σ13 [100]
● (115)/(100)

**Fig.1** The fabrication process of bicrystal test lines.

**Fig.2** Failure distributions of bicrystal lines with different types of grain boundaries: Σ13 [100] and (100)/(115).

RESULTS

Effects of The Grain Boundary Type

Because of the many processing steps, we performed checks for reproducibility by testing several patterns of each type of boundary. We tested 8 (115)/(100) and 10 Σ13 [100] test structures each containing 22 lines. We obtained MTTF's of 300 ± 30 hours and 480 ± 40 hrs, and DTTF's of 0.48 ± 0.05 and 0.52 ± 0.06 for the (115)/(100) and Σ13 [100] boundaries respectively. The failure distributions of both types of boundary are shown in figure 2.

The MTTF's of the bicrystal lines were compared to the MTTF's reported for various types of microstructures ranging from polycrystals [7], bamboo structures [7,8], and single crystals [9,10]. The values obtained here appear to be of the appropriate order of magnitude; smaller than those of single crystals (4,500h [9] and >222h [10]), but higher than those with bamboo structures (165h [7] and 57h [8]), and orders of magnitude higher than those of grain sizes smaller than the line width (0.2h and 1.7h [7]). The long lifetimes of the bicrystal lines and the good reproducibility of the results, seem to confirm their integrity. These results also demonstrate that this sample preparation technique is a viable means of producing single crystal as well as bicrystal interconnects. The sequence of line failures appeared to be random with respect to their position in the multiline pattern. This suggests that the temperatures of the lines were approximately the same.

Effects of The Grain Boundary Orientation

To study the dependence of lifetimes on the orientation of grain boundaries with respect to the current direction, we patterned test structures varying the angle θ between the current direction and the grain boundary. Four different θ values were examined, 90°, 85°, 75°, and 60°. The alignment error was approximately ± 1 °. Lines were tested at 250°C.

The failure distributions as a function of θ, are shown in figure 3 . While the DTTF's range only from 0.45 to 0.49, the MTTF's vary strongly with θ. The MTTF reduces more than 50% when the normal to the grain boundary plane varies 5° from the direction perpendicular to the flow of the current (θ = 85°). A 15° variation (θ = 75°) results in an 80% lifetime reduction, and a 30° variation reduces the MTTF by about 92%.

Temperature Dependence

The lines used in this study were patterned so that θ is 85°. This angle was chosen for several reasons. First, at θ = 90°, the MTTF is very high, thus requiring very long test periods. Secondly, at 90°, very large variations in MTTF may result from any slight misalignment. Lastly, of the three remaining angles, only the 85° position allows us to obtain the maximum number of test patterns (4 to 5) per bicrystal film.

The lines were patterned from films with Σ13 [100] boundaries, and were tested at 5 different temperatures (250°C, 265°C, 280°C, 300°C, and 320°C). Figure 4 shows the temperature dependence of MTTF. The $E_a$ of 0.94eV obtained from this plot is much higher than the 0.5eV obtained for polycrystalline Al films [7,11,12].

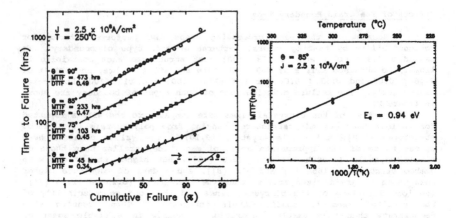

**Fig. 3** Failure distributions of
Σ13[100] bicrystal lines as a function
of the angle θ between the current
direction and the grain boundary plane.

**Fig. 4** Log of the MTTF vs 1/T
for Σ13[100] bicrystal lines
with θ = 85°

DISCUSSION

## The Failure Distribution of Lines

It is widely accepted that the times to electromigration-induced
failure of lines are lognormally (LN) distributed [13,14]. However, some
authors have suggested different types of distributions for different
microstructures. In particular, computer simulations by Schoen [15] and
experimental results from Towner [16] suggest either a LN or a logarithmic
extreme value (LEV) distribution depending on the line width w and grain
size d. In the failure unit model mentioned above [4], the cdf for the
failure times of a line, G(t) is a MLN distribution if the cdf for the
failure times of an unit, F(t), is assumed to be that of a LN distribution
function. The failure times of bicrystal lines are well fit by a LN
distribution as shown in Fig. 2 and 3. This appears to agree with the LN
assumption of Cho and Thompson's failure unit model [4]. Lloyd and Kitchin
[17] pointed out that the MLN distribution is experimentally very similar to
the single LN. The two distributions are nearly indistinguishable without
large numbers of samples. However, for very early failures, the MLN
distribution gives a much shorter TTF than does the LN. Lloyd and Kitchin
showed that if the data is treated as LN, the extrapolated time to 1%
failures is approximately an order of magnitude optimistic as compared to
predictions of the MLN model. This shows the importance of knowing the
correct distribution.

## Failure Mechanism

The reproducibility of our results shows that the measured difference
in the MTTF's of the two types of boundaries is real. If we assume that
electromigration-induced diffusion along the grain boundaries does play a
role even when the boundaries are normal to the flux direction, θ = 90°, we
might expect a correlation between MTTF and the grain boundary diffusivity.
Friedel et al [18] found that in Al, the energy of a symmetric grain

boundary is about 75% of the energy of a boundary lying far from any plane of symmetry. Przybylowicz [19] showed that the higher the grain boundary energy, the lower the $E_a$ for grain boundary diffusion. Thus, we expect that diffusion along the asymmetric (115)/(100) grain boundary is faster than along the symmetric Σ13 [100] boundary. An alternative explanation for the difference in the electromigration resistance of the two types of grain boundaries, regardless of the failure mechanisms, is that grain boundaries are sources and sinks of vacancies, and the source/sink efficiency tends to be lower in certain special low Σ, low energy boundaries, than in more general, high energy boundaries [20,21].

If grain boundary diffusion was the only failure mechanism, we would not expect any failures since the diffusion path (i.e., the line width x sinθ) is less than three microns for all θ, which is much shorter than the Blech length [22]. Optical and Scanning Electron Microscopy (SEM) examinations indicated that failure always occurred at the grain boundaries. Line narrowing was sometimes observed at the grain boundary in stressed lines that had not yet failed. The line narrowing was often observed to be asymmetric about the boundary plane with the region adjacent to the grain boundary on the cathode side having more mass depletion than the region on the anode side of the grain boundary. Grooving was not observed in the unstressed lines. In addition, the position of the grain boundary in the stressed lines did not change relative to the position of the unstressed lines. The appearance and location of the failure sites indicate that the failure mechanism is likely to involve an accelerated grain boundary grooving induced by electromigration. This mechanism which might involve surface and lattice diffusion was proposed by Ohring [23] to explain the film thinning and damage observed in stressed films [24, 25]. Ho and Glowsinki [26] showed that in polycrystalline Al films, the void shape gave indications that grain boundary grooving caused by surface electromigration may be an important void growth mechanism. Lloyd and Nakahara [27] also reported electromigration-induced damage in the form of grain boundary grooving and void nucleation and growth. These results therefore suggest that diffusion at the $Al/Al_2O_3$ interface plays an important role in the failure of these lines.

If the primary diffusion paths are along the $Al/Al_2O_3$ interfaces, we would still expect the line intercept of the grain boundary and the interface to be the site of a divergence of the point defect flux. When θ is 90°, we would expect little or no diffusion along the grain boundary. However, as θ is reduced, more diffusion along the grain boundary would be expected, leading to a greater flux divergence, and a shorter time to failure. Moreover, the view that the grain boundary functions only to divert transport along the interface would explain why θ = 90° does not appear to be a singularity, in that the MTTF at θ = 90°, and the 90° value extrapolated from the results at other angles, appears to be finite.

The $E_a$ of 0.94eV obtained here clearly indicates that grain boundary diffusion which has an $E_a$ of 0.5eV [7,11,12] is not the only or dominant mechanism. The measured activation energy has to lie between that of the two operating mechanisms. That it is significantly lower than that of lattice diffusion, 1.4 to 1.5 eV [12,28], suggests that interfacial diffusion may contribute to failures in bicrystal lines.


CONCLUSIONS

In summary, we have studied electromigration in large populations of Al lines with identical, single grain boundaries of controlled type, location, and orientation. Our results show that 1) the failure times for lines with single identical grain boundaries are lognormally distributed; 2) the MTTF's depend more strongly on the boundary orientation (with respect to the current direction) than on the type of grain boundaries (for bamboo structures); and 3) both grain boundary diffusion and $Al/Al_2O_3$ interfacial diffusion contribute to failure of bicrystal lines (and likely true and near-true bamboo lines as well). This indicates that suppression of interfacial diffusion, and the associated grain boundary grooving, should lead to improved reliability of bamboo lines.

## ACKNOWLEDGEMENTS

We would like to thank T.S. Hsieh for his instruction on the preparation of NaCl bicrystals and Professor Robert Baluffi for the use of his laboratory. This work was supported by the M.I.T. Joint Services Electronics Program Contract No. DAAG/29/83/K003, and by the Semiconductor Research Corporation through Contract No. 90-SP-080.

## REFERENCES

1. F. M. d'Heurle and P. S. Ho, p. 243 in **Thin Film - Interdiffusion and Reactions**, ed. by J. Poate, K. Tu and J. Mayer, Electrochemical Society Inc., John Wiley and Sons, New York (1978).
2. S. Vaidya and A. K. Sinha, Thin Solid Films 75, 253 (1981)
3. S. Vaidya, T. T. Sheng, and A. K. Sinha, Appl. Phys. Lett. 36(6), 464 (1980)
4. J. Cho and C. V. Thompson, Appl. Phys. Lett. 54(25), 2577 (1989)
5. Hai P. Longworth and C. V. Thompson, Appl. Phys. Lett. 60(18), 2219 (1992).
6. Hai P. Longworth, Sc. D. Thesis, MIT, February 1992.
7. Jaeshin Cho, Ph. D. Thesis, MIT, September 1990.
8. J. M. Pierce and M. E. Thomas, Appl. Phys. Lett. 39, 165 (1981).
9. A. Gangulee and F. M. d'Heurle, Thin Solid Films 16, 227 (1973).
10. F. M. d'Heurle and I. Ames, App. Phys. Lett. 16, 80 (1970).
11 J. R. Black, IEEE Trans. on Electron Devices ED-16(4), 338 (1969).
12. H. U. Schreiber, Solid State Electronics 24, 583 (1981).
13. M. J. Attardo, R. Rutledge, and R. C. Jack, J. Appl. Phys. 42(11), 4343 (1971)
14. D. J. LaCombe and E. L. Parks, 24th IRPS/IEEE, 1 (1986).
15. J. M. Schoen, J. Appl. Phys. 51(1), 513 (1980)
16. Janet M. Towner, IEEE/IRPS 100 (1990)
17. J. R. Lloyd and J. Kitchin, J. Appl. Phys. 69(4), 2117 (1991)
18. J. Friedel, B. D. Cullity, and C. Crussard, Acta Met. 1 , 79 (1953).
19. K. Przybylowicz, p. 422 in **Diffusion in Metals and Alloys**, Proc. Int'l Conf., Tihany, Hungary, Sept. 1982 (Trans. Tech., Switzerland).
20. R. W. Baluffi, Met. Trans. B, 13B, 527 (1982)
21. W. Jaeger and H. Gleiter, Scripta Met. 12, 675 (1978).
22. I. A. Blech, J. Appl. Phys. 47(4), 1203 (1976).
23. M. Ohring, J. Appl. Phys. 42(7), 2653 (1971).
24. R. Rosenberg and M. Ohring, J. Appl. Phys. 42(13), 5671 (1971).
25. R. Rosenberg, J. Vac. Sci. Tech. 9(1), 263 (1971).
26. P. S. Ho and L. D. Glowinski, J. Naturforsch. 26a, 32 (1971)
27. J. R. Lloyd and S. Nakahara, Thin Solid Films 64, 163 (1979).
28. A. T. English and E. Kinsbron, J. Appl. Phys. 54(1), 268 (1983).

# A MICROSCOPICAL AND STATISTICAL STUDY OF ELECTROMIGRATION DAMAGE AND FAILURE IN Al-4wt.%Cu TRACKS

W.C. SHIH*, T.C. DENTON** and A.L. GREER*,
*University of Cambridge, Department of Materials Science and Metallurgy, Pembroke Street, Cambridge CB2 3QZ, United Kingdom
**BNR Europe Limited, London Road, Harlow, Essex, CM17 9NA, United Kingdom

## ABSTRACT

We report a statistical analysis of electromigration (EM) induced damage and failure in unpassivated 1.4 mm long Al-4wt%Cu interconnects. The populations of damage sites of various types observed microscopically in life-tested samples are correlated with track width (1.1, 1.5 and 2.1 $\mu$m), temperature (200, 230, 260 and 300 °C), current density (1, 2 and 3 × $10^{10}$ A m$^{-2}$) and lifetest data. Some implications for damage and failure mechanisms are discussed.

## INTRODUCTION

Accelerated life-testing and measurement of the cumulative failure distribution have been commonly used as a way of correlating the performance of interconnects with alloy composition, heat treatment (affecting the microstructure) and track width. Useful information on reliability can be obtained in a reasonable time, but there is uncertainty about whether the mechanisms of EM damage and failure are the same under test and under service conditions, and therefore about whether extrapolations from test conditions are valid. In this work, which is part of a larger study (see companion paper [1]), we aim to define the mechanisms more clearly and thereby to increase the confidence which can be placed on estimates of reliability. We attempt a thorough characterization of the types and populations of damage sites on life-tested samples, permitting a statistical correlation of EM damage with median time to failure (MTF) under various test conditions. The information obtained should be useful not only for assessing mechanisms and whether they change with testing conditions (in particular, temperature) but also in setting up realistic computer modelling of failure.

## EXPERIMENTAL

The metallization is unpassivated Al-4wt.%Cu, 1 $\mu$m thick, with a nominal sheet resistance of 32.5 m$\Omega$/sq. which is patterned and then stress-relief annealed at 435 °C under nitrogen gas for 30 min. The tracks which are the focus of this study are 1.4 mm long, with nominal widths of 1.4, 1.8 and 2.4 $\mu$m. Actual track width was estimated from the variation of resistance with width in unstressed tracks, and with an etching undercut of 0.3 $\mu$m, the actual widths are found to be 1.1, 1.5 and 2.1 $\mu$m. Each track has two 4.5 $\mu$m wide guard rails designed for detecting short-circuiting due to hillocking, but this test is not performed in the current study.

Life-testing is performed as a function of temperature (200, 230, 260 and 300 °C) and current density (1, 2, and 3 × $10^{10}$ A m$^{-2}$). The time to fail is taken to be when there is an open circuit or when the resistance has increased by $\geq$ 20 %. The tracks are cooled to room temperature for resistance measurement, the intervals between measurements depending on the temperature and the test-time so far (giving roughly equal increments of ln{time}). The total current supplied to each test batch was also monitored during the life-test, and when this changed, the current to each individual track was measured and any open circuits noted. The MTF for each set of tests (variation of track width, temperature or current density, with other factors held constant) was obtained from a group of 20 tracks or more. The temperature rise due to Joule heating by the EM-stressing current was also estimated and used in the analysis of results.

Failed tracks from the life-tests were examined using scanning electron microscopy (SEM) in both secondary and backscattered electron imaging modes. The latter is useful to detect $Al_2Cu$ grains in the microstructure. As is shown in detail in the companion paper [1], there is a strong correlation between the $Al_2Cu$ grains and the development of damage. The counting of damage sites was carried out using a SEM (Camscan S2) which has an integrated storage frame producing digital images with a high signal-to-noise ratio, enabling slight EM damage to be detected.

## STATISTICAL ANALYSIS

For the statistical study here, the choice of tracks to be examined was determined by the need to have comparisons enabling the isolation of various effects — track width, temperature and current density. For each set of conditions, four tracks each 1.4 mm in length were examined.

### Classification of EM damage types

While more detailed classifications are possible, in the present work we separate the types of damage into 'hillocks', 'voids' and 'grooves'. The widely defined categories of 'hillocks' and 'voids' represent the types of damage associated with compressive stress (track thickening, top hillocks and side hillocks) and those associated with tensile stress (track thinning and voids), respectively. Examples are shown in Figs. 1 and 2. 'Grooves' are considered to be grain boundary features not obviously associated with either compressive or tensile stress.

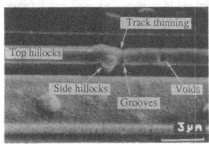

*Figure 1   A secondary electron SEM image, showing types of EM damage. The EM stressed track is central, between two 4.5 μm unstressed tracks. [1.5 μm track width, 260 °C oven temperature, $1 \times 10^{10}$ A m$^{-2}$ current density, 640 hr test time, electron flow left to right]*

*Figure 2   A secondary electron SEM image, showing types of EM damage. A failure site appears to have developed from a void without extensive hillocking [1.5 μm track width, 260 °C oven temperature, $1 \times 10^{10}$ A m$^{-2}$ current density, 640 hr test time, electron flow left to right]*

Time to failure

From examination of the tracks tested under identical conditions, but showing different lifetimes, we found that the total damage population, as well as the individual populations of hillocks, grooves and voids, increase linearly with electromigration stressing time (Fig. 3). The data are from 1.1 $\mu$m wide tracks tested to failure at 260 °C at a current density of $1 \times 10^{10}$ A m$^{-2}$. The data shown span the range from the shortest to the longest recorded lifetime for these testing conditions, at which the MTF is 400 hrs. Since all the populations of different damage types are proportional to time, the fractions of different types of damage do not vary with time. Also, the linearity of the increase of total damage population with stressing time allows us to compare the damage rates under different stressing conditions.

Effect of track width on MTF at different temperatures

Figure 4 shows the variation of MTF with track width for test temperatures of 200, 230, 260 and 300 °C. At 300 and 260 °C, the MTF decreases slightly as the track width is reduced. However, at 230 and 200 °C , the MTF increases as the track width is reduced. Although the value given for the MTF of 1.1 $\mu$m tracks at 200 °C must be regarded as a preliminary estimate (test still in progress as only 35% of samples failed so far), there is a clear trend with temperature in Fig. 4. The activation energies for MTF are consequently dependent on track width, being ~ 1.07, 0.96 and 0.86 eV atom$^{-1}$ for 1.1, 1.5 and 2.1 $\mu$m tracks respectively.

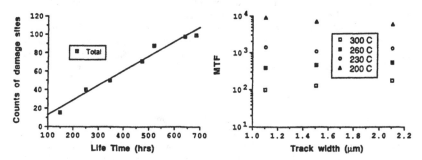

*Figure 3   Population of EM damage as a function of lifetime.*
*Figure 4   MTF as a function of track width at different temperatures.*

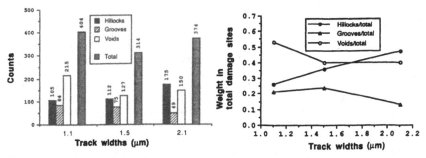

*Figure 5   Populations of various types of damage for different track widths.*
*Figure 6   Fractions of different damage types as a function of track width, derived from Fig. 5.*

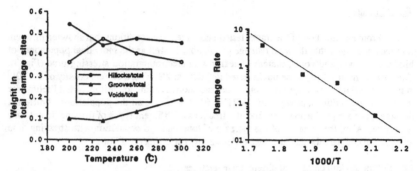

*Figure 7   Fractions of damage sites of different types as a function of temperature.*
*Figure 8   Damage rates (for four 1.4 mm track lengths per hour) as a function of temperature,*
*giving an activation energy of 0.85 eV atom⁻¹.*

### Effect of track width on EM damage

Data on 1.1, 1.5 and 2.1 μm tracks tested at 260 °C for 640 hrs at a current density of 1 $\times 10^{10}$ A m$^{-2}$ are shown in Fig. 5. We can see that the damage rate increases slightly as the track width is reduced, consistent with the trend in MTF (Fig. 4) for tracks of the same width. (In contrast, but also consistent with the trend in MTF, the damage rate at 200 °C decreases slightly as the track width is reduced.) A comparison of the fractions of different types of damage as a function of track width (Fig. 6) shows a higher relative population of hillocks in wider tracks. This could arise from greater compressive stress in wider tracks or from a greater population of grain boundary triple junctions at which hillocks seem to be favoured [1].

### Effect of temperature on EM damage

Figure 7 shows the fractions of different types of damage sites for 2.1 μm tracks tested at $1 \times 10^{10}$ A m$^{-2}$ as a function of temperature (200, 230, 260 and 300 °C). The fractions are not strongly temperature-dependent, but there seem to be an increase in the relative groove population and a decrease in the relative hillock population at higher temperature, the fraction of voids remaining roughly constant. The damage rate increases as the test temperature increases, indicating (Fig. 8) an activation energy of 0.85 eV atom$^{-1}$, very close to the value of 0.86 eV atom$^{-1}$ found from the temperature dependence of the MTF for the same tracks (Fig. 9).

### Effect of current density on EM damage

Comparison of the damage populations on 2.1 μm wide tracks tested at an oven temperature of 260 °C and at current densities of 1, 2 and $3 \times 10^{10}$ A m$^{-2}$ shows (Fig. 10) that the balance of damage types is largely independent of current density, suggesting no change in mechanism of electromigration damage and failure within the range of test conditions. Comparison of damage rates (sites per hour for 5.6 mm total track length) to extract the current density dependence is complicated by the need to standardize temperature. Although the oven temperature is the same in each case, the degree of Joule heating due to the electromigration stressing is different at the different current densities. The maximum increases of track temperature above the oven temperature are found to be 2.4 K, 13.0 K and 31.0 K for the three current densities used. The activation energy determined in this work, 0.85 eV atom$^{-1}$ is used to calculate the damage rates that would be expected at a constant temperature of 260 °C.

Figure 11 shows the data for damage rates (both uncorrected for Joule heating, and corrected as described) plotted logarithmically against current density. For the corrected data the current density exponent is ~ 1.9. In comparison, for the same track width and temperature (2.1 μm and 260 °C), the exponent for MTF (Fig. 12) is ~ 2.2. Thus for the development of damage and for the MTF, both the activation energies and the current density exponents seem to be closely similar. This may be taken to suggest that they have similar underlying mechanisms.

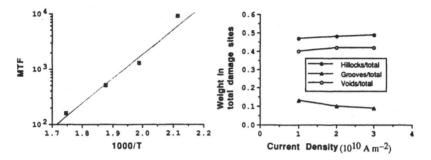

*Figure 9   MTF as a function of temperature, giving an activation energy of 0.86 eV atom$^{-1}$.*
*Figure 10   Fractions of damage sites of different types as a function of current density.*

## DISCUSSION

The effect of track width on MTF at different temperatures, shown in figure 4, is significant. At 230 and 200 °C, the increase of MTF with decreasing track width suggests a transition from near-bamboo towards bamboo structures. On the other hand, at 260 and 300 °C, the effect of the increased bamboo nature is not seen, and there is the more general decrease in MTF with decreasing width. The change of activation energy from 0.86 eV atom$^{-1}$ for 2.1 μm to 1.07 eV atom$^{-1}$ for 1.1 μm, suggests changes in the balance of different types of atomic transport paths, with transport being generally more difficult in the narrower tracks where the grain structure is closer to bamboo.

Given that damage populations rise linearly with time (Fig. 3), they are therefore measures of time elapsed. Since, for a given track width, damage rates (per unit length of track) and the MTF scale inversely to each other as a function of temperature and current density, there is (to a reasonable approximation) a constant damage population at the MTF (~9 × 10$^4$ m$^{-1}$ for 2.1 μm tracks). Thus damage populations during a test can be used as measures of the fraction of the MTF expended. However, there is a spread in the damage population at failure analogous to the deviation in failure times, and determining the damage population in a given track does not assist in predicting when that track will fail.

On the other hand, it seems clear that failure happens at sites which are part of the distribution of damage sites [1]. The scale of damage at a given site (which could be quantified as the the volume of material added to or taken from the track) will increase with time, and this should lead to an increased probability of failure. In this way the probability of failure of a track would depend mainly on the damage sites that were first to appear. This weighting in importance towards early sites would be increased if, as might be the case, early damage sites grow relatively faster (this being the reason for their earlier appearance). If damage developing early in electromigration stressing is particularly important for failure, then it would be of interest to correlate SEM observations after a short stressing time with the lifetime of the tracks on subsequent testing.

106

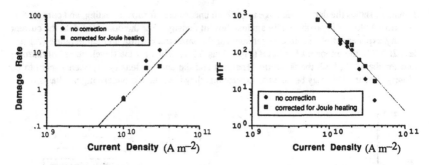

*Figure 11  Damage rates (for four 1.4 mm track lengths per hour) as a function of current density. The current density exponent is 1.9 for the data corrected for Joule heating.*
*Figure 12  MTF as a function of current density, giving a current density exponent of 2.2 for the data corrected for Joule heating.*

## CONCLUSIONS

During EM stressing the populations of all types of damage rise linearly with time. The damage rate increases non-proportionally with track width, slightly with temperature and sharply with current density.

Within the total population of damage sites, the fractions of different damage types are: (1) strongly dependent on track width; (2) independent of current density; (3) slightly dependent on temperature.

The temperature dependence of the damage rate (in 2.1 $\mu$m tracks) is given by an activation energy of 0.85 eV atom$^{-1}$, and the temperature dependence of the MTF gives an activation energy of 0.86 eV atom$^{-1}$. The rate of damage (number of sites appearing per time) is proportional to the square of the current density, and the MTF is inversely proportional to the square of the current density.

SEM examination during electromigration stressing reveals the development of damage. For standard metallization, this is not useful in predicting residual lifetime. Damage populations would reveal the fraction of MTF so far expended, but not the fraction of the lifetime of the particular track.

There appears to be a close link between the development of damage and failure. SEM examination of electromigration stressed tracks potentially can help in understanding the mechanisms of damage development, and thus in predicting MTF and its deviation.

## ACKNOWLEDGEMENTS

Samples were provided by GEC Plessey Semiconductors Ltd, and financial support (as part of a larger project [1]) was from the Science and Engineering Research Council (UK) and the Department of Trade and Industry (UK).

## REFERENCES

1.    A.L. Greer and W.C. Shih, "Microstructure and the development of electromigration damage in narrow interconnects", in this volume.

# TEXTURE EFFECTS ON THE ELECTROMIGRATION BEHAVIOR
## OF LAYERED Ti/AlCu/Ti FILMS.

K.P. Rodbell, D.B. Knorr[A] and D.P. Tracy[A]
IBM Research Division, Yorktown Heights, NY 10598
[A]Materials Engineering Department, Rensselaer
Polytechnic Institute, Troy, NY 12180-3590.

ABSTRACT

A strong correlation was found between film texture, quantified as fiber plots of (111) intensity versus tilt angle from the normal direction, and the resulting electromigration behavior of layered Ti/AlCu/Ti films. Superior electromigration behavior was found for those films which had a low volume fraction of randomly oriented grains, strong and sharp texture. Film microstructure and electromigration lifetime data on DC magnetron sputtered and electron gun evaporated Al-Cu and layered fine lines, fabricated and tested in the same laboratory, are included for a direct comparison. Outstanding electromigration lifetimes were measured for sputtered, layered, submicron films with copper concentrations between 0.12 - 2wt.%Cu. In contrast the electromigration lifetimes of evaporated layered films were found to degrade rapidly at < 2wt.%Cu. This anomalous electromigration behavior was attributed to both film texture and subtle structural differences in the Ti-Al intermetallics formed.

INTRODUCTION

Many factors are known to influence the electromigration lifetimes of aluminum films such as effects of chemistry, film geometry, and microstructure [1]. An effective reduction in hillock formation and enhanced electromigration performance is known to occur for Al-alloys, e.g. Al-Cu [1,2], and multilayered structures of Al (or Al-alloys) with a layer (or layers) of a refractory metal [3-8]. In Al-Cu alloys the electromigration lifetime is a function of Cu concentration, increasing as the Cu content increases from 0 to ~4 wt.%Cu [1,2]. The role of Cu in increasing the electromigration lifetimes in Al-Cu alloys is not well understood. It has been proposed that Cu atoms in solid solution preferentially migrate before Al in an electric field, thereby improving the electromigration behavior; it is supposed that Al atoms will not electromigrate until Cu is depleted [9]. The formation of $\theta$-phase $Al_2Cu$ precipitates at Al grain boundaries inhibits void growth and coalescence, limits vacancy migration and stabilizes the Al grain boundaries [1]. Annealing conditions are known to greatly influence the shape and distribution of precipitates in AlCu alloys [10,11] and to control the film microstructure. Film microstructure has a strong influence on electromigration behavior [1], with grain size, grain size distribution and film texture known to control the reliability of pure Al [12, 13] and AlCu [14] films. The motivation for this work was to characterize the microstructure of AlCu and layered Ti/AlCu/Ti films obtained from different deposition tools and to access the role of grain size, grain size distribution and film texture on electromigration lifetimes.

AlCu alloys and layered Ti/AlCu/Ti/cap (~25nm Ti/950nm Al-x wt.%Cu/25nm Ti/40nm cap) films were deposited onto oxidized silicon wafers by both sputtering and evaporation. The cap layer consisted of either an AlCu alloy or reactively sputtered TiN (25-50nm thick). Evaporated films were deposited in a Temescal 1800FC (planetary) evaporator onto oxidized silicon substrates and wafers with previously built liftoff masks. Sputtered films were deposited in Perkin Elmer 4450 (planetary), MRC 662 (parallel scanning) and Varian M2000 (single wafer) sputtering systems, see Table I for details. Following blanket metal deposition fine lines were fabricated using both optical and e-beam lithography and reactive ion etching (in $BCl_3$-$Cl_2$-$CHCl_3$) forming lines 0.5 - 2.5$\mu$m wide. Metal films that had been evaporated onto liftoff masks were submersed into a suitable solvent and agitated ultrasonically producing similar lines. Patterned wafers were annealed at 400°C for 1h in forming gas (90% $N_2$ - 10% $H_2$) and coated with ~2$\mu$m of sputtered $SiO_2$. Contact holes were opened in the oxide, the wafers diced and the chips mounted in 28 pin ceramic packages with Al-1wt.%Si wire ultrasonically wirebonded to them. Electromigration lifetests were performed in air ambients at 250°C at a current density of $2.5 \times 10^6$ A/cm². The sample temperature rise due to joule heating was typically < 3°C at these conditions. Both electrical open and short failures were recorded. The lognormal life distribution was used to calculate the median lifetime ($t_{50}$) and $\sigma$, the shape parameter of the lognormal distribution. X-Ray Fluorescence (XRF) was used to determine the average copper content in blanket films, 2 MeV $He^{2+}$ Rutherford

Mat. Res. Soc. Symp. Proc. Vol. 265. ©1992 Materials Research Society

Backscattering Spectroscopy (RBS) was used to determine the copper distribution and XRF, RBS and sheet resistivity were used to measure film thickness. Film microstructure was characterized by transmission electron microscopy (TEM) and by X-ray diffraction (XRD). Texture was measured on blanket films by Bragg diffraction ($\theta/2\theta$) scans and by the pole figure technique. Intensity measurements from Al (111), (200) and (220) peaks and from Ti (0002) were taken as integrated areas from Bragg scans. Since thin films have axial symmetry, the corrected and normalized (random = 1) pole figure data could be plotted as (111) intensity versus tilt angle, $\phi$, from the sample normal direction.

**TABLE I.** Film deposition parameters.

| Film | Power | Pressure (Torr) | Rate (nm/s) |
|---|---|---|---|
| **TEMESCAL FC1800 EVAPORATOR** | | | |
| Al, AlCu | -- | $2\times10^{-6}$ | 2 |
| Ti | -- | $5\times10^{-7}$ | 2 |
| **PERKIN ELMER 4450** | | | |
| Al, AlCu | 8kW DC | $12\times10^{-3}$ Ar | 2.5 |
| Ti | 1kW RF | $12\times10^{-3}$ Ar | 0.1 |
| **MRC 662 PARALLEL SCAN**[A] | | | |
| Al, AlCu | 8kW DC | $10\times10^{-3}$ Ar | 7 cm/s |
| Ti | 6kW DC | $10\times10^{-3}$ Ar | 243 cm/s |
| TiN | 6kW DC, 100V DC Bias | $10\times10^{-3}$ N$_2$/Ar | 40 cm/s |
| **VARIAN M2000**[B] | | | |
| Al, AlCu | 11kW DC | $7\times10^{-3}$ Ar | 17 |
| Ti | 2.4kW DC | $4\times10^{-3}$ Ar | 1.5 |
| TiN | 6kW DC, 100V RF Bias | $4\times10^{-3}$ N$_2$/Ar | 1.1 |

[A]Deposition rate controlled by varying the scan rate.
[B]Substrate heated to 100°C.

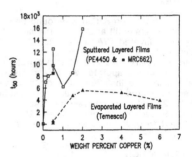

**Fig. 1.** $t_{50}$ (at 250°C, $2.5\times10^6$ A/cm$^2$) as a function of Cu solute concentration for annealed, 2.2$\mu$m wide, 25nm Ti/ 950nm Al-Cu/ 25nm Ti/ 50nm AlCu; sputtered versus evaporated films.

## RESULTS

Sputtered (PE4450 and MRC 662) layered films had electromigration lifetimes > 6000h at Cu concentrations between 0.12 and 2wt.%Cu, whereas evaporated (Temescal) films had decreased lifetimes below 2wt.%Cu (Fig. 1) [7]. The electromigration results and film characteristics are summarized in Tables II (evaporated films) and III (sputtered films). The average grain size and grain size distribution were determined by plotting log grain size (from TEM) versus cumulative probability (see Figures 2 & 3). The texture was quantified as the maximum (111) intensity ($I_{max}$), the volume fraction of (111) and random grains, and the half width of the (111) component ($\omega$). Figure 4 shows a higher $I_{max}$ (41 vs. 5.3), a lower random volume fraction (0.15 vs. 0.2) and tighter $\omega$ (6° vs. 29°) for as-deposited sputtered Al-2wt.%Cu versus evaporated Al-1.9wt.%Cu. With annealing the texture differences between these films narrowed, although $I_{max}$ was greater for the sputtered film (300 vs. 9.5) and $\omega$ was smaller (8° vs. 25°), see Tables II and III. In general a higher $I_{max}$, a smaller random volume fraction and a tighter (111) texture (low $\omega$) were obtained for the PE4450 sputtered films compared to the Temescal evaporated films, both as-deposited and annealed. Furthermore for the sputter deposition conditions used (see Table I), the PE4450 films were more highly textured than the M2000 films.

The electromigration lifetimes of evaporated Al-0.67wt.%Cu and Al-1.9wt.%Cu films were considerably improved with annealing (Table II), which caused not only grain growth and a lower film resistivity but also improvements in film texture. Only slight differences were measured in the lifetimes of annealed Al-0.5wt.%Cu and Al-2wt.%Cu sputtered films (Table III) versus annealed Al-0.67wt.%Cu and Al-1.9wt.%Cu evaporated films (Table II), respectively. Evaporated Al-0.67wt.%Cu had a slightly better $t_{50}$ (50h vs. 36h) than sputtered (M2000) Al-0.5wt.%Cu, although $\sigma$ was higher (1.01 vs. 0.43). The poor electromigration results for sputtered (M2000) Al-0.5wt.%Cu films were unexpected since these films were narrower than the evaporated Al-0.67wt.%Cu films (0.8$\mu$m vs. 2.5$\mu$m, which would tend to increase $t_{50}$), had a lower volume fraction of random grains (0.07 vs. 0.29), but had a larger $\omega$ (22° vs. 17°). The large volume fraction of random grains (0.29) and the high $\omega$ (= 17°) for the evaporated Al-0.67wt.%Cu films could account for the large $\sigma$ measured. Random grains and large misorientation angles between adjacent (111) grains (i.e., $\omega$) will act as flux divergences, which lead to early electromigration failures and, consequently, a high lognormal $\sigma$. For example the highly textured sputtered (PE4450) Al-2wt.%Cu had a longer $t_{50}$ (197.7h vs. 79.1h) and a lower $\sigma$ (0.43 vs. 0.96) than the poorly textured evaporated Al - 1.9wt.%Cu films.

**TABLE II**. Electromigration results and film characteristics for evaporated AlCu.

| Film (thickness) (sample width) | $t_{50}$ (h) | $\sigma$ | $I_{max}$ | Volume Fraction (111) | Volume Fraction (random) | $\omega$ (°) | Grain Size ($\mu$m) | $\rho$ ($\mu\Omega$cm) |
|---|---|---|---|---|---|---|---|---|
| Al-0.67Cu (400nm) | 4.1 | 0.29 | -- | -- | -- | -- | 0.45 | 3.4 |
| annealed (2.5$\mu$m) | 50.1 | 1.01 | 16.2 | 0.71 | 0.29 | 17 | 0.80 | 3.1 |
| Al-0.67Cu OUC[A] | -- | -- | 159 | 0.76 | 0.24 | 20 | 0.23 | 2.9 |
| annealed (2.2$\mu$m) | 155.2 | 0.29 | 303 | 0.78 | 0.22 | 7 | 0.79 | 3.5 |
| Al-1.9Cu (400nm) | 4.7 | 0.23 | 5.3 | 0.8 | 0.2 | 29 | -- | 3.7 |
| annealed (2.5$\mu$m) | 79.1 | 0.96 | 9.5 | 0.92 | 0.08 | 25 | -- | 3.3 |

[A]OUC = 25nm Ti/950nm Al-0.67wt.%Cu/25nm Ti/50nm Al

**TABLE III**. Electromigration results and film characteristics for sputtered AlCu.

| Film (thickness) (sample width) | $t_{50}$ (h) | $\sigma$ | $I_{max}$ | Volume Fraction (111) | Volume Fraction (random) | $\omega$ (°) | Grain Size ($\mu$m) | $\rho$ ($\mu\Omega$cm) |
|---|---|---|---|---|---|---|---|---|
| Al OUC[A] | -- | -- | -- | -- | -- | -- | -- | 2.8 |
| annealed (2.2$\mu$m) | 48.8 | 0.60 | 126 | 0.81 | 0.19 | 9 | -- | 3.3 |
| Al-0.5Cu OUC[A] | -- | -- | 165 | 0.8 | 0.2 | 4 | 0.3 | 3.1 |
| annealed (2.2$\mu$m) | 8600 | 0.98 | 102 | 0.9 | 0.1 | 10 | 0.8 | 3.4 |
| Al-0.5Cu OUC[B] | -- | -- | 75 | 0.95 | 0.05 | 8 | -- | 3.3 |
| annealed (1.3$\mu$m) | > 8000 | -- | 250 | 0.97 | 0.03 | 7 | -- | 3.7 |
| Al-0.5Cu (1$\mu$m)[C] | -- | -- | 18 | 0.92 | 0.08 | 31 | 0.2 | 3.0 |
| annealed (0.8$\mu$m) | 35.6 | 0.43 | 14 | 0.93 | 0.07 | 22 | 1.4 | 2.8 |
| Al-0.5Cu OUC[C] | -- | -- | -- | -- | -- | -- | ** | 3.2 |
| annealed (0.8$\mu$m) | ** | ** | 305 | 0.93 | 0.07 | 7 | ** | 3.7 |
| Al-2Cu (1$\mu$m)[A] | -- | -- | 41 | 0.85 | 0.15 | 6 | -- | 3.2 |
| annealed (2.2$\mu$m) | 197.7 | 0.43 | 300 | 0.94 | 0.06 | 8 | -- | 2.9 |
| Al-2Cu OUC[A] | -- | -- | 85 | 0.94 | 0.06 | 11 | -- | 3.0 |
| annealed (2.2$\mu$m) | > 15000 | -- | 234 | 0.95 | 0.05 | 5 | -- | 3.6 |
| Al-2Cu (1$\mu$m)[C] | -- | -- | -- | -- | -- | -- | 0.17 | 3.3 |
| annealed (0.8$\mu$m) | 288.5 | 0.38 | 7 | 0.95 | 0.05 | 15 | 1.3 | 2.9 |
| Al-2Cu OUC[C] | -- | -- | -- | -- | -- | -- | ** | 3.6 |
| annealed (0.8$\mu$m) | ** | ** | 55 | 0.85 | 0.15 | 14 | ** | 3.7 |

[A]Perkin Elmer 4450: OUC = 25nm Ti/950nm Al-x wt.%Cu/25nm Ti/50nm Al
[B]MRC 662 Parallel Scan: OUC = 25nm Ti/950nm Al-x wt.%Cu/25nm Ti/40nm TiN
[C]Varian M2000: OUC = 25nm Ti/950nm Al-x wt.%Cu/25nm Ti/40nm TiN
** to be determined

The microstructure of sputtered and evaporated layered films with < 2wt.%Cu were compared in an effort to explain the order of magnitude difference in their electromigration lifetimes. Within experimental error, the grain size, grain size distributions and film resistivity were identical for these films both before and after annealing. The Al (111) texture, however, was different; the random volume fraction tended to be lower for the sputtered films both before (Fig. 5) and after annealing, independent of the Cu content. It was found that film texture was not noticeably altered by solute additions (< 2wt.%Cu) but was influenced by the sputtering tool used and was a strong function of the deposition parameters for a particular tool.

The texture of AlCu films deposited on Ti was generally improved independent of the deposition technique. Figure 6 shows a higher $I_{max}$ (305 vs. 14), an identical random volume fraction (0.07) and a tighter $\omega$ (7° vs. 22°) for sputtered (Varian M2000), annealed, layered films versus Al-0.5wt.%Cu. The texture of thin Ti films (30 - 100nm) was found to be (0002) oriented and slightly dependent on the deposition conditions (Fig. 7 and Table IV). The Ti texture was quantified as the maximum (0002) intensity ($I_{max}$) and the half width of the (0002) component ($\omega$). AlCu films deposited on (0002) textured Ti layers had stronger $I_{max}$ and smaller $\omega$ texture components. The AlCu random volume fraction was less affected by the presence or absence of a Ti layer, but was influenced by the AlCu deposition process.

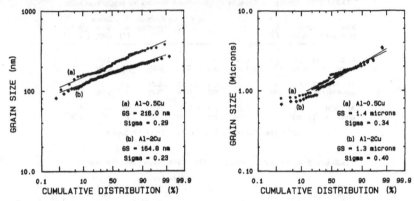

**Fig. 2.** Grain size distributions as cumulative probability plots for 1$\mu$m thick sputtered (M2000) Al-0.5wt.%Cu and Al-2wt.%Cu as-deposited (left) and annealed for 1h at 400°C in 90% $N_2$ - 10% $H_2$ (right).

**Fig. 3.** TEM micrographs of sputtered (M2000) Al - 2wt.%Cu (top) as-deposited and (bottom) annealed (1h at 400°C in 90% $N_2$ - 10% $H_2$).

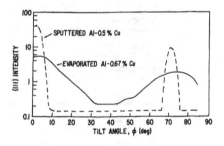

**Fig. 4.** (111) pole intensity versus tilt angle, $\phi$, from the sample normal direction, for as-deposited 1$\mu$m sputtered (PE4450) Al - 2wt.%Cu versus 400nm evaporated Al - 1.9wt.%Cu.

## DISCUSSION

All of the sputtered layered samples investigated showed a high degree of AlCu (111) fiber texture and large average grain sizes with a low variance, independent of the deposition tool. These factors alone would predict enhanced electromigration lifetimes, compared to a randomly oriented AlCu film with a small average grain size and/or a large grain size variance. In general, the evaporated layered films had higher random volume fractions, both before and after annealing, and lower lifetimes. These results are consistent with previous results in which stronger film texture resulted in improved electromigration lifetimes [12-14]. A perfect (111) fiber texture produces only tilt boundaries, a (111) spread, i.e. $\omega >$ 0, produces twist components while random grains can have boundaries of any orientation. If flux divergences are more likely to occur at boundaries of widely different orientations, due to inherent differences in transport properties, then, on average, films with higher random volume fractions and / or large misorientation angles between adjacent (111) grains will tend to electromigrate faster due to a larger density of flux divergence sites (or weak-links [15]).

The deposition process was found to directly affect the microstructure and, therefore, to control the number and distribution of weak links in a film. Varying the deposition parameters to consistently obtain a highly (111) textured Al film will improve those thin film reliability phenomena which depend on grain boundary transport, such as interdiffusion reactions, electromigration and stress-induced voiding [16]. Texture measurements are required to fully characterize film microstructure, in addition to grain size measurements, since films with identical grain sizes and grain size distributions may or may not have the same film texture [12, 13]. The annealing conditions, solute concentration, solute distribution and substrate will also affect film texture, although in this study the Al texture was not noticeably altered by small amounts (< 2wt.%) of Cu solute additions.

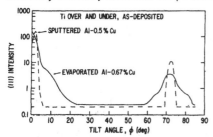

**Fig. 5.** (111) pole intensity versus tilt angle, $\phi$, from the sample normal direction, for as-deposited sputtered (PE4450) 25nm Ti/ 950nm Al-0.5wt.%Cu/ 25nm Ti/ 50nm AlCu versus evaporated 25nm Ti/ 950nm Al - 0.67wt.%Cu/ 25nm Ti/ 50nm AlCu.

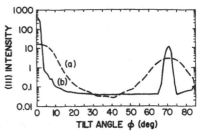

**Fig. 6.** (111) pole intensity versus tilt angle, $\phi$, from the sample normal direction for sputtered (Varian M2000), annealed (a) Al-0.5wt.%Cu (1$\mu$m thick) and (b) 25nm Ti/ 950nm Al-0.5wt.%Cu/ 25nm Ti/ 40nm TiN.

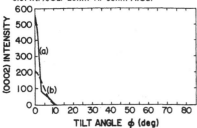

**Fig. 7.** (0002) pole intensity versus tilt angle, $\phi$, from the sample normal direction for (a) sputtered (Varian M2000) as-deposited 30nm Ti (condition (B) in Table IV) and (b) evaporated (Temescal), as-deposited Ti (100nm thick).

**TABLE IV.** Titanium (as-deposited) texture.

| Film | $I_{max}$ | $\omega(°)$ |
|------|-----------|-------------|
| 100nm Ti[A] | 220 | 13 |
| 30nm Ti[B] | 550 | 10 |
| 30nm Ti[C] | 300 | 10 |
| 30nm Ti[D] | 300 | 10 |

[A]Evaporated (FC1800): 5x10$^{-7}$ Torr, 2nm/s, no heat

[B]Sputtered (M2000): 2.4kW DC, 4mTorr Ar, 1.5nm/s, 100°C

[C]Sputtered (M2000): 2.4kW DC, 17mTorr Ar, 1.5nm/s, 100°C

[D]Sputtered (M2000): 2.4kW DC, 4mTorr Ar, 1.5nm/s, 300°C

Previously a different Ti-Al intermetallic structure was found in sputtered ($TiAl_3$) versus evaporated ($Ti_9Al_{23}$) layered films [7]. It was argued that the presence of $Ti_9Al_{23}$ in evaporated layered films was responsible for the decreased electromigration lifetimes. The formation of superlattice structures, such as ($Ti_9Al_{23}$) in the Ti-Al system, result in Ti rich unit cells, i.e. Al deficient structures. Such intermetallics tend to form large unit cells due to slight shifts in the atomic positions of the Ti and Al atoms from the symmetric positions found in the closely related line compound $TiAl_3$. A larger point defect density, which more readily allows for atomic diffusion, would also be expected in Ti-Al superlattices than in $TiAl_3$ stoichiometric compounds. This would result in faster Al and Cu diffusion through the superlattice redundant layer(s), than through similar $TiAl_3$ layers, resulting in shorter lifetimes. Film texture would affect both the center AlCu conductor and the Ti-Al intermetallic layers, while the type of intermetallic formed would only affect the reliability of the redundant layers. Although we do not know the relative magnitude of each of these effects (since they occur in series) both certainly influence the electromigration reliability of layered Ti/AlCu/Ti films.

## CONCLUSIONS

A strong correlation was found between film texture, quantified as fiber plots of (111) intensity versus tilt angle from the normal direction, and the resulting electromigration behavior of layered Ti/AlCu/Ti films. For films with an identical grain size and grain size distribution, superior lifetimes were found for those films which had low volume fractions of randomly oriented grains, strong and sharp texture. The importance of obtaining good texture, and therefore reliable films, suggests that it be routinely measured in addition to other parameters (grain size, grain size distribution, resistivity, stress, etc.) in order to quantify film quality, since films which appear to have similar microstuctures do not necessarily have similar texture components.

## ACKNOWLEDGEMENTS

We would like to thank S. Brodsky, J.D. Mis and S. Sanchez for their help with film deposition, TEM grain size measurements and TEM sample preparation, respectively.

## REFERENCES

1. T. Kwok and P.S. Ho in Diffusion Phenomena in Thin Films and Microelectronic Materials edited by D. Gupta and P.S. Ho (Noyes Publications, Park Ridge, NJ 1988), p. 369.

2. I. Ames, F.M. d'Heurle and R.E. Horstmann, IBM J. Res. Develop. *14*, 461 (1970).

3. J.K. Howard and P.S. Ho, U.S. Patent No. 4 017 890 (April 12, 1977).

4. J.K. Howard, J.F. White and P.S. Ho, J. Appl. Phys., *49*, 4083 (1978).

5. D.S. Gardner, T.L. Michalka, K.C. Saraswat, T.W. Barbee, J.P. McVittie and J.D. Meindl, IEEE Electron Devices, *ED-32*, (1985).

6. D.S. Gardner, K. Saraswat, T.W. Barbee, U.S. Patent No. 4 673 623 (June 16, 1987).

7. K.P. Rodbell, P.W. Dehaven and J.D. Mis, in Proceedings of the First MRS Symposium on Materials Reliability Issues in Microelectronics, (MRS Publications, Pittsburgh, 1991), *225*, p.91.

8. K.P. Rodbell, P.A. Totta and J.F. White, U.S. Patent No. 5 071 714 (December 10, 1991).

9. R. Rosenberg, J. Vac. Sci. Technol., *9*, 263 (1972).

10. D.R. Frear, J.E. Sanchez, A.D. Romig, J.W. Morris, Jr., Metall. Trans., *21A*, 2449 (1990).

11. D.R. Frear, J.R. Michael, C. Kim, A.D. Romig, Jr. and J.W. Morris, Jr., in Proceedings of SPIE Conference on Metallization: Performance and Reliability Issues for VLSI and ULSI, San Jose, CA (IEEE, New York 1991).

12. D.B. Knorr, K.P. Rodbell and D.P. Tracy, in Proceedings of the First MRS Symposium on Materials Reliability Issues in Microelectronics, (MRS Publications, Pittsburgh, 1991), *225*, p.21.

13. D.B. Knorr, D.P. Tracy and K.P. Rodbell, Appl. Phys. Lett., *59*, 3241 (1991).

14. S. Vaidya and A.K. Sinha, Thin Solid Films *75*, 253 (1981).

15. D.B. Knorr and K.P. Rodbell, this symposium.

16. S.A. Lytle and A.S. Oats, J. Appl. Phys., *71*, 174 (1992).

### EFFECTS OF FILM TEXTURE ON ELECTROMIGRATION LIFETIME PREDICTIONS

D.B. Knorr* and K.P. Rodbell**
*Materials Engineering Department and Center for Integrated Electronics,
Rensselaer Polytechnic Institute, Troy, New York 12180-3590.
**IBM Research Division, T.J. Watson Research Center, Yorktown Heights,
New York 10598.

ABSTRACT

Pure aluminum film was deposited on oxidized silicon to develop three
different conditions for study. Grain size and grain size distribution were
the same for all cases isolating texture as the major microstructural variable. Electromigration tests were done on 1.8μm thick lines at $1x10^6$ A/cm$^2$
at several temperatures between 150°C and 250°C. Over the entire temperature
range a strong (111) texture improves the median time to failure. The
applicability of a log normal fit to the failure distribution is questioned
because the sites that are susceptible to electromigration damage vary in a
statistical fashion, sometimes giving what appears to be a bimodal failure
distribution. Activation energy is also a function of texture but depends on
the applicability of the log normal failure distribution to give values that
can be correlated with transport processes.

### INTRODUCTION

The influence of microstructure on the electromigration (EM) behavior
of aluminum-based metallurgies is well established. Three microstructural
attributes are important: 1) grain size, 2) grain size distribution, and
3) texture [1]. the microstructure influences the flux divergence at local
sites in a line. Local differences in current - induced transport results
in depletion of material (voids) or accumulation of material (hillocks) by
grain boundary transport. A small grain sized material has many more sites
to nucleate damage than a large grain sized material. The proximity of small
and large grains induces material transport away from the vicinity of small
grains only to accumulate at large grains which block diffusional transport.
The presence of a texture means that specific boundaries have different
orientations and, therefore, different boundary diffusivities.
An ongoing study is considering the effect of texture on electromigration behavior [2,3]. Pure aluminum films with three distinct textures but
essentially the same grain size and grain size distribution have received
extensive microstructural characterization [4,5]. Electromigration results
are reported at 225°C [3] and, to a lesser extent, at 150°C [2]. Additional
data are reported here for tests at 150, 175, 200, and 250°C. The activation
energy is an important quantity that is calculated from an Arrhenius plot of
accelerated test data which is then used in extrapolations to predict the
reliability at operating conditions. An analysis of data from the three
textures considers the applicability and validity of the Arrhenius plots to
correlate the failure data and to make lifetime predictions.

### EXPERIMENTAL PROCEDURES

Films of 1μm thick pure aluminum were deposited on oxidized silicon by
a self-ion assisted technique (partially ionized beam, PIB [6]) or by
sputtering. Two PIB conditions (PIB-2/1 and PIB-2/2) and one sputtered
condition (Sp-2) were patterned into electromigration test structures by
standard silicon processing techniques followed by a forming gas (90% $N_2$ -
10 $H_2$) anneal at 400° for 1 hour. Both passivated (200°C) and unpassivated
(all other temperatures) lines of 1.8μm width were tested at a current

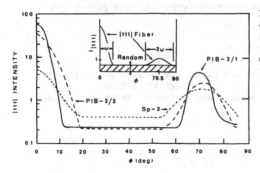

Figure 1. Fiber texture plot of (111) pole intensity as a function of tilt angle for three aluminum film conditions ω gives the half width of the (111) component.

## Table I

### SUMMARY OF MICROSTRUCTURE DATA

| Deposition Condition | Medium Grain Size (μm) | Standard Deviation | Component Volume Fraction Random | (111) | ω (deg) |
|---|---|---|---|---|---|
| PIB-2/1 | 0.75 | 0.48 | 0.23 | 0.77 | 11 |
| PIB-2/2 | 0.64 | 0.49 | 0.22 | 0.78 | 17 |
| Sp-2 | 0.75 | 0.52 | 0.42 | 0.58 | 17 |

density of $1\times10^6$ A/cm$^2$. From 10 to 19 lines per condition were tested at temperatures of 150, 175, 200, 225, and 250°C. Specimens from all conditions were exposed simultaneously in the same oven. Microstructural and failure analyses included determination of grain size and grain size distribution by transmission electron microscopy, texture by X-ray diffraction pole figure analyses, and post-test examination of failed lines by scanning electron microscopy (SEM) and focused ion beam (FIB) microscopy.

## RESULTS

The microstructural information is summarized in Table I. The grain size data were obtained by TEM of the pad area. Figure 1 plots the texture as log (111) intensity versus tilt angle $\phi$ showing the two components, (111) grains and random grains. The tilt angle is measured from the specimen

## TABLE II

### SUMMARY OF ELECTROMIGRATION FAILURE PARAMETERS

| Test Temp. (°C) | PIB-2/1 $t_{50}$ | σ | PIB-2/2 $t_{50}$ | σ | Sp-2 $t_{50}$ | σ |
|---|---|---|---|---|---|---|
| 150 | 7946 | 0.37 | 3922 | 1.01 | 1833 | 1.34 |
| 175 | - | - | - | - | 86 | 0.41 |
| 200 | - | - | 206 | 1.29 | 78 | 0.8 |
| 225 | 772 | 0.28 | 235 | 1.23 | 29 | 1.6 |
| 250 | 240.2 | 0.45 | 26.1 | 0.66 | - | - |

normal direction toward the plane of the specimen. These fiber textures demonstrate axial symmetry so there is no preferred orientation by rotation about the normal axis. Two metrics characterize the texture: $\omega$ which defines the width of the (111) component distribution and the volume fraction of random grains; both are listed in Table I for the three materials.

The electromigration failure results are compiled in Table II. It is very important to understand how the $t_{50}$ and $\sigma$ values are obtained. A data set is analyzed statistically to produce a best fit as log t versus cumulative failure probability. Both $t_{50}$ and $\sigma$ ( $= \ln t_{50}/t_{16}$) are calculated from the regression fit through the entire data set. Selected data are presented as log t versus cumulative probability: 150°C in Figure 2 for all material, 175°C in Figure 3 for Sp-2, and 250°C in Figure 4 for PIB-2/1 and PIB-2/2. The data at 225°C were published elsewhere [2,3].

The activation energy for electromigration is calculated from these data. Figure 5 shows $\ln t_{50}$ plotted against $1/kT$ with the slope equal to the activation energy. A full statistical analysis of the data yields the following activation energies:

- PIB-2/1:  0.61 ± 0.10 eV          - PIB-2/2:  0.81 ± 0.08 eV
- Sp-2:        0.90 ± 0.10 eV

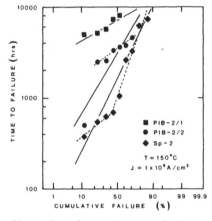

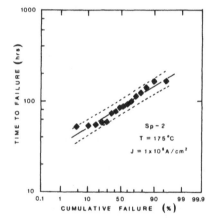

Figure 2. Electromigration failure data at 150°C. Solid lines are regression fits assuming a log normal failure distribution. Dashed lines indicate a bimodal distribution.

Figure 3. Electromigration failure data at 175°C. Solid line gives regression fit and dashed lines give 95% confidence bounds.

## DISCUSSION

The data confirm that texture has a very strong effect on lifetime over the temperature range 150°C to 250°C. The distinguishing microstructural attribute is the texture where $t_{50}$ always ranks in the order PIB-2/1, PIB-2/2, and Sp-2. Experience with other aluminum alloy metallurgies ]7] shows that a strongly textured sample has a tight (111) distribution with $\omega < 10°$ and a volume fraction of random grains $\leq 0.10$, the electromigratoin behavior orders in a similar manner with texture, and deposition conditions control the texture. Annealing modifies the texture slightly but is much less important than deposition parameters ]2,3]. Modification of deposition conditions to obtain only "strong link" texture components will enhance electromigration reliability.

The $t_{50}$ and $\sigma$ values in Tables II are derived from regression fits to

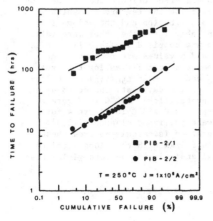

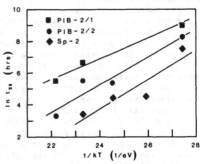

Figure 5. Arrhenius plot of $t_{50}$ data for three conditions. Lines are regression fits where the slope represents the activation energy.

Figure 4. Electromigration failure data at 250°C.

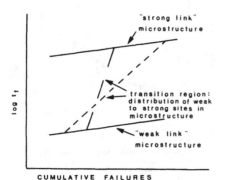

Figure 6. Schematic representation of failure data indicating the effects of microstructure on $t_{50}$ and $\sigma$.

the probability plot of the log time-to-failure data such as Figures 2-4. A straight line fit presumes a log normal distribution of failure times and the presence of only one failure mode. The $\sigma$ values vary quite dramatically for the three material conditions. The PIB-2/1 data are distinctive because the strongest texture results in consistently low $\sigma$ values. The remaining two textures show a wide range but generally higher magnitudes for $\sigma$. The data for Sp-2 at both 150°C (Figure 2) and 225°C [2,3] are most interesting because the failure distributions are bimodal with a low slope for early fails ($\sigma$ values of ∿ 0.7 and ∿ 0.4, respectively) and a very high slope for later fails. The last fail for all three conditions occurs at a comparable time at 225°C.

The presumption of a monomodal failure distribution implies that all failure sites are log - normally distributed. If all sites are "weak" (susceptible to EM damage) or all sites are "strong" (resistant to EM damage), this condition is satisfied. If the distribution of sites ranges between weak and strong, a broader failure distribution results with a correspondingly greater $\sigma$. This behavior is shown in the schematic plot of potential failure distributions in Figure 6. The data reported in Figures 2-4 and References [2,3] fit this interpretation. PIB-2/1 is a "strong link" microstructure which is resistant to damage. Sp-2, the weakest texture, shows behavior in the "weak link" and transition region while PIB-2/2 is entirely in the transition region.

The previous discussion concerns a statistical interpretation, but the local microstructure defines the "weak link" and "strong link" components. Examination of failure sites by SEM and FIB presents several important observations:

1) a small grain adjacent to larger grains is often the voiding site;
2) several damage locations are present in each 254µm line;
3) a void and an extrusion are paired in close proximity;
4) a large grain, the width of the line, is often the extrusion site.

The ranking of the three materials clearly and unambiguously demonstrates the influence of texture while microscopic examination of the failure sites shows the importance of local grain structure to the damage process.

Local microstructural inhomogeneities (grain structure and texture) influence the local flux divergence during electromigration. The effect of grain structure is reported from tests on lines with a bimodal grain size distribution [8,9] where damage accumulates at triple points coincident with an abrupt change in grain size. The cumulative flux divergence at a triple point depends on the geometry (boundary inclination with respect to the direction of current flow) and on the diffusivity of each boundary connected to the triple point [10,11]. On this basis the effect of texture can be postulated. The diffusion coefficient is a strong function of boundary orientation [12,13]. The (111) texture has (111) boundaries that are a complex mixture of symmetric, non-symmetric, tilt, and twist components. The random texture component can produce a boundary of almost any orientation when a randomly oriented grain contacts either a (111) component grain or, less likely, a random component grain. The strong texture effect on electromigration implies that the magnitude of orientation - induced flux divergence is minimized by (111) boundaries with a low misorientation. Two effects could be operative here: 1) low boundary diffusion coefficient which slows the transport of matter to extrusions or from voids, and/or 2) near isotropy of boundary diffusivity regardless of magnitude which minimizes the flux divergence along grain boundaries connected to a triple point.

The median grain size in the lines (< 0.75µm) and the 254µm line length means that there are on the order of 800 boundaries along the length of each line. The number of potential damage sites is much lower being on the order of 10's of sites per line. Thus, a texture which produces a spectrum of weak to strong sites coupled with the restricted number of damage sites per line results in a broad failure distribution (high σ). Alternately, potential damage sites of nearly equal strength, high or low, produce a tight (low σ) failure distribution. This behavior indicates that testing a large number of lines is required to fully delineate the failure distribution. Moderately textured samples might be expected to show a sigmoidal distribution or some combination of "strong link" + transition region or "weak link" + transition regions depicted in Figure 6.

A surprising result is the apparent texture effect on the calculated activation energy for electromigration. Most studies report a value of 0.4 to 0.8 eV [14] with typical values of 0.5 to 0.6 eV. Attardo and Rosenburg [8] found that a sharper (111) texture gave a higher activation energy although grain size was substantially different between their two materials. The results reported here show the opposite effect where sharper (111) texture correlates with lower activation energy. The tendency for the weaker textures to have a bimodal failure distribution or high σ indicates that damage sites of various strengths are involved so the use of a log normal distribution is questionable. The variation in activation energy also affects the separation of the upper and lower curves in Figure 6. The dependence of activation energy on texture and the sigmoidal shape of the failure curves in Figure 6 substantially complicates reliability predictions. Further research is required to understand the local aspects of the texture/

grain structure that maximize $t_{50}$, give a constant activation energy, and give failure data that do not a deviate from the curve in Figure 6.

## CONCLUSIONS

1. A strong texture maximizes the time to failure over the temperature range of 150 to 250°C while minimizing the distribution of failure times.
2. The log normal distribution may not be appropriate for correlating electromigration failure data where damage sites of various strengths are present.
3. The activation energy for electromigration depends on texture, but the correlation assumes that the log normal fit to the failure data applies.

## REFERENCES

1. S. Vaidya and A.K. Sinha, Thin Solid Films, 75, 253 (1981).

2. D.B. Knorr, K.P. Rodbell, and D.P. Tracy, in Materials Reliability Issues in Microelectronics, edited by J.R. Lloyd, F.G. Yost, and P.S. Ho (Mater. Res. Soc. Proc. 225, Pittsburgh, PA 1991), pp.21-26.

3. D.B. Knorr, D.P. Tracy, and K.P. Rodbell, Appl. Phys. Lett., 59, 3241 (1991).

4. D.B. Knorr, D.P. Tracy, and T.-M. Lu, in Evolution of Thin Film and Surface Microstructures, edited by C.V. Thompson, J.Y. Tsao, and D.J. Srolovitz (Mater. Res. Soc. Proc. 202, Pittsburgh, PA 1991) pp.199-204.

5. D.B. Knorr, D.P. Tracy, and T.-M. Lu, Textures and Microstructures, 14-18, 543 (1991).

6. S.-N. Mei and T.-M. Lu, J. Vac. Sci. Tech. A, 6, 9 (1988).

7. K.P. Rodbell, D.B. Knorr, and D.P. Tracy, this symposium.

8. M.J. Attardo and R. Rosenburg, J. Appl. Phys., 41, 2381 (1970).

9. J. Cho and C.V. Thompson, Appl. Phys. Lett., 54, 2577 (1989).

10. M. Genut, Z. Li, C.L. Bauer, S. Mahajun, P.F. Tang, and A.G. Milnes, Appl. Phys. Lett., 58, 2354 (1991).

11. J.E. Sanchez, Jr., and J. W. Morris, Jr., in Materials Reliability Issues in Microelectronics, edited by J.R. Lloyd, F.G. Yost, and P.S. Ho (Mater. Res. Soc. Proc. 225, Pittsburgh, PA 1991) pp.53-58.

12. I. Kauer, W. Gust, and L. Kozma, in Handbook of Grain and Interphase Boundary Diffusion Data, Vol. 1 (Ziegler, Stuttgart, 1989), pp.118-143.

13. D. Turnbull and R.E. Hoffman, Acta Metall., 2, 419 (1954).

14. T. Kwok and P.S. Ho, in Diffusion Phenomena in Thin Films and Microelectronic Materials, edited by D. Gupta and P.S. Ho (Noyes, Park Ridge, NJ, 1988), p.369.

# QUANTITATIVE ANALYSIS OF ELECTROMIGRATION-INDUCED DAMAGE IN Al-BASED INTERCONNECTS.

O. KRAFT, J.E. SANCHEZ, JR., and E. ARZT
Max-Planck Institut für Metallforschung, and Institut für Metallkunde, University of Stuttgart, D-7000 Stuttgart, Germany

## ABSTRACT

Electromigration in metal interconnect lines produces sites of damage, such as voids, hillocks and whiskers, which by definition are the sites of flux divergence in the lines. Detailed observations of damage volume and morphology, especially in relation to the local microstructure, may yield vital information about the processes which produce the damage and ultimate failure in the interconnects. We present fractographic measurements of void volumes and the spacing between voids and corresponding hillocks in Al and Al-2% Cu interconnects which have been electromigration tested until failure. It is shown that the void density as well as the shape of failure voids depend on the current density. Further it is found that the distribution of the spacings between voids and corresponding hillocks changes as a function of current density.

## INTRODUCTION

Electromigration (EM) is generally described as the net drift of metal ions due to the momentum transfer from conduction electrons which move in an applied electric field. This drift can be observed and the average drift velocity measured using, for example, the Blech edge displacement technique [1]. The effect has been attributed to mass depletion and developing tensile stresses at the cathode end and mass accumulation and developing compressive stress at the anode end of the stripe [2,3]. This stress gradient leads to a diffusional flux that counterbalances the EM mass flux. If this effect is incorporated in Black's [4] equation the following expression for the drift velocity results [1]:

$$v = \frac{D}{kT} eZ^* \rho \, (j - j_c) \qquad \text{where} \qquad j_c = \frac{\sigma^* \, \Omega}{eZ^* \, \rho \ell} \qquad (1)$$

Here D is the diffusivity (for the normally low homologous temperatures and large grains D is identical to the grain boundary diffusivity), $eZ^*$ is the effective charge of the atom, $\rho$ the electrical resistivity, j the current density, $j_c$ the threshold current density, $\Omega$ the atomic volume, $\ell$ the length of the stripe, and $\sigma^*$ the maximum hydrostatic stress the metallization is able to sustain. From equation (1) follows that below the threshold product

$$(j \cdot \ell)_c = j_c \cdot \ell = j \cdot \ell_c \leq \frac{\sigma^* \, \Omega}{eZ^* \, \rho} \qquad (2)$$

no EM or stripe drift occurs.

Mat. Res. Soc. Symp. Proc. Vol. 265. ©1992 Materials Research Society

It is common to apply this approach in understanding the effects of EM to continuous lines with near-bamboo microstructure, e.g. [5-7]. The assumed situation there is shown in Fig. 1, where the EM flux, concentrated in the grain boundaries, produces damage (i.e. voids and hillocks) at sites of flux divergence. Equation (1) suggests a direct relation between $\ell$ and the magnitude of the net flux along the line which produces the observable damage. It follows (equation 2) that below a certain spacing $\ell_c$ for a given current density no voiding should occur. Furthermore a correlation between the product $(j\ell)_c$ and the maximum stress $(\sigma^*)$ is suggested, hence strengthening the film may improve EM resistance [6,8].

In this paper we apply methods of fractography, as used in the analysis of damage and failure in structural or mechanical test components, to the microstructure of EM damage in order to achieve a better understanding of the failure mechanism. We present a quantitative analysis of EM induced void size, void density and the spacing between corresponding voids and hillock in Al and Al-Cu interconnects of near bamboo microstructure. We also present the first attempts to correlate these results to existing mechanisms and models for electromigration reliability.

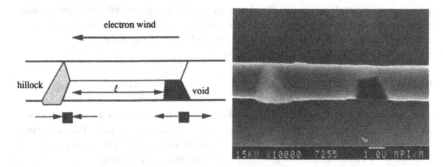

Fig. 1: Schematic illustration of the EM induced damage morphology in a continuous line, as a result of the divergences in the EM-flux.

Fig. 2: SEM micrograph of a void and the corresponding hillock on an Al line.

## EXPERIMENTAL PROCEDURE

Al and Al-2wt.%Cu films 0.5 μm thick were magnetron sputter deposited onto thermally oxidized (100) oriented 100 mm diameter silicon substrates. The deposition rate was approximately 2 nm/s at an Ar pressure of $2\text{-}3 \cdot 10^{-3}$ torr and the sputter system base pressure was better than $4.0 \cdot 10^{-7}$ torr. Films were annealed in forming gas in a hot wall furnace at 400°C for 45 minutes (Al) and at 425°C for 30 minutes (Al-Cu). After annealing films were patterned into parallel line arrays (PLA) of 20 lines 1 mm long (Al) or 25 lines each 1.5 mm long (Al-Cu) using standard lithographic and etch processes. Such PLA structures allow the simultaneous testing of large numbers of interconnects at a constant applied voltage with a single power supply. Failure sites will not be destroyed by arcing as the line produces an open circuit with this testing technique [9]. This method is suitable for quantitative characterization of EM induced damage.

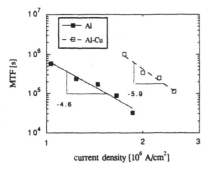

Fig. 3: A comparison of the Median Time to Failure (MTF) for the Al and the Al-Cu films as a function of current density.

The EM tests of unpassivated lines were performed on diced wafer chips placed on a hot chuck of a manual probe station. Testing temperature was 230°C, and the interconnects were 2.0 μm (Al) and 1.3 μm (Al-Cu) wide; current densities were in the ranges $1.0\text{-}1.9\cdot10^6$ A/cm$^2$ (Al) and $1.75\text{-}2.5\cdot10^6$ A/cm$^2$ (Al-Cu). The PLA resistance was constant during a stepwise increase of the current to the test condition, indicating that the line temperature was not measurably increased above the hot chuck temperature by Joule heating. After EM testing samples were observed in a scanning electron microscope (SEM). Fig. 2 shows the typical morphology observed, here a void (appearing black) and a corresponding hillock indicate the direction of current and mass flux. (In all figures the direction of the electrons is from right to left.)

The entire length of each line in the PLA was examined in the SEM for damage. A micrograph was taken of each void, and the distance to the next (downstream) hillock was measured. We assumed that the void area thus measured is directly proportional to the void volume since typically the sidewalls of the voids were perpendicular to the film surface.

## RESULTS

Results of EM testing for the two films are shown in fig. 3: the dependence of Median Time to Failure (MTF) on the current density is given by the well established relation MTF $\sim$ j$^{-n}$, where the current exponents (n) are 4.6 (Al) and 5.9 (Al-Cu). These values are higher than the often reported value of 2.

The SEM observations of the tested structures clearly reveal differences in the extent of EM damage between films and current densities. Void density, defined as the total number of voids divided by the entire length of interconnects in the PLA circuit, is shown in figure 4a. The increase of void density with current density is due to a greater number of voids at all spacings, to be discussed below.

The „length" (measured in the direction along the line) of voids responsible for failure is determined as follows. By definition a fatal void must cross the linewidth (W). In order to compare fatal void geometries or sizes in lines of different widths the measured void area ($A_v$) is normalized by W, which is 2.0 μm and 1.3 μm in the Al and Al-Cu interconnects, respectively. The resulting measure of the fatal void size is shown in figure 5a, where each plotted value is an average of 20 (Al) or 25 (Al-Cu) fatal voids. We note a possible trend of decreasing void length above approximately $1.8\cdot10^6$ A/cm$^2$ for the Al and above $2.3\cdot10^6$ A/cm$^2$ for the Al-Cu interconnects. Such a trend would suggest a change in the void growth mechanism, which is particularly obvious when the Al and Al-Cu data are combined.

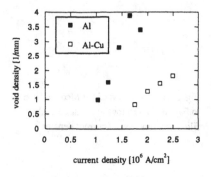

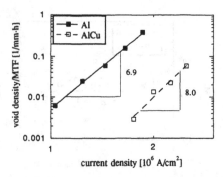

Fig. 4a: Void densities of both films Al and Al-Cu as a function of applied current density.

Fig. 4b: Average void nucleation rate as determined by dividing void density by MTF of each test.

For both films and all test conditions the measured spacings are lognormally distributed, as shown in figure 6a for the Al film at two extreme current densities. It is seen that the median as well as the minimum spacings decrease as the applied current increases. Preliminary analysis suggests that the minimum spacing $\ell_{min}$ decreases proportional to $j^{-3}$, which is a stronger dependency than it would be expected from equation (2). The threshold product $(j\ell)_c$ is shown in figure 6b as a function of the current density. Here the minimum spacing $\ell_{min}$ found in each test was used to calculate the threshold product: $(j\ell)_c = j \cdot \ell_{min}$. The threshold product is shown to decrease with j for both Al and Al-Cu, whereas the absolute values of the threshold products are higher for the Al-Cu film compared to the Al film.

DISCUSSION

The dependence of MTF on the current density shown in figure 3 is stronger than expected. Such high current exponents often occur when a high current density produces Joule heating in the interconnect. However, we can practically rule out this possibility, since a resistance increase was not detected and voids and hillocks were randomly distributed over the entire line length. The reason for the high current exponent is at present unclear.

The void density, as measured after failure of the last line, increases characteristically with current density (fig. 4a). By analyzing the data of spacings between voids and hillocks it turns out that this is due to the decrease of the minimum spacing $\ell_{min}$ as well as an increase of the number of void-hillock pairs with larger spacings. Because the test duration decreased with increasing current density, it makes sense to divide the void density by the MTF at that particular current density; in this way an approximate „average void nucleation rate" is obtained (fig. 4b), which exhibits a strong power-law dependence on j. (Note that even the somewhat inconsistent last data point for Al in fig. 4a now conforms to the straight line fig. 4b). It is also instructive that void nucleation is approximately two orders of magnitude slower in Al-Cu.

A similiar normalization with respect to MTF can be carried out for the average void area (fig. 5b). Because of the expected continuous void nucleation, however, the resulting „average void growth rates" must be considered as rough approximations. It is seen that again a power-law

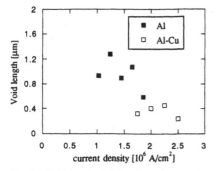

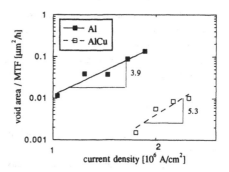

Fig. 5a: Void length defined as measured void area $A_v$ divided by the width W of the line.

Fig. 5b: Estimate of the average void growth rate obtained by dividing the measured void area $A_v$ by MTF of each test.

behaviour obtains and that Cu slows down void growth by two orders of magnitude. The current exponent, while greater than 1 (the value expected on the basis of equation (1)), is characteristically smaller than that for the void nucleation rate.

The determined threshold products $(j\ell)_c$ for both films have the same order of magnitude as found by others using the edge displacement method [1, 10]. However, it is inconsistent with the simple Blech model that the threshold product depends on the current density. Further the results suggest that the Al-Cu film is about six times stronger than the Al film. This is not consistent with recent results by R. Venkatraman [11], who has shown that continuous Al and Al-Cu films have about the same mechanical properties as determined using the wafer curvature technique. Measurements of the strength of interconnects are necessary to test this result for isolated lines.

An observation which is significant for future modelling of the damage accumulation is contained in fig. 5a. The decrease in „length" of the fatal void characterizes a trend which has been qualitatively observed on micrographs: with increasing current density, less material has to be moved by EM and the failure void becomes increasingly more „slit-like". There is an interesting analogy to this situation in creep-failure of high-temperature materials, e.g. [12]. Grain boundary

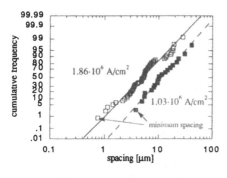

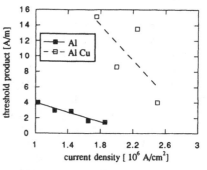

Fig. 6a: Distributions of spacings between voids and corresponding hillocks after EM testing of the Al film at two different current densities.

Fig. 6b: Dependence of the threshold product $(jl)_c$ on the current density for both films.

voids which grow under the action of a tensile stress maintain a spherical equilibrium shape only if surface diffusion is more rapid than grain-boundary diffusion; if on the other hand void growth is controlled by (slow) surface diffusion, voids grow in a crack-like manner. Further analysis and additional tests at higher current densities are presently being performed to verify this mechanism in electromigration.

## CONCLUSIONS

As a result of quantitative characterization of EM induced damage the following conclusions can be made:

1. The void density increases with current density due to both a decrease in the minimum void-hillock spacing and an increase in number of voids at all spacings. When the void density is normalized by MTF, an approximate „void nucleation rate" can be obtained which varies strongly with current density.

2. We find an apparent void size decrease above a current density of $1.8 \cdot 10^6$ A/cm$^2$ for Al and $2.3 \cdot 10^6$ A/cm$^2$ for Al-Cu. This corresponds to a qualitative trend to a greater frequency of slit like voids at higher current densities. A possible reason for this transition is a change of the rate-controlling diffusion mechanism.

3. The threshold product $(j\ell)_c$ has been found to lie in the range between 200 and 1200 A/cm, roughly the same order of magnitude as in other studies. However we find that the critical product decreases with increasing current for both Al and Al-Cu.

4. Careful observation of EM induced damage may yield information about microscopic mechanisms for void and hillock formation which should be helpful for further modelling.

## ACKNOWLEDGEMENTS

We acknowledge helpful discussions with Dr. J. Lloyd. We are also grateful to Dr. S. Bader and Prof. W.D. Nix for an ongoing collaboration Stanford-Stuttgart, which is supported by the Humboldt Foundation and the Max-Planck-Society in the form of a Max-Planck Research Award.

## REFERENCES

1. I.A. Blech, J. Appl. Phys. 47, 1203 (1976)
2. I.A. Blech and C. Herring, Appl.Phys. Lett. 29, 131 (1976)
3. I.A. Blech and K.L. Tai, Appl.Phys. Lett 30, 387 (1977)
4. J.R. Black, IEEE Trans. Electr. Dev. 16, 338 (1969)
5. M. Shatzkes and J.R. Lloyd, J. Appl. Phys. 59, 3890 (1986)
6. E. Arzt and W.D. Nix, J. Mat. Res. 6, 731 (1991)
7. C.A. Ross, Mat. Res. Soc. Proc. 225, 35 (1991)
8. E. Arzt, O. Kraft, J. Sanchez, S. Bader and W.D. Nix, Mat. Res. Soc. Proc."Thin Films: Stresses and Mecanical Properties III" (1991)
9. J.E. Sanchez Jr., L.T. McKnelly and J.W. Morris Jr., J. El. Mat. 19, 1213 (1990)
10. C.A. Ross, J.S. Drewery, R.E. Somekh and J.E. Evett, J. Appl. Phys. 66, 2349 (1989)
11. R. Venkatraman and J.C. Bravman, Mat. Res. Soc. Proc."Thin Films: Stresses and Mecanical Properties III" (1991)
12. A.C.F. Cocks and M.F. Ashby, Progr. Mat. Sci.17, 189 (1982)

# ELECTROMIGRATION PERFORMANCE OF FLUORINATED ALUMINUM FILMS FOR VLSI APPLICATIONS

K.P. MacWilliams, W.E. Yamada, S. Brown, G.K. Yabiku,
*L. Lowry, and M. Isaac

The Aerospace Corporation
P.O. Box 92957 M2-244 Los Angeles, CA 90009
* JPL / Cal Tech
4800 Oak Grove Dr. MS158-103 Pasadena, CA 91109

## ABSTRACT

We have previously shown greatly enhanced resistance to stress-induced hillock formation through fluorine incorporation in aluminum films. Utilizing relatively low F incorporation (<0.1 atomic %), hillock formation density is reduced ~10x over pure or similarly Cu-doped aluminum films. Electromigration tests were performed on a matrix of structures with varying topology (step heights and slopes) and fluorine incorporation dose. We find that although F improves the stress-induced hillock formation by an order of magnitude, the electromigration performance of flat structures is only slightly improved with F incorporation. Analyzing various step heights and step slopes, the non-fluorinated Al experienced a decreasing electromigration lifetime with increasing step height. However, the optimally implanted F samples showed almost no lifetime reduction with step coverages over a similar regime. In addition, scanning electron micrographs of the failed samples revealed that the failures of the fluorinated samples differ markedly from the non-fluorinated samples. Finally, SIMS profiles taken on F and Cu (for comparison) implanted samples reveal the fundamentally different nature of the two beneficial components: Cu redistributes relatively easily throughout the Al film to segregate to grain boundaries. In contrast, the F profile is extremely stable with similar anneals and provides its beneficial effect by forming a distributed refractory metal-like structure within the interconnect.

## INTRODUCTION

Electromigration and stress-induced migration are the two primary reliability concerns in integrated circuit interconnects. Electromigration and stress-migration can induce both voids and hillocks in the interconnects. Hillock protrusions can be induced after high temperature anneals due to thermal expansion coefficient mismatches between the silicon substrate, the surrounding insulator and the aluminum. Hillock formation can lead to cracking of the interlevel dielectric, interlevel shorts, and increased electromigration susceptibility. Electromigration opens or shorts may result from atomic motion caused by electron momentum transfer in the presence of high current density. As VLSI aluminum interconnect line widths have been reduced below ~4μm, electromigration and stress migration have been the key concerns in continuing to scale integrated circuits to increase performance. In addition, these failure modes have become especially critical as the number of metalization layers has increased and the topology has become more harsh.

Electromigration, hillock formation, and stress induced voiding result from physically or electronically induced motion of aluminum atoms. Past efforts to reduce these failure modes in aluminum have involved the intentional addition of impurities - typically Cu or refractory metals in the 1% to 4% range [1,2]. Other techniques involve the use of complex and expensive multilayer or encapsulated refractory metal schemes [3,4]. Each of these techniques not only increases resistivity of the interconnect per unit area, but also greatly increases the etch difficulty. We have previously presented a technique which utilizes ultralow level F incorporation to reduce hillock formation density by approximately an order of magnitude over pure Al, or Al films with similar Cu doping levels [5]. Our goal of this work is to determine if the fluorine incorporation technique also has a beneficial effect on the electromigration failure mode, and to gain additional insight into its already proven positive influence on aluminum interconnect reliability.

## TEST STRUCTURE FABRICATIONS

A 1000Å $SiO_2$ layer was thermally grown on 75mm diameter <100> silicon substrates. Interconnect test structures were made both with and without topology. The structures with topology were formed by depositing CVD polysilicon to thickness of 0.9µm or 0.5µm over the thermally grown $SiO_2$. The flat topology samples had no deposited poly silicon. The topology samples were then patterned using conventional photoresist procedures into 2µm wide steps with 4µm wide spaces. The polysilicon on the topology samples was etched using two different reactive ion etch conditions to produce vertical (~88°) and sloped (~50°) sidewalls on the 0.9µm and 0.5µm poly silicon. Next, the topology samples had 1200Å CVD $SiO_2$ deposited and reflowed at 950°C to provide electrical insulation from the polysilicon and conform to the underlying topology. All samples then had 8000Å Al-1%Si sputter-deposited at a rate of ~2nm/sec. Several wafers were then implanted with varying doses of fluorine to a depth of ~1200Å. To ascertain the chemical influence of fluorine on aluminum, neon was also implanted as a control to separate out the physical effects of the implants. All implanted and non-implanted samples were then patterned into standard, 800µm long, Kelvin-connected, electromigration structures and annealed in forming gas (4% $H_2$ in $N_2$) at 400°C for 20 minutes to complete processing. All samples were then diced, packaged and wire-bonded for electromigration testing.

## RESULTS AND DISCUSSION

We previously showed that relatively low levels of F incorporation in Al (<0.1 atomic %) can reduce hillock formation density about ten times over pure or similarly Cu-doped aluminum films [5]. Extensive SEM and optical microscopy inspections were used to study the hillock formation dynamics as a function of implant dose, species, and anneal temperature. The results of the study for hillocks greater than $1\mu m$ in diameter are summarized in Figure 1. The error bars on the F 300°C data are typical of the relative errors for the other 5 data sets shown. Note that the relatively low levels of Cu provide only a slight reduction in hillock formation, whereas above $\sim 5(10)^{14}/cm^2$ the comparable F doping provides a drastic reduction in hillock formation. Considering the increase in resistivity with F addition (similar to the Al film resistance increase with Cu addition) and the hillock formation behavior, it appears that the optimal F incorporation for hillock reduction is $\sim 1(10)^{15}/cm^2$.

Using the completed JEDEC electromigration structure shown in figure 2, we determined the electromigration performance of the standard and fluorinated aluminum films. Note that the $800\mu m$ test line affords about 130 $2\mu m$ wide steps for the topology samples (with 0.5 and $0.9\mu m$ high steps). The electromigration measurements were taken using conventional current and temperature acceleration techniques. All samples were stressed at $1.0(10)^6 A/cm^2$ calculated using the flat, unthinned portion of the interconnect. The ambient temperature was 220°C with an 8°C rise due to Joule heating. The line width of the test line was $6\mu m$. For each sample type, 27 or 54 lines were stressed to open failure. Although an extrusion monitor was present on the electromigration structure, no failures were observed across the $4\mu m$ extrusion monitor spacing.

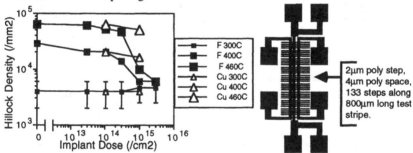

Figure 1. Hillock density greater than one micron as a function of implant dose, species, and anneal temperature.

Figure 2. Layout of electromigration structure showing optional topography steps.

Figure 3 shows a plot of median time to failure (MTF or $t_{50}$) versus implant dose for the flat topology samples. Note that although F improves the stress-induced hillock formation by an order of magnitude, the electromigration performance of flat structures is only slightly improved with F incorporation. There does appear to be a relative maximum in the MTF vs F dose which provides an optimal MTF with $1(10)^{15}/cm^2$ F implant dose, although admittedly this is not appreciable. F doses in excess of $1(10)^{15}/cm^2$ actually appear to have a detrimental effect on MTF.

Figure 4 shows a plot of median time to failure versus step coverage for both the $1(10)^{15}/cm^2$ F implanted ("optimally implanted") Al-1%Si film and the non-implanted, standard Al-1%Si film. The 14% and 20% step coverage data points resulted from the 0.9μm step and the 28% and 35% step coverages resulted from the 0.5μm vertical and sloped steps respectively. Analyzing various step heights and step slopes, the non-fluorinated Al experienced a decreasing electromigration lifetime with reduced step coverage (or increased step height). This trend is consistent with the decreased electromigration lifetime with increased step height of [6]. However we do not see the same magnitude of lifetime reduction with a 0.9μm step presumably due to the greater number of steps in [6] (1600 vs our 130). The $1(10)^{15}/cm^2$ Ne implanted sample behaved very similarly to the standard Al film, and the standard deviations shown in Figure 3 were also experienced in Figure 4. The primary point in Figure 4 is the behavior of the optimally implanted F samples at decreased step coverages. Note that, in contrast to the standard Al films, the F implanted samples show almost no lifetime reduction with step coverages over a similar regime. Comparing the behavior of the fluorinated and standard Al films for 14% step coverage (0.9μm step), we see that fluorination can provide a 4-5X improvement in MTF.

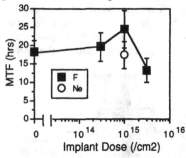

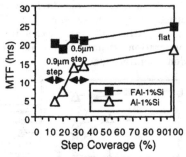

Figure 3. Electromigration MTF as a function of F dose into Al-1%Si film at 1E6A/cm2 and 220C.

Figure 4. Electromigration MTF as a function of metalization step coverage for Al-1%Si film and 1E15/cm2 F implanted Al-1%Si film.

To gain insight into why the fluorination can provide a 4-5X improvement in MTF with low step coverage, scanning electron micrographs were taken of the failed samples. The SEM pictures of Figure 5 show that the failure dynamics of the fluorinated samples differ markedly from the non-fluorinated samples. The fluorinated samples consistently appeared to have a more distributed region of damage before failure, whereas the standard Al films tended to fail over a more abrupt distance. The distance of the damage along the F-Al interconnect before failure was typically ~4x that of the standard Al interconnect. Possibly the more distributed failure characteristics of the fluorinated films allowed a bridging to take place to the next topology step thus providing an enhanced lifetime. Since the failure dynamics of corners in interconnects involve current crowding over a limited distance, a similar advantage may also be realized with fluorination due to bends in an interconnect.

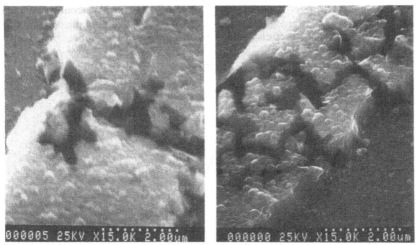

Figure 5.  Scanning electron micrographs of open failures due to accelerated test conditions.  Left SEM shows relatively abrupt failure of Al-1%Si while right SEM reveals more distributed region of damage before failure in the fluorinated sample.

Finally, SIMS profiles were taken on F and Cu (for comparison) implanted samples to reveal the fundamentally different nature of the two beneficial components.  Figure 6 shows how Cu redistributes relatively easily throughout the Al film to segregate to grain boundaries during annealing.  In contrast, the F profiles shown in Figure 7 are extremely stable after similar anneals.  We have previously shown that fluorinated aluminum films have approximately the same film stress as standard Al films or Cu-doped Al films [5].  However, the amount of stress relaxation with time or anneal is markedly less for the fluorinated aluminum.  This lack of stress relaxation, which can result from creep mechanisms, provides the superior stress-induced hillock reduction with fluorination.

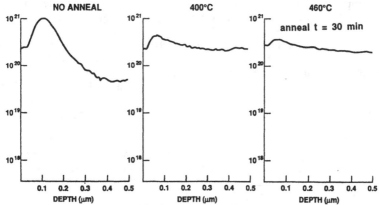

Figure 6.  SIMS sputter profile distribution of Cu in Al-1%Si as implanted, and after 400°C and 460°C anneals.  The ease of redistribution with anneal is evident.

The physically stronger fluorinated aluminum film results from the much stronger Al-F bond relative to Al-Cu or Al-Al. Thus the implantation of F into Al provides its beneficial effect by forming a very stable, distributed, refractory metal-like structure within the interconnect.

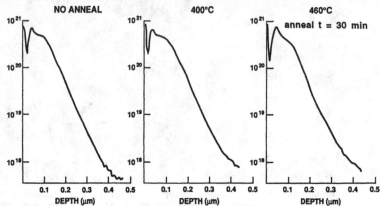

Figure 7. SIMS sputter profile distribution of F in Al-1%Si as implanted, after a 400°C anneal and after a 460°C anneal. Note the stability of F in Al with anneal.

## CONCLUSIONS

Although F improves the stress-induced hillock formation by an order of magnitude, the electromigration performance of flat structures is only slightly improved with F incorporation. The primary electromigration advantage of F implanted over standard Al films is the 4-5X improvement in MTF realized over harsh topologies. Scanning electron micrographs revealed that this improvement is due to a difference in the failure dynamics of the F implanted versus standard Al films. The F implanted films had a more distributed region of damage over the interconnect length before failure, whereas the standard Al films tended to fail over a more abrupt distance (~1/4 that of the F-Al films). SIMS profiles taken on F and Cu implanted samples demonstrated the strength of the Al-F bond by showing the stability of its distribution with anneal. The extremely stable F profile provides its beneficial effect by forming a distributed refractory metal-like structure within the interconnect. Based on the results of this study and the wide body of data on Cu-doped Al films, there may be a doping scheme involving the incorporation of both Cu and F that will reduce hillock formation, minimize electromigration and stress migration without sacrificing etch ease and interconnect conductivity.

References
[1]  A. Rey, P. Noel and P. Jeuch, Proc. 1984 VLSI MIC Conf., Vol.139, 1984.
[2]  T. Hosoda, et al., Proc. 27th IEEE Int. Reliab. Phys. Symp., p.206, 1989.
[3]  J.K. Howard, J.F. White and P.S. Ho, J. Appl. Phys., Vol.49, p.4083, 1978.
[4]  T.J. Faith, J. Appl. Phys., Vol.52, p.4630, 1981.
[5]  K.P. MacWilliams, et al., IEEE Symp on VLSI Tech., p.33, 1990.
[6]  L. Kisselgof, et al., Mat. Res. Soc. Symp. Proc., Vol. 225, p.107, 1991.

# MICROSTRUCTURAL ASPECTS OF INTERCONNECT FAILURE

J. SANCHEZ, E. ARZT
Max-Planck Institut für Metallforschung, and Institut für Metallkunde
University of Stuttgart, 92 Seestrasse, 7000 Stuttgart, Germany

## ABSTRACT

The range of microstructural effects on thin film and interconnect properties is briefly described, and the improvement of interconnect reliability with increased strength is reviewed. We show that the strengthening effect of dispersed second phases depends on their resistance to coarsening during thermal treatments. The rapid coarsening of $\Theta$ phases during annealing and accelerated electromigration testing is reviewed, leading to a discussion of metallurgical factors which determine the coarsening behavior. We describe alloy systems expected to have reduced coarsening rates. We suggest that the recently reported increased reliability of Al-Sc interconnects is due to finely dispersed coherent phases which are particularly resistant to coarsening. The range of electromigration failure morphologies is illustrated with particular emphasis on transgranular slit failures. The failures are discussed in terms of diffusion pathways and models for failure.

## INTRODUCTION

The reliability of Al alloy interconnects in integrated circuit (IC) devices is of paramount interest for IC manufacturers and users. Interconnect failures are due to electromigration and stress induced voiding processes. The processes which limit reliability are determined by the interconnect microstructure, which in turn may be influenced by the interconnect width and thickness. Interconnect microstructure includes grain size, grain size distribution, and grain crystallographic texture as well as the effects of alloying additions to the Al film. However the interconnect microstructure may evolve during annealing, reliability testing and device service, and the potential effect of microstructure changes on the interconnect performance may be profound. We shall briefly review the range of microstructural effects on film properties and interconnect lifetime.

Since metallizations are typically Al alloyed with Cu [1], Ti [2], Pd [3] and Sc [4], we shall include those microstructural aspects associated with alloying such as solute distributions and the distributions of dispersed second phases. We focus on potential effects of dispersed second phases on film and interconnect properties, and show that the potential beneficial effects of dispersed second phases, such as film strengthening, may depend on the relative stability of dispersed phases. Here we define "particle stability" as the resistance to coarsening during treatments such as annealing and accelerated electromigration testing. Using Al-Cu thin films and interconnects as an illustrative system for microstructural evolution we review the coarsening [5] of $Al_2Cu$ $\Theta$ phases during annealing and the enhancement of $\Theta$ coarsening [6] during electromigration testing. Examples of electromigration failures located along coarsened $\Theta$ phases will be shown [6-8], which suggest a correlation between $\Theta$ coarsening and processes which limit interconnect lifetime. This leads to the proposal that increased interconnect reliability may depend in part upon improved particle stability. We review the metallurgical factors which determine the rate of particle coarsening, and illustrate that alloy systems such as Al-Pd, Al-Ti and Al-Sc alloys may provide more stable second phases and stronger films and interconnects. In particular the coherent $Al_3Sc$ phases produced in the Al-Sc system are especially resistant to coarsening. We note that interconnects fabricated from these alloys have been shown to provide electromigration resistance comparable to or greater than the widely used Al-Cu alloys.

Finally, we illustrate the range [7] of morphology of electromigration damage and failures in Al and Al-Cu interconnects, including hillocks, whiskers, large equiaxed voids and transgranular slit failures. The morphology of these features will be discussed in terms of mechanisms for failure.

## EFFECTS OF MICROSTRUCTURE-STRENGTH CORRELATIONS FOR FILMS AND INTERCONNECTS

Grain size (d) is the most obvious microstructural feature to affect interconnect reliability. Since electromigration diffusion processes occur primarily via boundary diffusion, increased grain size reduces the amount of diffusive flux along interconnects. This generally leads to an increase in the electromigration lifetime. Interconnect line width (W) also determines the grain structure. When

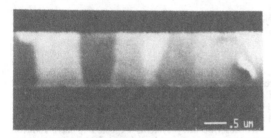

Fig. 1. Plan view focussed ion beam (FIB) micrograph of a bamboo grained Al-2% Cu interconnect 1.3 μm wide and 0.5 μm thick, showing contrast between different grains.

W/d ≈ 1, large segments of the interconnect are composed of bamboo grains as illustrated in figure 1 for a 1.3 μm wide Al-2% Cu interconnect. Obviously no boundary transport is possible in these line segments, and interconnects with W/d ≈1 and below show a significant increase in electromigration lifetime [9,10]. However the microstructure of narrow interconnects is not expected to be entirely bamboo [11]. Often electromigration damage occurs at those line segments containing a boundary path along the interconnect length, figure 2, which shows correlated void and hillock damage in a 1.8 μm wide Al interconnect. The effect of W/d on failure times has been modelled recently on the basis of grain boundary statistics in a line [20].

Film strength is also determined in part by grain size in an effect analogous to the "Hall-Petch" effect in bulk Al. Recent work [12] has shown that the yield strength ($\sigma_f$) of Al and Al-Cu films is the sum of film thickness (h) dependent and grain size dependent components, $\sigma_h$ and $\sigma_d$, respectively,

$$\sigma_f = \sigma_h + \sigma_d = m/h + k/d^n \qquad (1),$$

where the coefficients (m) and (k) and the exponent ($n \approx 1/2$) have been measured experimentally [12]. However it is not obvious that grain size strengthening occurs in unpassivated bamboo interconnects. Recent work [13] has demonstrated that the stress levels (or yield stresses) in narrow unpassivated Al-0.5% Cu interconnects were substantially below those in continuous films of the same alloy. The interconnects sustained relatively small uniaxial stresses (less than ≈ 80 MPa at temperatures above 200°C) aligned along the interconnect length, and were virtually stress free across the line width. Continuous films typically have uniform biaxial stresses and can sustain stresses of several hundred MPa. We conclude that grain size strengthening in continuous films does not directly translate to the strengthening of narrow interconnects. The possibility exists however that grain size may have some effect on the strength of passivated interconnects, since boundaries may play a role in stress relaxation processes which are limited by the passivation.

Interconnect crystallographic texture also helps to determine electromigration reliability. It has been shown that interconnects patterned from films with more uniform {111} texture have improved reliability [9,14,15]. It has been proposed that this effect is due to the tilt boundaries

Fig. 2 FIB micrograph of electromigration damage in a laser reflowed Al interconnect 1.8 μm wide and 0.5 μm thick. Void and hillock are connected by a boundary parallel to line. (Electron flux was from right to left.)

formed between the {111} fiber oriented grains, where the boundary tilt axis is normal to the film plane. These boundaries [15] are expected to have a more uniform distribution in specific boundary energies and a (presumed) more uniform distribution in boundary diffusivities ($D_{gb}$). This microstructure will lead to a decrease in mass flux divergences at, for example, triple points. Randomly oriented grains will in general produce random boundaries with an increased distribution in $D_{gb}$, which is more likely to produce larger flux divergences and earlier failures.

Texture may also determine in part the strength levels in continuous films. The effect of crystal orientation on dislocation motion in films is shown in figure 3 [16]. We consider single phase films, capped with a surface (oxide) layer, on a rigid substrate. Dislocation glide is impeded by the top and bottom interfaces and by grain boundaries, as shown schematically in figure 3. The flow stress ($\sigma_f$) required for dislocations to glide on the plane indicated there is dependent on the film thickness as described by

$$\sigma_h = \frac{\sin \phi}{\cos \phi \, \cos \lambda} \frac{G^* b}{h} = C_{ijk} \frac{G^* b}{h} \qquad (2).$$

Here $\phi$ and $\lambda$ are the included angles between the glide plane normal direction and Burgers vector and the film normal direction, respectively, and b is the Burgers vector. The constant $C_{ijk}$ is defined for simplicity as $= \sin \phi / \cos \phi \, \cos \lambda$. The dislocation motion leaves trailing dislocation segments at the top and bottom of the oxide/film and film/substrate interfaces, respectively, which leads to an effective shear modulus $G^*$ which is a detailed function [16] of the thicknesses and shear moduli of the film, oxide cap and substrate. The film thickness determines the width of the glide plane ($L_h$) for a given grain orientation by the factor $L_h = h / \sin \phi$, and the grain orientation also results in the resolved stress on the glide planes that is reduced from the applied stress by a factor ($\cos \phi \, \cos \lambda$). $C_{ijk}$ thus helps determine the dependence of the film flow stress on grain orientation (ijk). We note that $C_{111}$ is $\approx 3.46$, while the average of the four lowest values of $C_{110}$ for the available {111} <110> slip systems in (110) oriented grains is $\approx 1.42$ [17]. Films of principally {111} textured grains will be stronger than {110} textured films, given equivalent grain sizes and film thicknesses. It is not known at present how variations in the strength level of variously textured grains along an interconnect will affect reliability. However interconnect stresses and stress relaxation processes have been proposed to affect electromigration lifetimes. Therefore the role of texture in determining reliability may in part be through the texture influence on the interconnect mechanical properties.

<u>Effect of Mechanical Strength on Interconnect Reliability</u>

We briefly review the proposed interactions between interconnect strength and electromigration reliability. On the basis of the "critical length" effect during electromigration in interrupted stripes, Blech [18] proposed that the electromigration mass flux may produce tensile and compressive stresses at regions of mass depletion and accumulation, respectively. The resulting stress gradients would induce mass fluxes in a direction opposite to the electromigration fluxes and would reduce the flux divergences and damage caused by the applied current. Since the stresses that may be produced are limited by the interconnect "yield" strength, stronger interconnects would be more

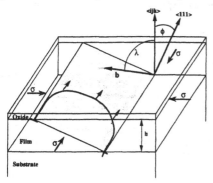

Fig. 3. Model for dislocation plasticity in thin films [16]. Here the dislocation bows between top and lower interfaces on glide plane (shaded). The film (grain) orientation helps to determine applied stress required for dislocation glide (see text).

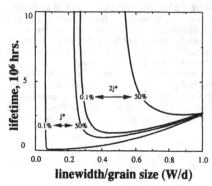

Fig. 4. Predicted effect of increased interconnect strength on the elecromigration lifetime for near bamboo interconnects as a function of linewidth/grain size ratio W/d. (After ref [20].)

resistant to electromigration damage since they would sustain larger stress gradients and induced back fluxes. This model has been extended to continuous lines under passivations [19]. In addition it has been shown [20] how an increase in strength of near bamboo interconnects may dramatically improve the electromigration reliability, figure 4. In this figure the critical current density j* is proportional to the interconnect strength, and it is shown that an increase in the strength (and j*) by a factor of two significantly increases the time for a 0.1% failure rate and for a 50% failure rate as the interconnect microstructure becomes more bamboo-like with decreasing ratio W/d.

Detailed understanding of the interactions between interconnect strength, large hydrostatic tensile stresses in passivated lines [21], stress relaxation processes and failure is lacking. However there is mounting evidence that mechanical strength should be a major consideration in the design of new metallization alloys. We may therefore describe possible methods for film and interconnect strengthening by the control of the microstructure and the proper choice of materials systems.

### Mechanisms For Film and Interconnect Strengthening

Metallizations are typically alloyed with additions such as Cu, Ti, Pd and Sc in order to improve the interconnection reliability. However because of the limited solubility of these elements in Al, phases such as $Al_2Cu$, $Al_3Ti$, $Al_4Pd$ and $Al_3Sc$ are formed. Film and interconnect properties may be sensitive functions of the phase size and distribution. We shall briefly describe several probable effects of particles on film and interconnect strength.

Traditionally, strengthening in crystalline materials involves impeding dislocation motion by presenting barriers to glide such as dispersed second phases and grain boundaries. In the context of plasticity in films, figure 3, a potential effect of dispersed phases of mean radius (r) and volume fraction ($f_v$) is to block dislocation motion on the glide plane as shown on figure 5a. The mean distance between blocking obstacles ($L_r$), for the regime of low volume fraction of second phase, is given by $L_r \approx r (f_v/2)^{-1/2}$. Thus particle strengthening is expected when $L_r < L_h/2$, or when

$$\frac{r}{(f_v / 2)^{1/2}} < \frac{h}{2 \sin \phi} \tag{3}.$$

Choosing nominal values of $\phi = 45°$ and $f_v = 0.05$, we calculate that dispersed particles of radius less than $r \approx 0.1 \ \mu m$ are expected to strengthen a 1.0 $\mu m$ film, while 0.5 $\mu m$ thick films require particles of size $r \approx 0.05 \ \mu m$ for strengthening by this mechanism.

An indirect mechanism of particle strengthening of films involves the particle pinning of boundary motion and the suppression of grain growth, leading to a retained small grain size. The grain size strengthening effect as described in equation 1 is illustrated in figure 5b. We note that such strengthening in Al films has also been observed in free standing films [22] and films deposited on Si substrates [23]. The effect of a dispersed phase of mean radius (r) and areal fraction ($f_a$) is to pin the grains at a size ($d_p$) given by [24]

$$d_p = (3.4 \ r) \ f_a^{-1/2} \tag{4}$$

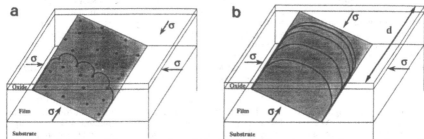

Fig. 5. Models for particle strengthening of thin films: a) a dispersion of small second phases will block dislocation motion, leading to strengthening; b) second phases (not shown) may maintain a small film grain size ($d_p$) leading to "Hall-Petch" type of strengthening.

The extent of this indirect strengthening may be calculated from the difference of the grain size dependent flow stress levels ($\sigma_d$) between single phase films (with grain size d) and films where particles have grains pinned at size $d_p$. Using equation 4 and $\sigma_d = k/d^n$ from equation 1 it can be shown that the fractional increase in film flow stress due to grain size pinning may be estimated as

$$(\sigma_{dp} - \sigma_d) / \sigma_d = (d / d_p)^{1/2} - 1 = (d\, f^{1/2} / 3.4\, r)^{1/2} - 1. \qquad (5).$$

We may estimate the magnitude of this effect for typical 1.0 μm thick films in which the annealed grain size is approximately twice [25,26] the film thickness, $d \approx 2.0$ μm. Assuming that $r \approx 0.05$ μm and $f_a \approx 0.05$, the fractional increase in film strength is significant, approximately ($\sigma_{dp} - \sigma_d$) / $\sigma_d \approx 0.62$. Again we note that the grain size strengthening effect in interconnects may not be operative to the same degree as in films. In fact reducing the interconnect grain size generally leads to reduced electromigration lifetimes. However the final effect of dispersed phases on reliability will be a combination of the strengthening and microstructural factors described above. A recent comparison [4] of the electromigration lifetimes of Al-Cu and Al-Sc alloys showed greater reliability of the Al-Sc interconnects in spite of a smaller grain size than in the Al-Cu interconnects.

We note that the beneficial effect of Cu additions on electromigration reliability is not due to a strengthening effect. Recent measurements found roughly equal strength levels in Al and Al-Cu films [12]. This is to be expected since the Θ phases found in the binary Al-Cu films reside along boundaries and triple points [27]. Considering the strengthening mechanisms described above, Θ phases with this morphology are expected to contribute little to the film or interconnect strength. The beneficial effect of copper additions in Al interconnects is generally agreed to be the reduction of grain boundary diffusivity due to Cu segregation to the boundaries.

## Θ COARSENING IN Al-2% Cu THIN FILMS AND INTERCONNECTS

Both mechanisms described above show that the efficiency of particle strengthening, at constant volume fraction, is increased for smaller mean particle radius. Thus a fine dispersion of particles *which remain small* during the thermal treatments of device fabrication is desirable. Particle size stability is determined by the rate of coarsening of the second phases. As an example, we review the kinetics of Θ particle size evolution during Ostwald coarsening in Al-2% Cu thin films during annealing. The enhancement of Θ coarsening during electromigration testing will also be presented. We then review those metallurgical factors which determine the coarsening rates of second phases and describe the characteristics of alloy systems which are expected to have improved particle stability and interconnect reliability. This leads to the suggestion that increased particle thermal stability, due to reduced second phase coarsening rates, may be a significant factor in the design of more reliable interconnect alloys.

### Experimental

Al-2% (wt.) Cu thin films 0.5 μm thick were sputter deposited onto thermally oxidized (100) 100 mm Si substrates at a rate of 120 nm/minute without applied heating or substrate bias. Base pressure of the sputter system was less than 2.0 10⁻⁵ mTorr. One wafer was sectioned into small (1 cm x 1 cm) sections which were annealed in forming gas in a hot wall tube furnace at 310°C for up to 40 minutes. Samples for transmission electron microscopy (TEM) were prepared with additional care taken to prevent inadvertant heating of the films. Measurements of Θ phase and Al grain size were obtained from TEM prints taken from several areas for each film condition. (See ref. [5] for details of sample preparation and phase size measurement.) In addition the films were examined in a scanning electron microscope (SEM) with an energy dispersive x-ray analysis (EDS) system. This allowed for the elemental characterization of local areas of interest in the film.

Other similar films were photolithographically patterned and etched to produce electromigration test circuits consisting of arrays of 25 interconnects in parallel. Line widths were from 1.0 μm to 6.0 μm, and line length was 1.5 mm. Prior to testing samples were annealed in forming gas at 425°C for 30 minutes. Wafer level accelerated electromigration testing on individually diced chips was performed as part of a larger study [7,8] of interconnect microstructure and electromigration failure analysis. Testing conditions were 1.75 10⁶ A/cm² to 2.5 10⁶ A/cm² and 175°C to 265°C. In addition pure Al interconnects were fabricated and tested. Their preparation is described in detail elsewhere [28,29], however several Al films were laser reflowed prior to patterning. The laser reflow process serves to locally anneal the metallization in order to improve film step coverage over device topology and to enlarge the film grain size. The grooves corresponding to the junction of grain boundaries and the film surface were readily apparent after the laser reflow process.

The microstructure of tested and untested but otherwise identically treated interconnects was observed with SEM, TEM and focussed ion beam (FIB) techniques. (The TEM results have been presented elsewhere [7,8,30].) The FIB induced secondary electron and ion yields are dependent upon the orientation of the target crystal with respect to the beam direction [31] and produce bright/dark contrast between grains of different crystallographic orientation in the film.

### Results: Θ Coarsening in Continuous Films

The as deposited Θ particles, about 200 nm in diameter, are uniformly distributed along grain boundaries and grain interiors [27]. Cross section TEM results on identical films show a relatively even dispersion of Θ through the film thickness ([27] and no evidence of the metastable Θ′ phases. Plan view TEM results [5] show the initiation of Θ phase coarsening at triple points after annealing for 5 minutes at 310°C, figure 6a. Annealing for 40 minutes at 310°C shows extensive Al growth and much larger Θ phases due to coarsening at triple points, figure 6b. The Θ particles remaining in the largest grain interiors, as shown in several grains in figure 6b, have also increased in size to about 0.06 μm for this anneal. Identical films annealed at 345°C have shown that coarsening has completely removed Θ phases in grain interiors after only 15 minutes [27]. It is also possible to detect the coarsened Θ phases in the SEM, figure 7a. Here the enhanced secondary electron emission provides bright contrast of the Θ phase for the same film condition as in figure 6b. These

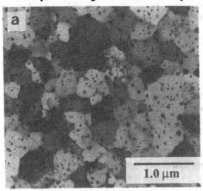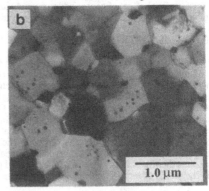

Fig. 6. Plan view TEM micrographs of 0.5 μm thick Al-2% Cu thin films after annealing for 5 minutes (a) and 40 minutes (b) at 310°C. Θ phases have black contrast and are shown to coarsen at triple points with continued annealing.

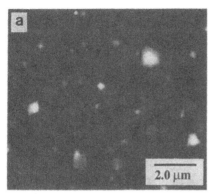 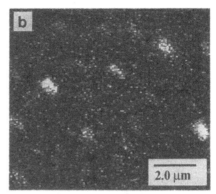

Fig. 7. a) SEM micrograph of 0.5 μm thick Al-2% Cu thin film after annealing for 40 minutes at 310°C (condition as in figure 6b). b) Energy analyzed x-rays from same area as in 7a showing enhanced Cu at bright areas, verifying these as Θ phases.

bright areas are further verified as the copper rich Θ phase by mapping of the copper x-rays generated by the incident electron beam and analyzed by the EDS. Figure 7b is an elemental copper mapping of the same area as in figure 7a, showing the correspondence between the bright areas in the secondary electron and copper x-ray mapping images. The bright contrast of Θ in the SEM will be useful for the subsequent correlation of electromigration damage with Θ phases after testing.

The coarsening rate of the triple point Θ is shown in figure 8. Note that the Θ size shows an initial period of rapid coarsening which decreases to steady state behavior after about 10 minutes. This initial period also corresponds to rapid initial grain growth of the aluminum, during which boundary migration may intersect additional small Θ phases previously in grain interiors, thereby increasing the Θ volume fraction and coarsening rate [5].

### Enhanced θ Coarsening During Electromigration

Θ coarsening is enhanced [6] during accelerated electromigration testing as clearly shown in figure 9 for a 5.0 μm wide Al-2% Cu interconnect. The testing conditions were 250°C and 1.75 $10^6$ A/cm$^2$. The median time to failure (MTF) for this array was approximately 24 hours, however the circuit remained on the hot chuck for approximately 40 hours to allow for the failure of all of the interconnects in the circuit. SEM figure 9a shows the Θ phase size and distribution in an untested interconnect after completion of the test (i.e., 40 hrs. at 250°C). Figure 9b shows the Θ phase morphology in a tested interconnect adjacent to that in figure 9a. The enhancement of the coarsening process is evident since the effect of electromigration has been to increase the size and decrease the number of Θ phases, figure 9b. This effect is in contrast to a simple "Cu accumulation" effect [32] where the average concentration of Cu increases near the positive

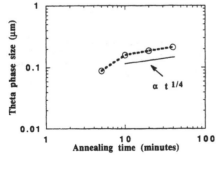

Fig. 8. Plot showing the increase of Θ phase size (radius) with annealing time at 310°C. After an initial period of rapid coarsening, a "steady state" coarsening rate proportional to $t^{1/4}$ is reached.

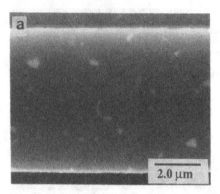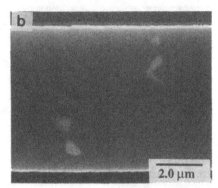

Fig. 9. Bright SEM contrast of Θ in an Al-2% Cu 5 μm wide interconnect after a) ≈ 40 hrs. at 250°C without current; b) identical interconnect and same thermal exposure as in figure 9a except with an applied current density of 1.75 10⁶ A/cm², showing the enhancement of Θ phase coarsening during electromigration.

terminal of the line. Such a process would increase the size and/or the number of Θ phases in this region. However Cu accumulation at the positive end and depletion at the negative end of interconnects occurs by a process in which the coarsening mechanism is modified by the directional electromigration flux of Cu along the line. A detailed discussion of the effect of electromigration on the kinetics of coarsening is outside the scope of this discussion, however we note that electromigration will modify and increase the rate of coarsening.

The Θ coarsening enhancement is significant since often electromigration induced voiding occurs along such phases [7,8,30]. Figure 10 shows an example of a large erosion [7] void near Θ phases in an interconnect from the same array as shown in figure 9. Examination of 1.3 μm wide Al-2% Cu interconnects tested at the same conditions as above clearly show a failure, figure 11a, and mass accumulation and depletion damage, figure 11b, along coarsened Θ phases. The Θ shown in figure 11b have coarsened along bamboo boundaries, effectively serving as obstacles to mass flux. (The direction of electron and mass flux is from right to left in the micrographs of tested interconnects shown here.) Such blocking Θ phases are formed only by enhanced coarsening during electromigration.

Morphology of Interconnect Failures

FIB and SEM microscopy were used to determine the morphology of failures in relation to local grain structures. The damage and failures associated with coarsened Θ phases have been shown. Of particular interest are the transgranular slit failures previously observed in Al-Cu-Si lines [7,30].

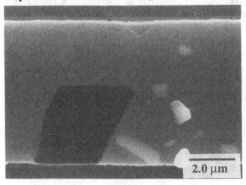

Fig. 10. SEM micrograph of electromigration tested Al-2% Cu 5.0 μm wide interconnect as in figure 9. Large erosion void is shown adjacent to coarsened Θ and hillocks. Current was from right to left.

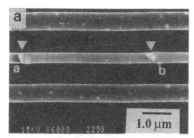

Fig. 11. Plan view SEM micrographs of electromigration damage at coarsened $\Theta$ (marked with arrows) in Al-2% Cu 1.3 µm wide interconnects. a) Shows voiding at region "a" and mass accumulation at "b". b) Shows a failure at the $\Theta$ phase (arrow). Current flux was right to left.

FIB figures 12a, b show a narrow failure in a laser reflowed Al interconnect in two orientations of tilt and rotation with respect to the ion beam. Each view provides different contrast in each grain, as described above. The relatively even contrast of material on either side of the void for both orientations suggests that the void is transgranular, but the evidence is not definitive. However the interesting change of contrast along the line indicated by the arrows does not correspond to a detectable surface groove. This may be explained as a low energy boundary or possibly a twin. A transgranular slit is clearly shown in figure 13a for a laser reflowed Al interconnect. The grain boundary is located along the line of contrast change, and the slit is shown by the arrows. Figure 13b is another view of the same slit, rotated 40° from the orientation in figure 13a to illustrate the even contrast across the slit. A final transgranular slit, figure 14, is shown to emanate from a small wedge void (at the black arrow), to traverse grain 2, and to cross the boundary between grains 2 and 3 while maintaining the same direction. The change in contrast between grains 1 and 2 indicates that the slit is *not* the site of the boundary that might have migrated into grain 2.

DISCUSSION

As shown in figure 8 the $\Theta$ coarsening kinetics at 310°C are described by the relation

$$r^4 - r_0^4 = Kt \qquad (5),$$

where $r_0$ is the initial phase radius, t is annealing time and K is a rate constant to be described below. This behavior is consistent with Ostwald coarsening controlled by solute diffusion along grain boundaries [33] and is described in more detail elsewhere [5].

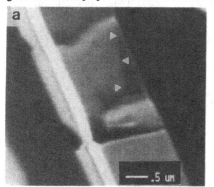
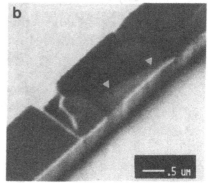

Fig. 12. FIB micrographs of slit failure in laser reflowed Al interconnects. a) Shows grain contrast, failure and a low angle or twin boundary (at arrows); sample tilted 45° normal to beam direction. b) Shows same sample tilted 35° from beam normal and rotated from position in 12a, showing change in grain contrast and the low angle or twin boundary (arrows).

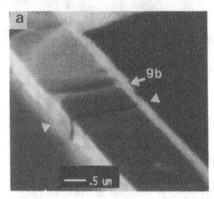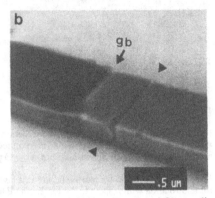

Fig. 13. FIB micrographs of transgranular slits in laser reflowed Al interconnects. a) Shows slit failure (at arrows) through the grain; sample tilted 45° from beam direction. b) Same sample tilted 55° from beam; even contrast across slit shows transgranular nature of slit.

K is given by [33]

$$K = \frac{4\,\gamma_i\,C_e\,D_{gb}\,\delta\,\Omega}{3\,A\,k\,T} \tag{6},$$

where $\gamma_i$ is the interphase (particle/matrix) boundary energy, $C_e$ is the equilibrium solute concentration, $D_{gb}$ is the solute boundary diffusivity, $\delta$ is the boundary thickness, $\Omega$ is the atomic volume, A is a geometrical constant, k is Boltzmann´s constant and T is the absolute temperature. (In this discussion we neglect the effect of second phase stochiometry on the coarsening behavior.) If we assume that second phase stability is desirable for improved interconnect reliability, then optimal alloying additions to Al would be those in which $\gamma_i$, $C_e$ and $D_{gb}$ are reduced.

The interphase energy $\gamma_i$ depends principally on the chemical and structural (i.e., crystal structure and crystallographic) differences between the matrix and second phase. However $\gamma_i$ is reduced by at least an order of magnitude [34] by coherency between the two phases. $C_e$ is obviously a minimum for insoluble alloying additions, while an a priori choice for a solute with decreased boundary diffusivity in Al is less obvious.

Al-Cu binary alloys typically form the incoherent Θ phase located at boundaries [27]. The rapid

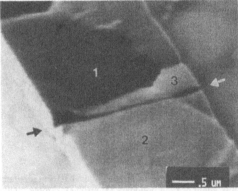

Fig. 14. FIB micrograph of transgranular slit failure (arrows) in Al interconnect, tested at 1.75 10⁶ A/cm², showing the slit to traverse grain 2 and cross the boundary between grains 2 and 3.

coarsening of $\Theta$ shown above is due to the high $\gamma_i$ between $\Theta$ and Al [35] and the rapid increase in $C_e$ (Cu in Al) with temperature. Among the additions to Al proposed for improved reliability, Ti, Pd and Sc are virtually insoluble in Al. The boundary diffusivities of these solutes and the coarsening behavior of the $Al_3Ti$, $Al_4Pd$ and $Al_3Sc$ precipitate structures have not been characterized for thin films. We note however that all three systems have shown substantially improved electromigration resistance. The reliability of Al-0.15 wt.% Sc interconnects was recently shown [4] to be significantly improved over that of similarly treated Al-0.5 wt.% Cu lines. TEM microscopy of the Al-Sc films revealed strain field contrast characteristic of small coherent phases. Other recent work [36] has characterized the microstructure and the efficient strengthening effects of $Al_3Sc$ phases in bulk Al-Sc alloys. There the $Al_3Sc$ phases were shown (by lattice imaging TEM [36]) to be coherent, less than 0.01 μm in diameter and to similarly provide the strain field contrast at lower magnifications as observed in the thin film-interconnect study. Thus we propose that improved reliability of the Al-Sc alloy is dependent on the finely dispersed coherent $Al_3Sc$ phases which retain their beneficial effect due to their resistance to coarsening. Further work is required in order to determine the effect of Ti, Pd and Sc additions on solute diffusivities, precipitate structures, film and interconnect properties and interconnect reliability.

Electromigration damage at coarsened $\Theta$ phases is inconsistent with models which require local Cu (and $\Theta$) depletion prior to flux divergences. Voiding at $\Theta$ is more likely to limit lifetimes for interconnects with W/d ≈1, since blocking $\Theta$ phases and associated damage easily produce failures. Reducing the $\Theta$ (or Cu) volume fraction may minimize this failure mechanism, while a more fundamental approach is to choose an alloy resistant to coarsening as described above.

The transgranular slit morphology is intriguing since the generally accepted electromigration diffusion processes are limited to boundaries. However careful examination of the damage shows that often other pathways must operate. The blocking $\Theta$ phases along bamboo boundaries (see also [6,7,8]) are examples of alternative pathway mass transport. Further it is not obvious that the small volume slit void growth is controlled by diffusive processes. Local stresses or stress concentrations may induce plasticity which induces a "crack-like opening" to occur. The restraint of the substrate to the adhering interconnect prevents a simple "fracture" interpretation of slits. However the strengthening effect of small coherent phases may act to reduce the effects of stress induced voiding. The reliability increase of $Al_3Sc$ interconnects was attributed in part to elimination of slit voiding [4]. It may be significant that transgranular slits appear to originate at wedge voids at the top edge-corner of the line, which may induce local stress concentrations. Similar wedge voids have been observed in passivated single crystal interconnects [37]. We note that slit voids (not always transgranular) are a common mode for stress induced failures.

TEM examination of a stress induced failure [37] showed that the parallel faces of the slit, which were vertical in the line, were both (111) type faces, consistent with a transgranular morphology. This orientation indicates that the material at the slit was not the typical ($1\bar{1}1$) fiber texture but with an orientation of the type ($1\bar{1}0$) or similarly normal to (111) planes. Work is in progress to determine possible relationships between interconnect texture and slit failures.

The range of failure morphologies generally observed for various interconnect materials under broad testing conditions suggests that a simple mechanistic analysis of failure time distributions is complicated at best. Hopefully future experimental work will focus on the determination of lifetimes of specific failure modes.

## CONCLUSIONS

(1)    Models for the effects of interconnect strength on reliability have been reviewed.

(2)    Mechanisms for second phase strengthening of films and (possibly) interconnects have been outlined.

(3)    Strengthening effects of dispersed phases have been shown to depend critically on their thermal stability, defined as the resistance to coarsening.

(4)    The coarsening kinetics of $\Theta$ phases in Al-2% Cu thin films during annealing were shown to be proportional to $t^{1/4}$, consistent with Ostwald ripening controlled by grain boundary diffusion.

(5)    Accelerated electromigration testing was shown to enhance $\Theta$ phase coarsening in Al-2% Cu interconnects.

(6)    Electromigration induced damage and failures occurred adjacent to coarsened $\Theta$ phases, inconsistent with Cu depletion models for interconnect reliability.

(7)    Transgranular slit failures were found in Al interconnects, inconsisistent with electromigration processes requiring grain boundary diffusion.

## ACKNOWLEDGEMENTS

We acknowledge the assistance of L.T.McKnelly and S. Bader with sample preparation, O.Kraft and R. Young (FEI Company, Cambridge, UK) with SEM and FIB microscopy. One author (JES) would like to acknowledge the support of the Max-Planck Society during his tenure in Stuttgart, Germany.

## REFERENCES

1. F. M. D'Heurle, Met. Trans., 2, 683 (1971).
2. F. Fischer, F. Neppl, in Proc. IEEE International Reliability Physics Symposium, (IEEE publishers, Las Vegas, 1984), vol. 1984, pp. 190-192.
3. Y. Koubuchi, J. Onuki, M. Suwa, S. Fukada, in Proc. VLSI Multilevel Interconnection Con. (IEEE publishers, Santa Clara, Ca., 1989), vol. TH-0259-2/89/0000-0419, pp. 419-425.
4. S.-I. Ogawa, H. Nishimura, in Proc. IEEE International Conference on Electron Devices and Materials (IEEE publishers, 1991), vol. IEEE-IEDM 1991, pp. 10.4.1-10.4.4.
5. J.E. Sanchez, Jr., L.T. McKnelly, J.W. Morris, Jr., in Phase Transformation Kinetics in Thin Films , edited by M. Chen ( Mater. Res. Soc. Proc. 230, Pittsburgh, PA, 1991) pp. 67-72
6. J.E. Sanchez, Jr., O. Kraft, E. Arzt, in Proc. First International Workshop on Stress Induced Phenomena in Metallizations (American Institute of Physics, Ithaca, NY, 1991)
7. J.E. Sanchez, Jr., L.T. McKnelly, J.W. Morris Jr., J. Electron Mater., 19, 1213 (1990).
8. J.E. Sanchez, Jr., J.W. Morris Jr., in Materials Reliability Issues in Microelectronics, edited by J.R. Lloyd et al (Mater. Res. Soc. Proc. 225, Pittsburgh, PA, 1991) pp. 53-58
9. S. Vaidya, A.K. Sinha, Thin Solid Films, 75, 253 (1981).
10. J. Cho, C.V. Thompson, Appl. Phys. Lett., 54 , 2577 (1989).
11. D.T. Walton, H.J. Frost, C.V. Thompson, in Materials Reliability Issues in Microelectronics, edited by J.R. Lloyd et al (Mater. Res. Soc. Proc. 225, Pittsburgh, PA, 1991) pp. 219-224
12. R. Venkatraman, J.C. Bravman, to appear in J. Mater. Res. (1992).
13. P.R. Besser, R. Venkatraman, S. Brennan, J.C. Bravman, in Thin Film Stresses and Mechanical Properties III, edited by W.D. Nix et al (Mat. Res. Soc. Proc. 239, Pitts., PA, 1991)
14. M.J. Attardo, R. Rosenberg, J. Appl. Phys., 41, 2381 (1970).
15. D.B. Knorr, D. Tracy, K. P. Rodbell, Appl. Phys. Lett., 59, 3241 (1991).
16. W.D. Nix, Metallurgical Transactions A, 20A, 2217 (1989).
17. J.E. Sanchez Jr., E. Arzt, to appear in Scripta Metallurgica et Materialia, (1992).
18. I.A. Blech, C. Herring, Appl. Phys. Lett., 29, 131 (1976).
19. C.A. Ross, J.S. Drewery, R.E. Somekh, J.E. Evetts, J. Appl. Phys., 66, 2349 (1989).
20. E. Arzt, W.D. Nix, J. Mater. Res., 6, 731 (1991).
21. B. Greenebaum, A.I. Sauter, P.A. Flinn, W.D. Nix, Appl. Phys. Lett., 58, 1845 (1991).
22. A.J. Griffin, F.R. Brotzen, C. Dunn, Scripta Metallurgica, 20, 1271 (1986).
23. M. Doerner, D.S. Gardner, W.D. Nix, J. Mater. Res., 1, 845 (1986).
24. D.J. Srolovitz, M.P. Anderson, G.S. Grest, P.S. Sahni, Acta Metallurgica, 32, 1429 (1984).
25. W.W. Mullins, Acta Metallurgica, 6, 414 (1958).
26. J.E. Sanchez, Jr., E. Arzt, Scripta Metallurgica et Materialia, 26, 1325-1330 (1992).
27. D.R. Frear, J.E. Sanchez, Jr., A.D. Romig, J.W. Morris, Met. Trans A, 21A, 2449 (1990).
28. E. Arzt, O. Kraft, J.E. Sanchez, Jr., S. Bader, W.D. Nix, in Thin Films: Stresses and Mechanical Properties III, edited by W.D. Nix (Mat. Res. Soc. Proc. 239, Boston, MA, 1991)
29. O. Kraft, J.E. Sanchez, Jr., E. Arzt, in Materials Reliability Issues in Microelectronics II, edited by J.R. Lloyd et al (Mater. Res. Soc. Proc. 265, Pittsburgh, PA, 1992)
30. J.E. Sanchez, Jr., L.T. McKnelly, J.W. Morris, Jr., submitted Appl. Phys. Lett., (1991).
31. P.H. La Marche, R. Levi-Setti, K. Lam, IEEE Trans. Nucl. Sci., NS-30, 1240 (1983).
32. A.J. Learn, J. Electron. Mater., 3, 531 (1974).
33. M V. Speight, Acta Metallurgica, 16, 133 (1968).
34. J.J. Hoyt, S. Spooner, Acta Metallurgica et Materialia, 39, 689 (1991).
35. H.B. Aaron, H.I. Aaronson, Acta Metallurgica, 18, 699 (1970).
36. R.R. Sawtell, Ph.D Thesis, University of California, Berkeley, Dept. of Mat. Sci., (1988).
37. M. Hasunuma, et al., in Proc. IEEE International Conference on Electron Devices and Materials (IEEE publishers, Washington, DC 1989), vol. IEEE-IEDM 1989, pp. 677-680.

# PART III

# Metallization Schemes for
# Interconnects, Wiring, and Packaging

# INTERCONNECTION CHALLENGES OF THE 1990's

MICHAEL A. FURY, IBM Corporation Zip 49X, Route 52, Hopewell Junction, NY 12533-6531, (914)894-3248, FAX (914)892-6399, fury@fshvmcc.vnet.ibm.com

## ABSTRACT

Silicon device performance has improved dramatically over the past two decades. During that time, the portion of total circuit delay attributable to interconnect wiring has increased to nearly half. Changes in the use of aluminum wiring and silicon oxide insulators must be considered to further improve performance.

The generation of ULSI circuitry now under development in the industry depends heavily on innovations in materials and processes to achieve significant improvements in wiring density and circuit performance. The subsequent generation places even greater demands on materials improvements for lower resistivity wiring and a lower dielectric constant insulator. Integrating materials and methods compatible with semiconductor processing while ensuring device reliability will create a significant financial challenge in addition to the technical challenge.

This presentation enumerates the interconnection challenges facing the materials scientist and the circuit designer. The ultimate limits in materials improvements are predictable and finite. Further improvements will be derived from design innovations such as treating interconnects as active rather than passive devices. Accurate physical modelling of MLM structures will complement even more complex electrical modelling. Manufacturing sciences will grow in sophistication if implementation is to succeed. The financial burden of these technical challenges will foster collaborations among semiconductor manufacturers, equipment vendors, and research universities.

## INTRODUCTION

The first transistor developed by John Bardeen and his associates in 1948 provided a relatively simple interconnect problem. It was solved with eighteen gauge wire and paper clips. Since then, things have gotten complicated. Device dimensions have decreased by five orders of magnitude. Device speed has increased by nine orders of magnitude. (Figure 1). The world has moved from transistors to circuits to integrated circuits to LSI to VLSI to ULSI. Throughout this period of dramatic evolution in electronics, the interconnect

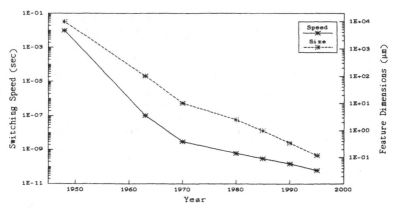

**Figure 1.** Advances in semiconductor switching speed and feature dimensions since the invention of the transistor.

technology used to integrate the myriad of available devices into functional circuits has maintained the identity of its forbearer -- the wire. The paper clips have been replaced by carefully engineered contact metallurgies; the plastic wire wrap has been replaced by silicon oxides, nitrides, polymers, or composites. Even the bulk copper of the wire has been replaced by aluminum and electromigration-resistant alloys and cladding. But for all of its high-technology sophistication, we still have a wire connecting discrete devices to make up a circuit.

There is no shame in that.

Circuit designers and process engineers and materials scientists have worked together for years to provide the function, cost and performance needed to drive our most sophisticated computers. Systems are measured in MIPS for desktop computers and GigaFLOPS for super-computers. We have entered a realm that demands new thinking, a realm in which the circuit delay caused by chip and packaging interconnects -- our wires -- is not only significant, but becomes predominant [1],[2]. (Figure 2).

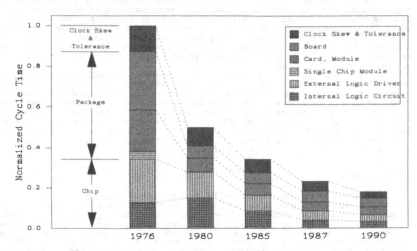

**Figure 2.**  Distribution of normalized circuit path delay components for various IBM mainframe computers (Davidson [1]).

One of the ways to mitigate these delays is to keep related circuit elements close to each other. "Don't leave the package until you have to." This designer's rule of thumb works so well that it has evolved to "don't leave the chip unless you have to." The process engineers and materials scientists groan a bit, but they continue to deliver. Single and double layers of on-chip metallization have given way to four level metal [3]. Global and local planarization techniques have enabled additional levels to be considered [4].

This complexity comes with a price. We have succeeded in bringing the circuit elements closer together, shortening the signal delay time between devices. Now the interconnects themselves are able to capacitively couple and drive the delay time back up. Consistent with Murphy's Laws, as the semi-conductor devices shrink, the device delays decrease inversely with scaling, while the RC delays can increase as scaling squared [5].

The circuit delays for device interconnects need to be manipulated to alter the shape, position, and the very nature of the delay curves.

## CURRENT TRENDS

The increase in on-chip circuit density has resulted in several key changes in the shape of multi-level metallization circuit elements and in the process strategy by which they are defined. (Table I). Lines of evaporated metal defined by wet etch or lift-off stencils are being replaced by sputtered

Table I.  Migration of Multilevel Metallization
         Process Technology Elements

|  | 1960s-1980s | 1980s-1990s |
|---|---|---|
| Levels | 1-2 | 3-4+ |
| Metal | Evaporated | Sputtered |
| Insulator | SiO2, Si3N4 | Plasma TEOS |
| Lines | Wet Etch | RIE |
| Vias | Wet Etch, Sloped | RIE, Vertical |
| Planarization | None; SOG | Etch Back; CMP |

metal defined by metal RIE. Wet etched vias with their sloping sidewalls, high area overhead, and non-planar structures, have given way to submicron [6], planar [7] vertical studs.  The myriad of established planarization techniques [8] is being challenged by the global planarization benefits of chemical-mechanical polishing in volume production [9].  Further extensions of these concepts are under development, such as processes to fill submicron spaces without voids and to planarize the surface for etchback. Extensions of current technology concepts to 1 Gbit devices have been proposed [10],[11], with the necessary speculation about what new inventions are required to realize that manufacturing capability.  (Table II).
A reduction in resistivity of the interconnects provides a linear

Table II.  Larrabee View of Technology Evolution

|  | 1989 | 1991 | 1994 | 1997 | 2000 |
|---|---|---|---|---|---|
| Minimum Feature ($\mu$m) | 0.8 | 0.6 | 0.35 | 0.25 | 0.15 |
| DRAM | 4M | 16M | 64M | 256M | 1024M |
| Litho | G Line | G Line I Line | DUV | DUV X-Ray | X-Ray |
| Chip Size (cm$^2$) | 0.9 | 1.4 | 2.1 | 4.0 | 7.5 |
| Killer Defect ($\mu$m) | 0.15 | 0.10 | 0.08 | 0.05 | 0.03 |
| Mask Levels | 18 | 20 | 25 | 28 | 32 |
| Tolerable Defects/cm$^2$ per Mask Level | 0.028 | 0.020 | 0.012 | 0.0029 | 0.0015 |

reduction in circuit delay. In practice, pure aluminum is not useful due to its electromigration failure modes, and is alloyed with copper or silicon. In high performance bipolar and CMOS devices tungsten is used [12] for vertical stud interconnects, trading resistivity for product reliability and process/tool availability over the numerous but short vertical runs. Activity is underway in the open literature on the use of copper [13] to replace aluminum, but copper is so subject to corrosion that implementation compromises will reduce the full leverage of its lower resistivity.

The dielectric constant of the interconnect insulator provides greater leverage for reducing circuit delays. Geffken [14] has shown that a reduction in dielectric constant from 4.0 of a typical silicon oxide to 2.8 of an organic-based insulator could provide a 15% improvement in CMOS performance, design rules remaining unchanged. Coupling this to a migration from aluminum to copper metal could net a 23% performance improvement for a multi-level interconnect scheme. Developing a dielectric material which satisfied all of the process compatibility and reliability constraints of ULSI is a key challenge facing the materials scientist in this decade. This challenge is especially critical since it is not at all evident that migrating to copper metal while maintaining silicon oxide as an insulator provides adequate performance leverage for the manufacturing investment required for implemen-

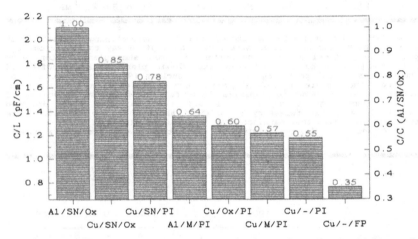

**Figure 3.** Comparison of three-dimensional total capacitances for high aspect ratio lines of pitch 2μm for different metal/etch-stop/passivation combinations. See end note (Edelstein [2]).

tation. A more convincing business case was made by Edelstein [2] for a 65% performance improvement by implementing copper metal with a fluoropolymer insulator with a dielectric constant of 1.8 (Figure 3), assuming such a material can be successfully developed.

This 65% improvement in performance may represent an upper limit to the performance gains that can be achieved through materials solutions alone. The existence of high speed super-computers today demonstrates that exceptional performance can be achieved with current technology. This performance comes with a design complexity and cost that today puts it in a market niche by itself. Systems will be built around massively parallel designs, with clever software to take advantage of it. These systems, though, will be designed around the materials limitations of current chip technologies, breathing new life into them, extending their life cycles, allowing them to mature and thus lowering their cost. The answers do not lie exclusively in making a faster chip. Herein lies another key challenge for interconnections in the '90's: breadth of focus.

## GROWING NEEDS

In order to sustain advances in performance for mainstream computers, it remains necessary, as always, to see beyond the materials and process technology horizon. There are three focal areas which I believe will be pivotal in this decade: design, modelling, and manufacturing sciences.

### Design

The trend to higher density is not without its practical bounds. More circuits on a chip will mean more levels of metallization built over larger chips. But in order to depart from the trend lines established through the 1980's, it will become necessary to design more system interconnects as active, rather than passive, circuit elements [15].

Packaging technology [16],[17],[18] is already migrating to the use of fine-line copper circuitry in a multi-level polyimide insulator, built atop a ceramic substrate metallized with refractory conductors or a glass ceramic with copper metallization (Figure 4). The active circuit element

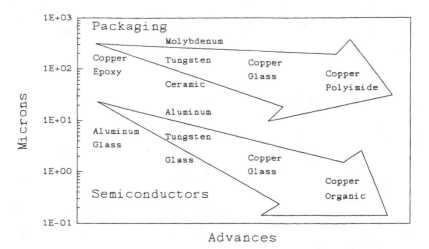

**Figure 4.** Convergence of packaging and semiconductor technologies in both materials and feature dimensions.

designs will need to be extended throughout the package, both in the thin film zone and in the ceramic layers. Furthermore, the speed advantages gained by active design will need to be preserved across the chip-to-package bonding elements, whatever method may be used.

This added layer of complexity will not stop with packaging interconnects. Implementation of transmission line concepts in a ULSI multi-level metallization scheme will be, first and foremost, a design challenge made achievable only by highly sophisticated computer design systems. Once the methodology has been established, the baton will again pass to the materials and process scientists to optimize the implementation of active interconnects in the subsequent generation of systems.

Design groundrules for the generation of ULSI DRAMs now under development are approaching 0.3 microns. Design groundrules for packaging are measured in tens of microns, if not mils. Yet, the gross dimensions of thin-film wiring for packages, which can measure several inches across, demand a defect density level which approaches that required for on-chip wiring. Even more demanding, then, is the defect density challenge facing semiconductor materials and processes to extend chip boundaries to wafer-scale integration while maintaining submicron design rules and without sacrificing yield and reliability objectives.

I've already alluded to supercomputer systems built around massively parallel architectures. I believe that we are still on the steep portion of the design learning curve in this area. The turning point will come when designers have exploited all of the available materials properties, and are again demanding new materials with which to build their systems.

The flurry of recent activity in high temperature superconductors is already starting to focus on processing methods compatible with thin film interconnects [19]. It is likely that these materials can provide significant performance leverage in specialty devices such as SQUIDS and Josephson junctions, and perhaps in packaging, but not likely that there will be applications in foreseeable silicon ULSI technologies.

Another design alternative is to depart entirely from metal and insulator constraints and introduce optoelectronic concepts [20],[21] to the module-to-module and chip-to-chip interconnect repertoire. This option would perhaps be the most exciting from a materials science perspective, as it requires the development, implementation, and integration of many device elements, each of which is new to systems applications for high density, high speed circuits. For instance, designers will require transmitters, receivers, conductors, reflectors, splitters, combiners, junctions, vertical and horizontal bends, terminators, and switches. Each of these elements will require a design solution in combination with a materials solution and a process integration methodology. Further extension to on-chip optical interconnects requires a significant re-thinking of our ULSI design strategies. Nonetheless, novel materials are already under development [22],[23],[24] to enable such technologies to flourish.

## Modelling

The use of computer modelling to simulate IC performance has long been standard procedure for circuit designers. Each generation of simulation programs increases in sophistication and in its approximation to reality. The modification of these systems to accommodate active device interconnect elements, while not trivial, will be within the realm of prior leaps in complexity.

Another area of active research has been process modelling, particularly physical and chemical deposition, light scattering in lithographic films, and dry etching. Many such models have been developed, and compilations [25] to identify them and methodologies [26],[27] to implement them have been published.

Of growing importance in the coming decade, however, will be the maturation of reliability, thermal, and mechanical models of multi-layer ULSI structures. In the current generation of VLSI structures, the metal grain size approximates the lateral line dimensions giving rise to new fields for study. The film thickness itself presents new challenges as the ratio of surface area to volume approaches and surpasses unity.

The mechanical stresses induced by mismatches in coefficient of thermal expansion and by intrinsic film stresses begin to manifest themselves in new failure mechanisms, such as the formation of thermal stress voids [28] in aluminum lines. The heat generated by operation of high speed devices will itself lead to unexpected phenomena, because the thermal conductivity behavior of thin films is not the same as the bulk [29].

Failure to understand these mechanisms is tantamount to failure of the design and process strategy for fabricating them. We must first understand the physics of these phenomena, then we must learn to model them effectively and predictively as the cost of running huge experimental matrices becomes prohibitive. As the models become more complex and expensive to run, even the selection of simulation parameters will need to be made using an efficient experimental design methodology.

## Manufacturing Sciences

Much has been written recently about the manufacturing competitiveness of U.S. factories, especially in relation to their Japanese counterparts. The race to establish and maintain world class semiconductor manufacturing leadership will parallel efforts in other industries, insofar as issues such as cycle time, inventory control, and cost containment are concerned. In other areas, the semiconductor industry will be faced with disproportionate technical challenges.

The ability to create and transfer submicron lithographic images is outstripping our ability to detect and quantify particulate contaminants, whose critical dimensions for causing device failure now measure in the hundredths of a micron. Further, it is no longer adequate to detect particles after the fact; they must be detected in the process of origin so that their very formation can be eliminated.

One method to reduce particle defects is to reduce wafer handling and exposure to the atmosphere, which has given rise to cluster tool concepts [30],[31]. A wafer may be processed through many steps without leaving the process vacuum. New sensors [32],[33] which enable in situ process monitoring can be coupled with in situ particle controls and in situ post-process measurements. The reliability of each component must be extremely high, however, if any productivity is to be achieved by the composite system.

The ability to qualify a new product technology and drive it to market will depend heavily on the successful implementation of rapid yield learning strategies [34], which are currently the object of university study as well as industrial activity.

The use of mature, real-time statistical process control techniques will be required to ensure the stability of the production floor. Each process must be understood well enough to enable the measured parameter to be related back to the correct parameter of origin in order to keep the product centered. In some cases, new statistics based on a combination of individual parameters will need to be created in order for the process control methodology to be sensitive enough to detect shifts, if the requirement is for real-time process control. When properly constructed, this hybrid statistic can signal even subtle fluctuations of significance in instances when the individual parameters indicate nothing. Such technology can be developed only when the process and tool is well characterized.

## EVOLVING PRACTICES

The financial burden of developing and implementing new semiconductor technologies is becoming too great for individual companies to bear. The Semiconductor Research Corporation (SRC) and Sematech are organizations that would have been unthinkable twenty years ago in the thriving infant semiconductor industry. Today they are necessities. The practice of pre-competitive collaboration among corporations and universities for strategic technological goals has been well documented in Japan. Bill Holton of SRC [35] and others have suggested that the practice be expanded here in the U.S. I happen to be one who agrees with them.

The opportunity to do creative work with someone else's money will always be an attractive one. Recent funding initiatives from NSF, individual states, and other agencies (Table III) have been so tailored as to encourage and even

---

Table III    Examples of Government Funding

✳ NSF Materials Synthesis and Processing Initiative

✳ NSF State/Industry University Cooperative Research

   Centers Initiative

✳ DARPA Advanced Materials and Processing Initiative

✳ NYS Research & Development Grants Program

✳ NYS Energy Research and Development Program

require collaborative efforts between universities and appropriate industrial partners. These partners may be semiconductor manufacturers, equipment vendors, or materials suppliers. The federal and state governments have recognized both the strategic importance of the semiconductor industry and the skyrocketing costs and complexities associated with it.

The offering of such grants pre-supposes that the semiconductor industry has looked far enough ahead to reliably predict the technology elements which will need to flow from the university laboratories to our development lines and factories. A scan of current research on semiconductors and derivative technologies indicates that this is a reasonable leap of faith. A review of corporate annual reports, however, indicates that the continuation of such visionary activity is at risk. In order to foster the continuation of relevant research in the universities, it is incumbent on the industry not only to invest in the research institutions themselves, but to establish and maintain dynamic technology transfer links. This requires the commitment of internal industrial resource to mentor and focus the external research, and to establish the internal customer for the deliverable result. It also requires the commitment of institutions and faculty to establish new paradigms which simultaneously satisfy needs for leadership in basic research, applied technology, academic excellence, and financial stability.

## SUMMARY

The role of interconnects in semiconductor and packaging technologies must undergo some fundamental changes in order to anticipate the demands for high performance and high reliability. Changes in materials will accommodate a finite, predictable portion of those demands, given today's design and structure concepts. The more significant leaps in performance will be derived from changes in the fundamental role of interconnects, from the increasing sophistication of modelling mechanical and thermal properties, and from educated attention to detail on the manufacturing floor. Finally, in order to make these technologies more affordable, we will need to develop their solid foundations in pre-competitive collaborations among the universities and involved industries.

## REFERENCES

1. E. Davidson, "Delay Factors for Mainframe Computers," IEEE 1991 Bipolar Circuits and Technology Meeting, September 9-10, 116 (1991).

2. D.C. Edelstein, "3-D Capacitance Modeling of Advanced Multilayer Interconnection Technologies," SPIE International Conference on Advances in Interconnection and Packaging, 1389, 352 (1990). In figure 3, Al = Al(Cu)-Ti, SN = $Si_3N_4$, Ox = $SiO_2$, PI = polyimide, FP = fluoropolymer, and M = refractory metal etch stop (removed from space between wires).

3. S.R. Wilson, J.L. Freeman Jr., C.J. Tracy, "A Four-Metal Layer, High Performance Interconnect System for Bipolar and BiCMOS Circuits," Solid State Technology, November, 67 (1991).

4. R.R. Uttecht, R.M. Geffken, "A Four-Level-Metal Fully Planarized Interconnect Technology for Dense High Performance Logic and SRAM Applications," VMIC Conference, June 11-12, 20 (1991).

5. M.B. Small, D.J. Pearson, "On-Chip Wiring for VLSI: Status and Directions," IBM J. Res. Develop. 34(6), 858 (1990).

6. "NTT Develops Multi-Layer Wiring Technique for ICs," Japan Industrial Journal, December 12, 5 (1991) (0.2 micron via hole technology announced).

7. W.J. Patrick, W.L. Guthrie, C.L. Standley, P.M. Schiable, "Application of Chemical Mechanical Polishing to the Fabrication of VLSI Circuit Interconnections," J. Electrochem. Soc., 138(6), 1778 (1991).

8. G.C. Schwartz, "Planarization Processes for Multilevel Metallization -- A Review," The Electrochemical Society Proceedings of the Symposia on Reliability of Semiconductor Devices and Interconnections and Multilevel Metallization Interconnection, and Contact Technologies, 89(6), 310 (1989).

9.  C.W. Kaanta, S.G. Bombardier, W.J. Cote, W.R. Hill, G. Kerszykowski, H.S. Landis, D.J. Poindexter, C.W. Pollard, G.H. Ross, J.G. Ryan, S. Wolff, J.E. Cronin, "Dual Damascene: A ULSI Wiring Technology," VMIC Conference, June 11-12, 144 (1991).

10. G. Larabee, P. Chatterjee, "DRAM Manufacturing in the '90s," Semiconductor International, May, 84 (1991).

11. H.J. Levinstein, "White Paper on IC Fabrication in the Year 2000", SRC Technical Report T90150, December (1990).

12. F.B. Kaufman, D.B. Thompson, R.E. Broadie, M.A. Jaso, W.L. Guthrie, D.J. Pearson, M.B. Small, "Chemical-Mechanical Polishing for Fabricating Patterned W Metal Features as Chip Interconnects," J. Electrochem. Soc., 138(11), 3460 (1991).

13. S.P. Murarka, "Overview of Copper Technology," Materials Research Society Conference Proceedings, VLSI VI, 179 (1991).

14. R.M. Geffken, "An Overview of Polyimide Use in Integrated Circuits and Packaging," The Electrochemical Society 179th Meeting, May 5-10, (1991).

15. S. Dabral, J.F. McDonald, "Design, Fabrication and Evaluation of a Parylene/Cu Test Vehicle," Current Research, Department of Electrical Engineering, Rensselaer Polytechnic Institute, Troy, NY (unpublished).

16. D.P. Seraphim, D.E. Barr, "Interconnect and Packaging Technology in the 90's," SPIE International Conference on Advances in Interconnection and Packaging, 1390, 39 (1990).

17. R.R. Tummala, "Electronic Packaging in the 1990's -- A Perspective from America," IEEE Transactions on Components, Hybrids, and Manufacturing Technology, 14(2), 262 (1991).

18. H. Wessley, O. Fritz, M. Horn, P. Klimke, W. Koshchnick, K.-H. Schmidt, "Electronic Packaging in the 1990's: The Perspective from Europe," IEEE Transactions on Components, Hybrids, and Manufacturing Technology, 14(2), 272 (1991).

19. A.E. Kaloyeros, A. Feng, J. Garhart, M. Holma, K. Brooks, W.S. Williams, "The CVD Route to High-Temperature Superconductors," in Superconductivity and Applications, edited by H. Kwok (Plenum Press, New York, 1990).

20. H. Markstein, "Optics Evolve as a Viable Interconnection Alternative," Electronic Packaging and Production, April, 50 (1991).

21. M. Osinski, "Vertical-Cavity Surface-Emitting Semiconductor Lasers for Optical Interconnections," Proceedings of the First International Workshop on Photonic Networks, Components and Applications, 70 (1991).

22. R.T. Chen, "Optical Interconnects: A Solution to Very High Speed Integrated Circuits and Systems," SPIE Integrated Optics and Optoelectronics II, 1374, 162 (1990).

23. J.F. McDonald, N.P. Vlannes, T.-M. Lu, G.E. Wnek, T.C. Nason, L. You, "Photonic Multichip Packaging (PMP) using Electro-Optic Organic Materials and Devices," SPIE International Conference on Advances in Interconnection and Packaging, 1390, 286 (1990).

24. I. Stambler, "Etched Silicon Pillars Glow," R&D Magazine, September, 22 (1991).

25. S.H. Chung, "Process and Device Simulation Programs Catalog," Semiconductor Research Corporation Publication #C91200, March (1991).

26. K. Rebab, T. Sanders, Y. Li, "Statistical Design Methodology for Process & Device Simulation," Semiconductor Research Corporation Publication #C91394, May (1991).

27. K.R. Green, J.G. Fossum, "A Pragmatic Approach to Integrated Process/Device/Circuit Simulation for IC Technology Development," Semiconductor Research Corporation Publication #C91409, May (1991).

28. S.K. Ghosh, K. Ramkumar, A.N. Saxena, "Stress Voiding in Al Interconnects: Effects of interlevel oxide films," submitted to MRS 1992 Spring Meeting.

29. A.N Saxena, New York State Sematech Center of Excellence in Multilevel Metallization (unpublished).

30. J.R. Hauser, S.A. Rizvi, "Cluster Tool Technology," Semiconductor Research Corporation Publication #C91732, September (1991).

31. J. Davies, J. Dunn, "The 90s Decade: Larger Wafers, Cluster Tools and Horizontal Partnering," Channel, September, 3 (1991).

32. K.D. Wise, K. Najafi, "Microfabrication Techniques for Integrated Sensors and Microsystems," Semiconductor Research Corporation Publication #C91750, October (1991).

33. P. Borden, "Installing in situ Sensors in Single-Wafer Plasma Etchers," Microcontamination, February, 43 (1991).

34. W. Maly, A.J. Strojwas, "Pennsylvania Sematech Center of Excellence for Rapid Yield Learning Annual Report," Semiconductor Research Corporation Technical Report #T91118, August (1991).

35. W.C. Holton, "Precompetitive Cooperative Research: The Culture of the '90s," SPIE Advanced Techniques for Integrated Circuit Processing, 1392, 27 (1990).

# RELIABLE METALLIZATION SYSTEM FOR
## FLIP-CHIP OPTOELECTRONIC INTEGRATED CIRCUITS

OSAMU WADA
Fujitsu Laboratories Ltd.
10-1 Morinosato-Wakamiya, Atsugi 243-01, Japan

## ABSTRACT

Thermal stability of evaporated Pd, Pt and Rh films as reaction barriers to Au-Sn solder was studied for the application to flip-chip optoelectronic integration. Sn in the solder diffused preferentially into a barrier metal uniformly to produce more stable intermetallic phases for all three metals. Pt and Rh exhibited sufficiently samll interdiffusion coefficients with high activation energies in the temperature range of device operation (Pt: 1.35 eV, Rh: 1.95 eV). This result demonstartes the usefulness of Pt and Rh in practical flip-chip integrated circuit fabrication. Aging test was conducted on flip-chip GaInAs/InP p-i-n photodiodes with Au-Sn/Pt metallization and no severe degradation was observed over 3400 h at 180 ° C. The same metallization techniques were applied in the fabrication of 10 Gbps optoelectronic integrated receivers as well as quad p-i-n photodiodes for coherent optical receivers.

## INTRODUCTION

Thermally stable metallization systems is a prerequisite for packaging in semiconductor electronic and optoelectronic devices and its importance is emphasized as device miniaturization and integration develop. Flip-chip integration is an attractive approach for optoelectronic integrated circuits with high performance in various optoelectronic functions, because chips made of different materials can be combined without introducing any parasitic reactances. In flip-chip integration where one chip is directly donded onto another, metallization system used must have high thermal stability.

Au-Sn solder is widely used in semiconductor packaging due to its high tensile strength, practical melting point and easy deposition technique such as evaporation. On the other hand, Sn was reported to deteriorate lasers through its persistent penetration into GaAs bulk.[1] Thus it is extremely important to develop metallization system by combining solder and a thin, stable barrier material. The purpose of this paper is to summarize the results on a study of the thermal stability of selected near noble metal barrier metals, e.i., Pd, Pt and Rh combined with Au-Sn solder.[2-4] We show that Pt and Rh have sufficient stability for practical application. Also the result of aging test of GaInAs/InP flip-chip p-i-n photodiodes[5, 6] and the fabrication of optoelectronic integrated receivers are described.[7, 8]

## THERMAL STABILITY OF Au-Sn/BARRIER SYSTEMS

### Experimental procedure

Samples used were formed on InP substrates with or without 50 nm thick Au/Zn/An p-type ohmic contacts.[3, 9] Metallization structures were formed by sequentially depositing a 100 nm thick adhesion layer and an approximately 300 nm thick barrier metal film by electron-beam evaporation, and then a 300 nm to 400 nm thick Au-Sn (30 wt% Sn) layer by resistance-heated evaporation. A SiN film was used to cap the surface to maintain planarity during heating. Heating was carried out in $N_2$ atmosphere. Metallurgical reaction was analyzed by Auger electron spectroscopy (AES), scanning electron microscopy (SEM) and x-ray diffractometry.

**Mat. Res. Soc. Symp. Proc. Vol. 265. ©1992 Materials Research Society**

## Au-Sn/barrier reaction

Figure 1 shows AES depth profiles observed on samples heated at a temperature well beyond the Au-Sn melting point (280 °C) for 10 min to 15 min as indicated in the figure. It is noticed that all three metals react preferentially with Sn in Au-Sn. Severer reaction is seen in Pd, Pt and Rh, in this sequence. For Pd, Sn penetrates deeply and consumes almost all Pd, leading to an appreciable reduction of Au concentration to leave a nearly pure Au layer at the surface. Similar behavior is observed in both Pt and Rh.

AES measurements were carried out on Pt and Rh samples heated at different temperatures from 140 °C to 400 °C. Essentially same composition profiles as Fig. 1 were collected at temperatures below the Au-Sn melting point. Figure 2 shows a SEM image of the cross section of a Rh sample heated at 226 °C for more than 600 hours. Reasonably uniform interfaces can be recognized for the reaction region, being consistent with AES data.

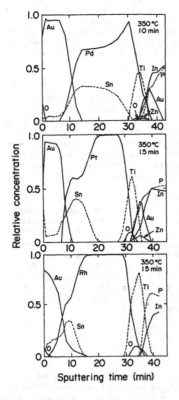

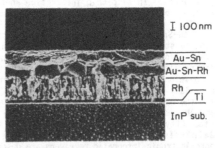

Fig. 2 SEM image of Au-Sn/Rh sample after heating at 226 °C for more than 600 h.

Fig. 1 AES depth profiles taken after heating at 350 °C.

The thermal stability of each system was studied by analyzing the temperature dependence of interdiffusion region thickness $\Delta x$, which was defined in AES plofile by taking the Sn front (at half of the peak concentration) and the 50% Au point coinciding with the barrier metal front. The time dependence of $\Delta x$ was confirmed to obey square root raw. We determined the effective interdiffusion coefficient $D_i$ by $(\Delta x)^2/4t$, where t is the heating time. A plot of $D_i$ versus the reciprocal heating temperature $1/T$ is shown in Fig. 3. A kink observed at least for Rh is attributed to the melting of Au-Sn at 280 °C. The value of $D_i$ is unambiguously determined below the melting point using the expression $D_i = D_0 \exp(-\Delta E/kT)$, where $D_0$ and k stand for the jump frequency and Boltzmann constant. The values determined are $D_0 = 4.41 \times 10^{-2}$ cm$^2$/s and $\Delta E = 1.35$ eV for Pt and $D_0 = 6.69 \times 10^3$ cm$^2$/s and $\Delta E = 1.95$ eV for Rh, both below 280 °C.

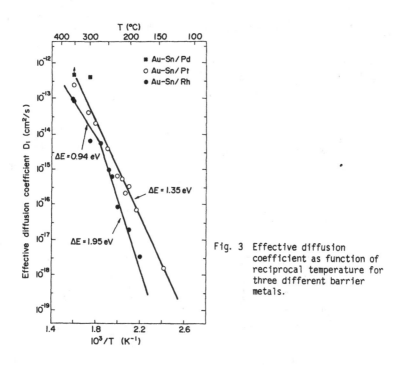

Fig. 3 Effective diffusion coefficient as function of reciprocal temperature for three different barrier metals.

## Interdiffusion characteristics

Figure 4 shows the result of $\theta - 2\theta$ scan x-ray diffraction measurements for Pt samples before and after heating at 350 ° C. As shown in as-deposited sample, Au-Sn solder consists of both AuSn and Au$_5$Sn phases. Interdiffusion takes place during the heating to produce an intermetallic compound layer with PtSn phase in the reaction region. This results in leaving a nearly pure Au layer at the Au-Sn surface, as is supported by Au diffraction peaks in Fig. 4. This behavior was observed to be enhanced as the heating temperature increased.

Same measurements were done on Rh samples in a heating temperature range between 250 °C and 450 °C, and the result is shown in Fig. 5. Data for as-deposited sample indicates the dominance of Au$_5$Sn and AuSn phases in Au-Sn layer and also implies the existence of Rh$_3$Sn$_2$ phase at the interface of Au-Sn and Rh. As the reaction is activated as the temperature increases, the

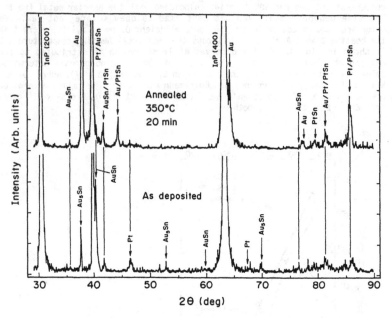

Fig. 4   X-ray diffraction patterns for AuSn/Pt samples with and without
heating at 350 °C.

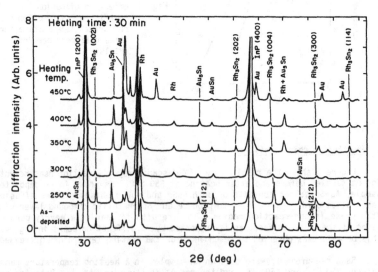

Fig. 5   X-ray diffraction patterns for Au-Sn/Rh samples before and after
heating at different temperatures.

formation of intermetallic phase $Rh_3Sn_2$ is enhanced and its lattice parameters seem to vary slightly, reflecting presumably the stress produced in the reaction region. The appearence of Au peaks at the highest temperature indicates the purification of Au in Au-Sn layer same as observed in Pt samples. The variation of peak intensity and diffraction angle observed in Rh samples is more difficult to interpret than in Pt samples, indicating more complicated reaction kinetics taking place in Au-Sn/Rh couple.

Based on the results described above, the formation of layered structures at different temperatures is illustrated for Pt in Fig. 6 and Rh in Fig. 7. It is common to both Pt and Rh that Sn preferentially reacts with the barrier. In Pt case, rather simple structure containing an intermediate layer dominated by PtSn phase is produced at all temperatures tested.

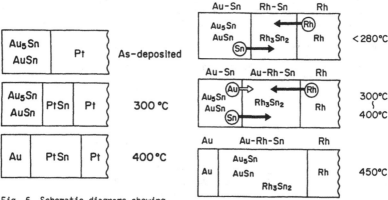

Fig. 6  Schematic diagrams showing Au-Sn/Pt reactions in different temperature ranges.

Fig. 7  Schematic diagrams showing Au-Sn/Rh reaction and diffusing species at differnt temperatures.

Rh-Sn reaction produces a little complicated structure. Rh reacts with Sn partially even at low temperature. As the reaction is enhanced at temperatures above the Au-Sn melting point, $Rh_3Sn_2$ phase is rigidly produced at the interface. This is associated with subsequent penetration of Au into the reaction region, eventually producing a layer composed of a mixture of $Au_5Sn$, AuSn and $Rh_3Sn_2$ phases. The initiation of Au penetration into Rh is observed by the tailing of Au into Rh as observed in AES data of Fig. 1. The penetration of Au is likely interpreted by grain boundary diffusion of Au in Rh-Sn compound layer. This interpretation is at least consistent with an analysis made on the Au concentration within Rh-Sn, in which Whipple model[10] has extensively been used for evaluating the grain boundary diffusion. The activation energy of Au diffusion into Rh-Sn layer was found from the data in Fig. 1 to be 0.94 eV beyond 280 °C. The activation energy determined from the analysis using above model was in agreement with this.

Thermal stability of Pt and Rh barriers

It is obvious that a certain amount of interdiffusion occurs between Au-Sn and barriers during the bonding process which is carried out typically in a couple of minutes at around 300 °C, but high activation energies obtained for Pt and Rh are advantageous in achieving long lives at device operation temperature.

The lifetime of the metallization structure is defined, rather pesimistically, as the time in which one-half of the barrier layer is consumed by the reaction. Assuming bonding conditions as indicated above, the lifetime defined in this way for a 300 nm thick Pt barrier using the result of Fig. 3 is in excess of $10^8$ h at the device operation temperature of 50 °C. For Rh barrier, a lifetime more than two orders magnitude longer than that for Pt is expected to be achieved due to its higher activation energy.

In order to confirm the stability of Pt barrier, an aging test was performed on GaInAs/InP filp-chip p-i-n photodiodes fabricated by the present metallization structure with a Pt barrier layer. The photodiode had back-illumination structure with a monolithic microlens as shown in Fig. 8. A starting epitaxial layer involved a 1.4 $\mu$m thick GaInAs photoabsorption layer. Au/Zn/Au and Au/AuGe films were deposited and alloyed for p- and n-type contacts, respectively. Metallization for bonding was formed with a 100 nm thick Ti adhesion layer, a 200 nm thick Pt barrier layer and a 4 $\mu$m thick Au-Sn solder layer. One such solder bump with a diameter of 15 $\mu$m and four 40 $\mu$m square bumps were prepared on the p- and n-type contacts, respectively. Fabricated photodiodes were flip-chip bonded on ceramic substrates having microstriplines and thick Au bonding pads. They showed a dark current less than 100 pA and a cutoff frequency as high as 21 Ghz at 10 V at room temperature.

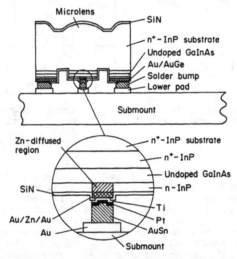

Fig. 8 Cross section of GaInAs/InP flip-chip p-i-n photodiode.

An aging test was performed on ten unscreened devices with the bias voltage of 10 V at an ambient temperature of 180 °C. Figure 9 shows the variation of the dark current during aging for 3400 h. The current level observed is low for this high aging temperature. Nine devices, except one device with defect, have exhibited extremely stable operation over the aging duration. High stability of these devices have been confirmed by their current-voltage characteristics monitored at room temperature before and after aging.[5] This result has demonstrated sufficient reliability of Pt barrier. It is expected that Rh extends the lifetime and enables the use of even thinner barrier layers.

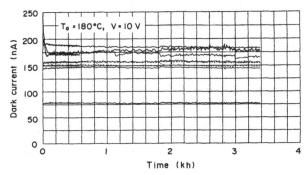

Fig. 9 Temporal variation of dark current under aging at 180 ° C.

## FLIP-CHIP OPTOELECTRONIC INTEGRATED RECEIVERS

The present metallization structures with Pt barrier layers were applied to fabricate flip-chip optoelectronic integrated receivers. Figure 10 shows the structure of an ultra-fast receiver consisting of a GaInAs/InP flip-chip APD (avalanche photodiode) and a Si preamplifier.[7] The APD had a gain bandwidth product of 80 GHz due to the reduction of the capacitance (70 fF) by the use of small junction with flip-chip structure. The Si preamplifier was fabricated by 28 GHz cutoff ESPER (emitter-base selfaligned structure with poly-silicon elelctrodes and resistor). The coupling of a slant-end siglemode fiber and a lensed APD provided a compact flat-package receiver. A receiver thus fabricated exhibited a bandwidth of 10 GHz with a transimpedance of 400$\Omega$. A receiver sensitivity evaluated for 10 Gbps, NRZ at an error bit rate of 10$^{-11}$ was extremely good, -23 dBm, at the 1.55 $\mu$m wavelength.

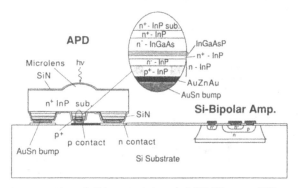

Fig. 10 Cross section of flip-chip integrated APD/Si-preamplifier receiver.

Figure 11 shows the structure of a GaInAs/InP quad p-i-n photodiodes for coherent optical receivers, which involves four lensed p-i-n photodiodes on a chip.[8] Six 40 $\mu$m diameter bumps were formed on the chip and interconnectins were made to provide two sets of balanced photodiode pairs. The chip was bonded on a ceramic substrate as shown in a SEM image in Fig. 12. The chip showed a cutoff frequency of 13 GHz for balanced pairs and a crosstalk of -40 dB between two pairs, sufficient for the application to polarization diversity coherent optical receivers.

These examples have demonstrated the usefulness of the present metallization technique in flip-chip optoelectronic integrated circuits.

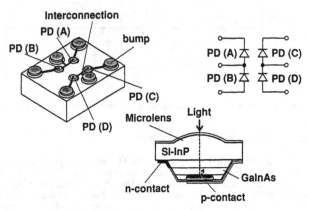

Fig. 11 Diagrams showing structure and circuit of quad p-i-n photodiodes.

Fig. 12   SEM image of flip-chip quad p-i-n photodiodes.

200 μm

SUMMARY

We have shown sufficient stability of near noble metals Pt and Rh when used as the barrier in a metallization system with Au-Sn solder.  Thermal interaction in these Au-Sn/barrier systems are characteristic in that Sn in the Au-Sn layer preferentially diffuses and reacts with the barrier metal to produce an intermediate layer containing intermetallic compounds such as PtSn and $Rh_3Sn_2$.  Thermal stability of Pt and Rh barriers has been found from activation analysis to be sufficiently high for practical application.  Adding to their small interdiffusion coefficient at the bonding temperature near 300 ° C, high activation energies in the temperature range for device operation are advantageous for achieving long lives.

Metallurgical lifetime exceeding $10^8$ h is expected for Au-Sn/Pt system and further more for Au-Sn/Rh system at a device operation temperature of 50 °C. Aging of flip-chip p-i-n photodiodes with Au-Sn/Pt metallization has shown no degradation over 3400 h even at 180  °C.  The present metallization structure has also been applied for fabricating APD/Si-preamplifier receivers for 10 Gbps systems and quad p-i-n photodiodes for coherent optical receivers.  This metallization structure will be important for reproducible and reliable flip-chip optoelectronic integrated circuit fabrication aiming for enhanced performance, functionality and integration scale.

ACKNOWLEDGMENT

The author would like to thank T. Kumai, M. Makiuchi, H. Hamaguchi and T. Mikawa for their useful discussions throughout the work.

REFERENCES

[1] K. Mizuishi, J. Appl. Phys., 55, 289 (1984)
[2] O. Wada and T. Kumai, Appl. Phys. Lett., 58, 908 (1991)
[3] O. Wada and O. Ueda, MRS Sympos. Proc., vol. 181, 273 (1990)
[4] O. Wada and T. Kumai, Jpn. J. Appl. Phys., 30, L1056 (1991)
[5] O. Wada, T. Kumai, H. Hamaguchi, M. Makiuchi, A. Kuramata and T. Mikawa, Electron. Lett., 26, 1484 (1990)
[6] O. Wada, M. Makiuchi, H. Hamaguchi, T, Kumai and T. Mikawa, IEEE J. Lightwave Technol., 9, 1200 (1991)
[7] H. Hamano, T. Yamamoto, Y. Nishizawa, Y. Oikawa, H. Kuwazuka, A. Tahara, K. Suzuki A. Nishimura, Electron. Lett., 18, 1602 (1991)
[8] M. Makiuchi, H. Hamaguchi, O. Wada and T. Mikawa, IEEE Photonics Technol. lett., 3, 535 (1991)
[9] O. Wada, J. Appl. Phys., 57, 1901 (1985)
[10] See for example J. M. Poate, K. N. Tu and J. W. Mayer, ed., "Thin Films — Interdiffusion and Reactions" (John Wiley, New York) 1978, Chap. 7 and Chap. 9

This article also appears in Mat. Res. Soc. Symp. Proc. Vol 260

# EFFECTS OF LINE WIDTH, THICKNESS AND ALLOY TEMPERATURE ON THE BREAKDOWN ENERGY OF THIN ALUMINUM LINES

Rahul Jairath, _Kamesh Gadepally_, Brad Bradford, James Shibley, Egil Castel & Rajeeva Lahri
Technology Development, National Semiconductor Corporation, Mail Stop E-120
2900 Semiconductor Drive, Santa Clara, CA 95052.

## ABSTRACT

Reliability issues have assumed greater importance as critical dimensions in microelectronic devices have entered the sub-micron regime. Wafer level reliability tests have been performed to study the breakdown energy of aluminum. 1500Å, 3500Å, 5500Å and 7500Å aluminum films were sputter deposited at 400°C on oxide-on-silicon substrates. Test structures with line widths varying from 1 μm to 10 μm were defined on the metal film. The aluminum lines were then passivated using a stacked layer of silicon dioxide and silicon nitride. Finally, the substrates were alloyed at temperatures from 400°C to 450°C. An attempt has been made to minimize process variability effects by processing the substrates through the same process conditions as far as possible. Failure was induced by stressing the metal lines with high DC current densities and at a constant slew rate. The samples were characterized using optical and scanning electron microscopy. Failure analysis of the samples processed under the different conditions will be discussed.

## INTRODUCTION

Wafer level reliability tests have gained popularity in the recent past, driven primarily by a strong need to reduce cycle time [1]. The breakdown energy test described by Hong & Crook [2] is known to provide a correlation between breakdown of metal lines and thickness, line width, step coverage and metallization defects. In most published studies conducted using this test methodology, line widths and metal thicknesses have focused within a fairly narrow range of values, while concentrating primarily on aluminum-silicon metallurgy. Here, we present results from an extensive study using Al-Cu(0.5%)-Si(1.0%) metallurgy in which the variables were line width, thickness and alloy temperature. The study made use of design of experiment (DOE) techniques which allowed us to leverage maximum information from the minimum number of experimental runs.

## EXPERIMENTAL PROCEDURE

The experiments performed as part of the study are listed in table 1. Line widths studied were 1.0 μm, 2.0 μm, 4.0 μm, 7.0 μm and 10.0 μm; aluminum thicknesses were 1500Å, 3500Å, 5500Å & 7500Å; and alloy temperatures studied were 400°C and 450°C. Substrates used in this study were commercial grade 150 mm P<100> single crystal silicon wafers deposited with 1500Å TEOS-based plasma oxide. The substrates were sputter deposited with varying thicknesses of Al-Cu(0.5%)-Si(1.0%) at 400°C. Test structures 1000 μm long were then defined on the metal film using standard mask and etch techniques. The test die contained structures varying in line width from 1.0 μm to 10.0 μm. Next, a stacked passivation layer of 3000Å silane

| Variable | Levels |
|---|---|
| Line Width (μm) | 1.0, 2.0, 4.0, 7.0 & 10.0 |
| Aluminum Thickness (Å) | 1500, 3500, 5500 & 7500 |
| Alloy Temperature (°C) | 400 & 450 |

Table 1: Factors investigated as part of the study, and their levels.

based plasma oxide and 11,000Å plasma silicon nitride was deposited on the substrates after which contact pads on the test die were opened. Finally, the substrates were alloyed at 400°C or 450°C for 30 minutes in a reducing ambient.

A Hewlett-Packard controller was used along with the appropriate software to control forcing current from a HP6634A DC power supply. The starting current for our tests was 50 mA with a constant slew rate of 400 μA/sec. Total energy to breakdown (TEB) was defined as the area under the V-I plot until failure was observed.

MODELING EFFORTS

The experimental design software, RS1, was used to design the experiment and analyze & interpret the results. The distributions for total energy to breakdown (TEB), time to breakdown ($t_b$), and current at breakdown ($J_b$) were found to be normally distributed. Thus, it was possible to model the responses using standard response surface methodology (RSM). Table 2 summarizes properties of the model developed from a total of 40 data points *each*. Each of the 40 data points represents the mean of five measurements made at similar locations on various substrates. The non-uniformity of the five measurements (=standard deviation/mean) were consistently less than 15%.

| Model | Response | # Of Terms | # Of Outliers Removed | Adjusted R-Sq |
|---|---|---|---|---|
| 1 | TEB | 6 | 4/40 | 0.996 |
| 2 | Current Density @ Breakdown | 6 | 2/40 | 0.920 |

Table 2: Summary of properties for the model developed.

For each of the two responses modeled, the adjusted regression coefficient (R-Sq) was greater than 0.9. This represents a good fit to experimental data and allowed for model predictions with high confidence levels.

RESULTS

Effect Of Alloy Temperature

Figure 1 is a main effects plot for TEB. It suggests that a change in alloy temperature from 400°C to 450°C does not have *any* effect on TEB. Plots for time to breakdown and current at breakdown follow similar trends and indicate that they, too, are independent of alloy temperature in the range 400 ℃ to 450 ℃. We have not included these plots here due to real estate restrictions.

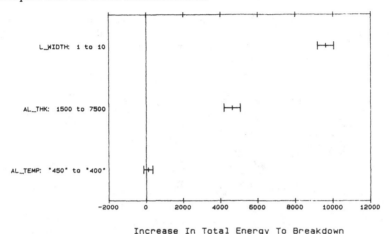

Increase In Total Energy To Breakdown

Figure 1: Main effects plot showing effects of line width, thickness and alloy temperature on the total energy to breakdown of aluminum lines.

## Scanning Electron Microscopy

Scanning electron microscopy (SEM) was used to investigate failures of the aluminum lines. Figures 2 and 3 represent micrographs of 10 μm wide aluminum lines, 7500Å and 1500Å thick respectively. Hillocks and extrusions from the metal line appears to be the predominant mode of failure in both cases. This is consistent with earlier findings of Scoggan and coworkers [3] who have observed similar failure

 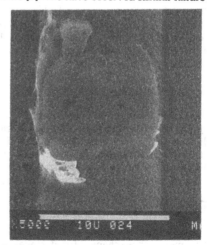

Figure 2: SEM of a 10 μm wide Al-Cu-Si film 7500Å thick.

Figure 3: SEM of a 10 μ m wide Al-Cu-Si film 1500Å thick.

mechanisms for 10 μm wide lines. Interestingly however, there does not appear to be any effect of thickness on the primary mode of failure at these large geometries.

Figure 4 is an SEM micrograph of a 4 μm wide, 5500Å thick aluminum line. In this case, two modes of failure are evident - (i) extrusions & hillocks, and (ii) fusing. Figure 5 is an SEM micrograph of a 2 μm wide, 7500Å thick line. The dominant failure mechanism appears to be fusing.

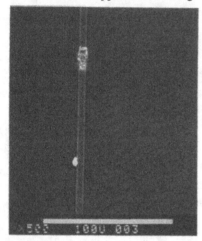

Figure 4: SEM of a 4 μm wide Al-Cu-Si        Figure 5: SEM of a 2 μ m wide Al-Cu-Si
         film 5500Å thick.                              film 7500Å thick.

### Total Energy to Breakdown (TEB)

Figure 6 is a contour plot which explains the dependance of TEB on aluminum thickness and line width. It may be observed that TEB changes rapidly at higher line widths and aluminum thicknesses (and therefore, at higher cross-sectional areas). However, at lower geometries, TEB appears to be fairly insensitive to changes in film thickness and line width.

### Current Density at Breakdown

Figure 7 is a contour plot explaining the dependance of current density at breakdown ($J_b$) on aluminum thickness and line width. At line widths greater than about 6 μm, $J_b$ is a fairly strong function of aluminum thickness, increasing linearly with decreasing thicknesses. However, at lower line widths this relationship is somewhat weakened. Also, this desensitization of $J_b$ at line widths less than 6 μm appears to increase with increasing aluminum thickness.

### DISCUSSION

Changes in resistance of a metal line are often indicative of damage in the line [4]. Commonly, these changes (or, damage) are due to (i) annealing the metal line under stress, (ii) joule heating, (iii) electromigration, (iv) fusing, and (v) environmental

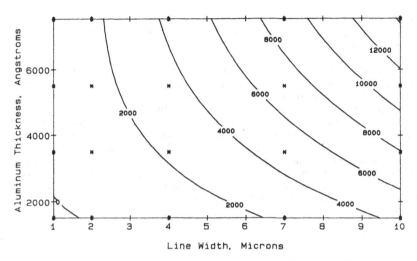

Figure 6: Contour plot showing effects of line width, thickness and alloy temperature on the total energy to breakdown of aluminum lines.

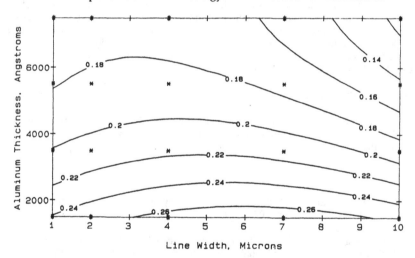

Figure 7: Contour plot showing effects of line width, thickness and alloy temperature on the current density at breakdown of aluminum lines.

conditions. In our study, all tests were conducted at room temperature, and therefore, the last reason has not been considered.

At the larger line widths, aluminum grain sizes are typically smaller than the line width. Under such conditions, damage may occur as a result of any or all, of the first four mechanisms described above. Figures 2 and 3 suggest that of these four mechanisms, failures due to electromigration are predominant. This is consistent with the fact that electromigration is primarily a grain boundary diffusion phenomenon. More importantly, at these larger line widths, changes in aluminum thickness, and hence, the cross-sectional area of the metal line do not appear to cause a change in the primary failure mechanism.

As line widths are reduced, aluminum grains begin to occupy a larger percentage of the metal line [5]. Therefore, grain boundary diffusion becomes increasingly reduced. This is consistent with earlier observations [3] that the activation energy for electromigration increases by about 20% when the line width is decreased from 10 $\mu$m to 5 $\mu$m, and by more than 40% when the line width is decreased from 5 $\mu$m to 0.8 $\mu$m. Under these conditions, joule heating assumes greater importance as a competing failure mechanism. This is further corroborated by figure 4 which shows failure by extrusion *and* fusing for 4 $\mu$m lines. At geometries less than 3 $\mu$m, aluminum grain size becomes larger than the line width, thus imposing severe constraints on grain boundary diffusion. At this point, joule heating (leading to fusing) becomes the predominant damage mechanism. Thus, failures are now a function of material properties of the metal line.

CONCLUSIONS

We have studied the breakdown energy of thin aluminum lines as a function of line width, thickness and alloy temperature. Alloy temperature did not appear to have any effect on TEB of the metal lines. TEB is more sensitive to changes at higher line widths and thicknesses when the primary mode of failure is electromigration related extrusions and hillock formation. At lower cross-sectional areas, however, the predominant mode of failure is fusing due to joule heating. Under these conditions, TEB appears to become more a function of material properties of the metal line, and less of the line width and thickness.

ACKNOWLEDGEMENTS

The authors gratefully acknowledge the help of Shirley Gouveia with data collection, Keith Daum & Walter Bachmann for SEM support, and Rory Headlong & Narahari Narasimha for their help with wafer processing.

REFERENCES

1. P. B. Ghate, " Industrial Perspective on Reliability of VLSI Devices", Mat. Res. Soc. Symp., Proc. Vol. 225., pp. 207-218, 1991.

2. C. C. Hong and D. L. Crook, " Breakdown Energy of Metal (BEM) - A New Technique for Monitoring Metallization Reliability at Wafer Level", IEEE/IRPS Proc., pp. 118-114, 1985.

3. G.A. Scoggan, B.N. Agarwala, P.P. Peressini and A. Brouillard, "Width Dependence of Electromigration Life in Al-Cu, Al-Cu-Si, and Ag Conductors", Proc. IEEE/IRPS, pp. 151-158, 1975.

4. B.J. Root and T. Turner, "Wafer Level Electromigration Tests for Production Monitoring, " Proc. IEEE/IRPS, pp. 100-107, 1985.

5. E. Kinsborn, Appl. Phys. Lett, 36, 968 (1980).

This article also appears in Mat. Res. Soc. Symp. Proc. Vol 260

## ELECTROMIGRATION in Cu/W STRUCTURE

C.-K. Hu, M.B. Small, and P.S. Ho[a)]

IBM Research Division, Thomas J. Watson Research Center,
Yorktown Heights, NY 10598

ABSTRACT

Mass transport by electromigration in sputtered Cu line segments on a continuous W line has been measured using the drift velocity technique at temperatures from 166 to 396 °C. The Ta/Cu/Ta line segments are patterned by dry etching techniques. Cu mass depletion (voids) at the cathode end and accumulation (hillocks) at the anode were measured as a function of time from scanning electron microscope micrographs. The edge displacement of Cu was found to increase linearly with time. The activation energy for Cu electromigration drift velocity, which relates to the product of effective charge number and diffusivity, $Z^*D$, is found to be 0.6 eV.

INTRODUCTION

Copper metallization is often used for packaging semiconductors where the copper films are used as a solderable material, or as a conductor in mutilayer copper-clad Invar boards, for example. Copper metallization for on-chip interconnects has received increased attention in recent years.[1,2] Use of Cu for advanced interconnection metallization is due to its better conductivity than Al alloys such as Al(Cu). One concern in using the Cu is its electromigration reliability. Electromigration[3] arises from the motion of atoms due to the influence of a high electric current density. As devices are continually scaled down to submicron dimensions, the electromigration reliability becomes increasingly important due to the shrinkage of conducting area. The mean failure lifetime of Cu, due to electromigration, in a conventional structure (a single level stripe) has been reported and an activation energy of 0.8 to 1.3 eV measured.[4,5] This method, however, uses a narrow stripe sample with large bonding pads on either side. These pads act as reservoirs resupplying material to the sample. Failure mechanisms have been attributed to variations of microstructure and temperature along the metal line.[6,8] The failure of a two-level Al line structure with W studs [9] has been shown to be quite different from the single level test. Material depletion or accumulation in these structures, which do not contain reservoirs, become important damage modes since either voids are created by depletion, or extrusions due to accumulation.

In this paper we report the electromigration mass transport rates in the sputtered pure Cu lines using the drift velocity technique[10] of isolated Cu line segments on a continuous W underlayer. This simulates electromigration mass transport at a Cu line/W stud interface without the reservoirs. This technique can also provide a the measurement of the Cu grain boundary(GB) self-diffusion activation energy with rather high precision.

EXPERIMENTAL PROCEDURE

The samples were prepared on thermally oxidized 5" silicon wafers. A plasma enhanced chemical vapor deposited(PECVD) trilayer of silicon nitride (100 nm)/SiO₂ (300 nm)/silicon nitride (100 nm) was deposited on these wafers. This was followed by a silicon nitride/SiO₂ trench etching process steps. After photo-resist striping and cleaning, the blanket sputtered Ti/TiN and

CVD W were deposited. A planarization process was performed to form the W lines.[11,12] The metallization used for the Cu metal line segments were magnetron sputtered Ta(30 nm)/Cu(400 nm)/Ta(80 nm) trilayer films. Followed by a 100 nm thick PECVD silicon nitride. The resistivity of sputtered Cu and Ta are 1.8 and 200 $\mu\Omega$-cm at room temperature, respectively. The top silicon nitride and Ta bilayer were patterned by reactive ion etching (RIE) in CF₄ plasma using the photo-resist stencil as an etch mask. Then the patterned images were transferred to Cu using N₂⁺/Ar⁺ ion milling with Ta as an etch mask and etch stop. The bottom Ta was etched in CF₄ RIE. The final thickness of top Ta layer, as determined by Auger electron microscopy with Ar+ sputtering, was found to be about 10 nm Ta. The length of Ta/Cu/Ta lines varied from 5 to 1000 $\mu m$ for 2.5 $\mu m$ wide line and 200 $\mu m$ long for 5.4 $\mu m$ wide lines. The 5.4 and 2.5 $\mu m$ wide Cu lines are connected in series through a continuous underlayer line of W, 5.4 $\mu m$ wide .25 $\mu m$ thick. All the samples were annealed at 400° for 1 hour in Helium before the testing. The sheet resistivities of Cu and W in the structure are 0.0045 and 0.048 $\Omega/\Box$, respectively. The median grain size of the Cu was about 0.4 $\mu m$. The samples were tested in a vacuum chamber at 50 torr of helium. The temperature variation across the test board was within 1 °C. The current densities used for testing were 1.6 and 3.1x10⁶ A/cm² for 5.4 $\mu m$ wide and for 2.5 $\mu m$ wide lines, respectively. The sample temperature was varied from 166 to 396 °C. The average sample temperature increase due to Joule heating was estimated to be 11°C from the measured resistance as a function of temperature and current. The edge displacement were measured using scanning electron microscopy (SEM). The values of the resistances of samples were stored periodically in a computer.

RESULTS

The electromigration flux Cu/W structure is given by

$$J = n V_e = n(DZ^*)e(\rho j/KT) \tag{1}$$

where J is the atomic flux, n the density of mobile metallic ions, j the current density, e the absolute values of the electronic charge, $Z^*$ the effective charge number, K the Boltzmann's constant and T the absolute temperature, $\rho$ the metallic resistivity. The diffusivity D is given by $D_o exp(-Q/KT)$ where Q is the activation energy and $D_o$ is the prefactor. Comparing the diffusivity of Cu in the Cu line segments to that in the W line,[13] mass transport in the latter would be very small. Thus this structure exemplifies the mass transport rate of Cu electromigration without a reservoir.

For a finite length of conductor line, the mass transport will deplete material at one end while accumulating it at the other. The material accumulation generates a compressive stress with a gradient along the direction of the mass transport. This gradient gives rise to a driving force for back-flow, retarding the electromigration,[14] leading to a drift velocity of

$$V_d = V_e - V_b = (D/kT)(Z^*e\rho j - \Delta\sigma\Omega_a/L) \tag{2}$$

where $V_b$ is the average velocity of the atoms caused by the back-flow stress $\Delta\sigma$, $\Omega_a$ the atomic volume, and L the line length.

According to eq.2, the effect of the back-flow stress in reducing the mass transport is inversely proportional to the line length. For sufficiently short lines, the stress can completely suppress the mass transport.

One can define a threshold value of the line length $L_c$, below which the drift velocity is zero and the mass transport vanishes. This condition is usually expressed in terms of a threshold j.L defined as $j.L_c = \Delta\sigma\Omega_a/Z^*e\rho$

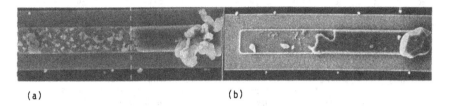

(a)                                        (b)

Figure 1. SEM micrographs of 20 $\mu$m long Cu line segments on 5.4 $\mu$m wide W line, stressed at (a) 230, (b) 407 °C. The dashed lines in (a) indicate the depletion region.

Figures 1 (a) and (b) show SEM micrographs illustrating depletion at the cathode and hillocks at the anode end of a 20 $\mu$m long and 2.5 $\mu$m wide Cu lines at 230°C for 427 hours and 407°C for 5 hours, respectively. The depletion clearly shows the flux of Cu atom follows the direction of electron flow. The threshold value of $(j.L)_c$ is determined according to Eq. (2) where $V_d=0$ from the measured 10 $\mu$m long segment. It is found to be weakly dependent on temperature and to have the value of $1.6 \times 10^3$ A/cm. The typical average edge displacements from the cathode as functions of time for lines longer than 100 $\mu$m are plotted in fig. 2. The linear behavior of these lines indicates that the drift velocity of Cu is independent of time. The lines intersect at a time close to zero, suggested no incubation time in Cu lines at these current densities. The electromigration lifetime in the Cu line/W stud structure (when the depletion volume in the Cu line reaches the critical failure volume) is the time required for the Cu line to be depleted from over the stud interface, this can be directly calculated from edge displacement data.

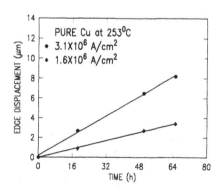

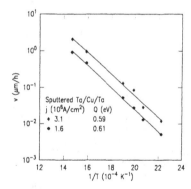

Figure 2. Plots of edge displacement as a function of time.

Figure 3. Plots of Cu drift velocity vs 1/T.

The values of drift velocity were obtained from the slopes of edge displacements with respect to time at each temperature and current density. The Arrhenius plot of drift velocities versus 1/T is shown in Fig. 3. From the slope of the best fitted line through these data the average Cu electromigration activation energy for drift velocity of 0.60 ±0.02 eV was obtained. An activation energy of D can be calculated, if the temperature dependence of $Z^*$ was known. Since the temperature dependence of $Z^*$ is unavailable in literature,. one can reasonably assume that $Z^*(T)$ in thin film would follow the same temperature dependence as that in bulk metal with a linear relation to $1/\rho(T)$[15]. (The experimental data for ultra-fast diffusers in bulk Pb shown $Z^*$ fitted to $1/\rho$ model very well.)[16] By this procedure one can calculate the temperature dependence of D according to eq. (1) using bulk values[15] $Z^*(T)$ of Cu. The resulting dependence as a function of 1/T is plotted in Fig. 4. The activation energy Q is found to be 0.66 eV which is a good agreement with published value of Ag GB diffusion in Cu of 0.62 eV.[17]

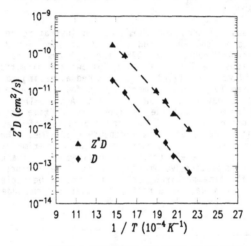

Figure 4. Temperature dependence of $Z^*D$ and D.

## DISCUSSION

Figures 5(a) and (b) show micrographs of the interfaces of samples stressed at 230°C using 2.5 and 20 KeV electron beam energies in an SEM. The surface morphology of the Cu line is clearly shown in Fig. 5(a). The secondary electron image from 20 keV provides a deeper penetration profile than 2.5 Kev electron beam energy. It clearly shows the mass depletion under the immobile thin Ta layer. One can conclude from figs. 5(a) and (b) that the contribution of free surface Cu diffusion in this structure is small. The dominant paths for Cu motion would be along GB planes and/or the Cu/Ta interface. The former motion causes voids along GB planes at depleted front edges, this is followed by the surface diffusion along the trailing surface. One can explain the formation of islands if Cu on the surfaces of grains has not completely moved out before the voids encircle the grain and the size of an island is less than the critical length. In such a case the Cu displacement is stopped due to electromigration driving force cancelled out by mechanical back flow as described in Eq. (2).

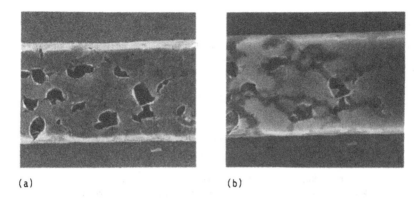

(a)                                        (b)

Figure 5. SEM micrographs of the electromigration stressed Cu lines with electron beam energy of (a) 2.5, (b) 20 KeV. The linewidth of Cu lines is 2.5 μm.

The observed value of 0.66 eV in Cu GB diffusion is much less than the published values of 1.1 eV[18] as measured from creep experiments in the temperature range of 560 to 819 °C. The discrepancy between these two is not clear. The atomistic mechanism of GB diffusion is not well understood. Many researchers have proposed it is the energy of vacancy migration [19,20], which determines the activation energy of GB diffusion. As with lattice diffusion, the activation energy of GB diffusion should be the sum of vacancy formation energy and the migration energy in the GB. One would expect the migration energy in grain boundaries (GBs) to be less than the lattice migration energy due to the open structure in GBs. The small vacancy formation energy in large angle GBs in fcc metals was proposed by assuming that the free volume is sufficiently large and every atom in the boundary plane is near a vacancy.[20] Thus the activation energy in GBs would be similar to vacancy migration energy. This observation of an activation energy of .66 eV in Cu GB self-diffusion (from electromigration drift velocity measurements) is in a close agreement with the lattice vacancy migration energy[21] of 0.71 eV. As such they seem to support the vacancy migration model in GBs.

CONCLUSIONS

The electromigration mass transport rate of sputtered 0.4μm thick pure Cu line been investigated. We conclude the following points:
a). The edge displacement increased linearly with time and a constant drift velocity is obtained for lines much larger than the critical length.
b). The electromigration activation energy of drift velocity and grain boundaries diffusion is found to be 0.6 and 0.66 eV, respectively. The value of the activation energy in grain boundaries is in a close agreement with the Cu lattice vacancy migration energy.

ACKNOWLEDGMENTS

We are grateful to Drs. R. Rosenberg, K.N Tu and D. Gupta for helpful discussions, and J. Doyle, N. Mazzeo, J. Kirleis, L.Cribb, E. Baran, and T. Ross for the technical assistance, and IBM Yorktown Si Facility for sample preparations.

REFERENCES

a)Permanent Address: Balcone Research Center, The University of Texas at Austin, Austin, TX 78758

1. C.-K. Hu, M.B. Small, F. Kaufman and D.J Pearson, Mat. Res. Soc. Symp. Proc. VLSI V, 369 (1990).

2. P.L. Pai and C.H. Ting, IEEE Electron Device Lett. 10,423 (1989).

3. H.B. Huntington, in "Diffusion in Solids: Recent Development", edited by A.S. Nowick and J.J. Burton (Academic, New York, 1974)Chap. 6.

4. C.W. Park and R.W. Vock, Appl Phys. Lett. 59, 175 (1991).

5. T. Nitta, T. Ohmi, M. Otsuki, T. Tokewaki and T. Shibata, J. Eleetrochem. Soc. 139, 922 (1992).

6. P.S. Ho and T. Kwok, Rep. Progr. Phys. 52, 301 (1989).

7. R.E. Hummel, in "Electro- and Thermo-transport in Metals and Alloys", edited by R. E. Hummel and H.B. Huntington, AIME, NY (1977), Chap. VI.

8. F. d'Heurle and R. Rosenberg, Phys. Thin Film, 7, 257 (1973).

9. C-K. Hu, P.S. Ho and M. Small, Mat. Res. Soc. Symp. Proc. 225, 99 (1991).

10. I.A. Blech and E. Kinsborn, Thin Solid Film, 25, 327 (1975).

11. D. Moy, M. Schadt, C-K. Hu, F. Kauman, A. Ray, N. Mazzeo, E. Baran, and D.J. Pearson, Proc. of the 6th International VLSI Mutilevel Interconnection Conf., (IEEE, NY, 1989) p.26

12. R. Geffken Proc. of the 8th International VLSI Mutilevel Interconnection Conf., (IEEE, NY, 1990) p.20

13. K. Vieregge, and D. Gupta,in "Tungsten and Tungsten Alloys-Recent Advances", edited by A. Crowson and E. S. Chen (The Minerals, Metal & Materials Soc., 1991) pp. 231-243

14. I.A. Blech, J. Appl. Phys. 47, 1203 (1976).

15. H. B. Huntington and A. R. Gone, J. Phy. Chem. Solids, 20, 76 (1961).

16. C.-K. Hu and H.B. Huntington, in "Diffusion Phenomena in Thin Films and Microelectronic Materials", edited by D. Gupta and P.S. Ho (Noyes Publications, Park Ridge, N.J. 1988) Chap. 10.

17. J.M. Schoen, J. M. Poate, C.J. Doherty and C. M. Melliar-Smith, J. Appl. Phys. 50, 6910 (1979).

18. B. Burton and G. W. Greenwood, J. Met. Sci., 4, 215 (1970).

19. R. Kirchheim, Acta metall. Matter., 40, 309 (1992).

20. D. Gupta, J. Appl. Phys., 44, 4455 (1973).

21. D. Gupta, in "Diffusion Phenomena in Thin Films and Microelectronic Materials", edited by D. Gupta and P.S. Ho (Noyes Publications, Park Ridge, N.J. 1988) Chap. 1.

# QUICK TESTS FOR ELECTROMIGRATION:
# USEFUL BUT NOT WITHOUT DANGER

J.R. Lloyd
Digital Equipment Corporation, 77 Reed Road, Hudson MA 01749-2895 USA
Max Planck Institut für Metallforschung, Institut für Werkstoffwissenschaft,
Seestraße 71, D-7000, Stuttgart 1, Germany

## INTRODUCTION

Traditional electromigration life testing is a time-consuming and often painful process, consuming many expensive samples and weeks of valuable time to collect enough data for a statistically significant conclusion. Annoying, to say the least, this often causes real problems in process development when cycle times of new products can often be measured in months. Therefore, there has been a strong incentive to search for alternatives.

Several potential quick electromigration tests have been developed with limited success (1-5), but more often than not the short cut misses the destination. Unfortunately, the unwary traveler is so happy that he has arrived earlier than she thought possible that the fact that where she is and where he wants to be are not the same often goes unnoticed.

This is not to say that conventional testing is not without its own set of problems. Since accelerated testing is a necessity, the desire to obtain rapid data often seduces the experimenter into employing test conditions where unimportant and invalid failure modes are excited. This is called "overstressing". In addition, one can cast suspicion that even the accelerations that we presently believe are appropriate may be too severe and the data so obtained may not be representative of the performance under realistic use conditions.

## CONVENTIONAL TESTING: TRIED, BUT TRUE?

To understand the problems with quick tests, one must first become familiar with conventional testing. Conventional testing is defined here as a test where the controllable variables of temperature and applied current density are increased to values where failure will be realized in an experimentally reasonable time yet the data can be extrapolated to realistic use conditions with some confidence. In order to accomplish this often tricky task, significant limitations must be placed on the accelerated conditions.

The task before us is to design tests at accelerated conditions without exciting invalid failure mechanisms. If this cannot be accomplished, we are faced with the highly unsatisfactory solution employed by the aircraft industry. Since they cannot trust their models completely and the consequences of failure are so great, the effect of landing an airplane over a period of several years is simulated by picking up an actual airplane and dropping it thousands of times! This is hardly efficient, and if the analogous solution were applied to electromigration testing it would mean the testing of thousands of samples for decades.

During operational use, integrated circuits are subjected to an environment which, although not benign, is not terribly severe. The temperatures experienced are usually on the order of 125C or less and the current densities on the order of a few hundred thousand amps/cm$^2$. What this means in terms of electromigration is that the major diffusion mechanism will be along a short circuit path, such as grain boundary, interfacial or dislocation core pipe diffusion rather than diffusion through the lattice.

Integrated circuit conductors are invariably covered by a passivation material which obviates the need to study surface diffusion. In addition, the metals of choice in the electronics industry, Al and its alloys, have a tough adherent natural oxide which effectively shuts off the surface as a diffusion path even if weaker passivations are applied. (It should be noted that these arguments pertain to the problem of electromigration in thin film conductors. If electromigration in solders or printed circuit board materials is of interest, then the preceding arguments about surface diffusion may not be appropriate.)

Traditionally the only temperature limitation considered has been to keep the temperature below that where lattice diffusion dominates. Experimentally, this has been shown to be on the order of 300C, somewhat higher than expected due to the finer grain size in thin films as compared to bulk specimens. (6)

This is, however, only the first of the limitations to conventional testing. It is an interesting feature of this limitation that as the grain size becomes larger, the maximum usable test temperature becomes lower. Since larger grain size has also been shown to be more reliable, we are faced with the annoying paradoxical requirement that more reliable material must be tested with smaller accelerations.

One area of concern not dealt with in the literature, is that the structure of the material at higher test temperatures can be very different from that at use conditions. For pure metals, this is not as much of an issue, since there are usually no phase changes except melting, which we obviously want to avoid. For alloys, however it can be a very different matter. A look at the Al/Cu phase diagram, for instance, shows that the solubility of Cu is appreciably greater at 300C than at 100C. (7) Therefore, if we were stressing a sample at 300C for a long time, the equilibrium microstructure would differ greatly from the same material kept at a lower use temperature. Most importantly, the distribution of the all-important intermetallic precipitates could be very different. Recent Transmission Electron Microscopy (TEM) studies of grain boundary segregation of Cu in Al/Cu alloys suggests just this. (8,9) The effect would not necessarily be revealed in experiments, since the temperature range experimentally accessible is relatively narrow and centered about traditional test temperatures. On this basis, we can cast doubt on our ability to extrapolate ANY test data from currently employed accelerated conditions even when the diffusion mechanisms are identical in both regimes. For Al/1%Cu, a commonly used integrated circuit alloy, we are probably limited to a test temperature on the order of 225C. Alloys with lower copper contents will be more restricted, whereas alloys containing more copper may allow higher temperature testing. Ternary alloys, with the lack of good phase diagrams limiting our ability to judge, are anybody's guess.

With the complexity of modern integrated circuits, extreme extrapolations of any reasonable experimental result are necessary. In order to obtain data in a reasonable time, we need to accelerate the stress conditions as much as possible. Unfortunately, the farther one accelerates the stress conditions the less representative is the material you are studying. Therefore, we need to be as close as possible to the use condition in order to minimize the certainty of error. This limitation on temperature obviously increases testing time, which is bad news to the reliability engineer asked to supply 1000 hour data by next week.

The other variable which can be used to accelerate test results is the current density. It has been established, experimentally and theoretically that electromigration lifetime is proportional to the inverse square of the current density. (10-13) Therefore, increasing the applied current will accelerate the failure process, but you must be careful. When an electric current is passed through a metal conductor, you get Joule heating and care must be taken so as not to allow the temperature rise to be significant. Long before the temperature rises to our previously determined limit, however, temperature gradients cause problems of a more severe nature. Electromigration is the biased diffusion of metal atoms and diffusion is thermally activated. A temperature gradient will produce a mass divergence flux which can be severe enough itself to induce failure. Since large temperature gradients are not met under use conditions, failure times obtained in a temperature gradient cannot be extrapolated to use conditions. One reason is that the median time to failure will be pessimistic. Temperature gradient-induced failure will occur before normal modes, such as those due to microstructure variations etc. On the other hand, since the thermal gradient will be relatively uniform from sample to sample, the deviation in the times to failure will probably be small. If the standard, but improper, lognormal failure statistics are used, the lower sigma will have a greater, and optimistic, impact on the calculated "failure rate" than will the lower median time to failure. The outcome could therefore be either pessimistic or optimistic, but will most certainly be wrong!!

The maximum current density that can be used before temperature gradient induced errors are introduced is a function of the choice of substrate, the type and condition of the metal, the amount of passivation and the design of the test structure. It can be easily seen when temperature gradient failure is dominating by the location of the failure site, which will be at the places where the highest temperature gradient lies. Normally this is at the end of a test structure, with voids forming upstream and hillocks forming downstream. (defined by electron flow) Unfortunately, even though this is a simple determination, even well published laboratories have missed the significance, leaving their conclusions suspect. With all this to complicate matters, it is best to determine the maximum current experimentally. A good rule of thumb, however, is that if Joule heating increases the temperature more than about 20C, you may be asking for trouble. In contacts and vias, Joule heating can be an even more insidious problem since it might not be detected in the normal TCR (Temperature Coefficient of Resistance) measurements. The contact area is very small and often highly resistive, which will produce inordinate local heating in accelerated tests. Contamination, oxidation or sometimes intentional coatings may be present which contribute to the increase in resistance. Since contacts and vias are necessarily connected by long metal lines, the relatively high resistance contact will produce Joule heating out of proportion to the calculated average of the whole test circuit, which is what is measured by the TCR. You might find that the average increase in temperature is about 10 degrees, and you think you're OK, but the contact itself is much hotter with extreme gradients. Failure might then occur at or near the contact giving a potentially false indication of a contact related problem. Lowering the current density and re-testing would indicate this. There is no reason to believe that contact and via failure shouldn't scale with current the same as stripe failure does, so if it doesn't scale as $1/j^2$, be suspicious.

We can see that to perform a standard electromigration life test, great care must be taken so as not to produce data which has little or no meaning in relation to reliability. Doubt can be cast on much of what has already been published. The reader may come away from this reading with the opinion that there is not much that we can do to accelerate testing any more than we already do, and that this may already be too much in some cases. The reader would be quite correct. To someone working in the aircraft industry where accelerations of 1 are common, the typical acceleration of a standard electromigration test, which can exceed 100,000, can be considered very much a "quick test" already.

The challenge is to provide data that does not suffer from overstressing errors which has perhaps only been partly met in the standard test, and in most "quick tests" has not been met at all. As it has been pointed out before, it's not nice to fool Mother Nature (14).

QUICK TESTS

Quick tests come in basically two flavors, those that are extremely accelerated to obtain early failure and those that measure some other parameter and relate this to electromigration lifetime. Both have their realms of usefulness and none of them are completely useless, but a test that can replace traditional lifetime testing has not yet been and may never be developed.

SWEAT

Perhaps the most well known of the "Quick Tests" or the "Wafer Level Tests" is is the "Standardized Wafer-level Electromigration Accelerated Test" or SWEAT developed by Root and Turner in 1985.(3) It has gained some favor in parts of the industry, but efforts to standardize the test have met with frustration and an inability to obtain reproducibility among laboratories.(15-17) Considerable effort has been spent trying to find a way out and use the technique, because, since it is so fast, it is very seductive.

In this technique, a sample is stressed with extremely high current density. No temperature acceleration is provided save for that arising from Joule heating. The assumption is made that the generalized Black equation,

$$t_{50} = Aj^{-n} \exp(\Delta H / kT) \qquad (1)$$

is valid throughout the experiment, regardless of the amount of Joule heating and the induced temperature gradients. The temperature is estimated by the TCR, and the current through the sample is computer controlled to maintain a "constant acceleration factor". It is claimed that this technique can approximate accelerations equivalent to 20 years in as few as 20 seconds!!

There are several serious problems with this approach. The most important is the belief that the Black equation is valid throughout the range of the accelerations calculated and that it is valid in the presence of substantial Joule heating temperature gradients. Furthermore, it is neglected that the temperature estimated from Joule heating is an average temperature of the sample and that at the high current density used in this test, a very non-uniform temperature profile exists where the actual peak temperature is substantially different from the measured temperature. Therefore we have little idea of what the actual temperature is and, even if eqn. (1) were valid, the accelerations are unknown.

Of more importance is the erroneous assumption, discussed above, that failure due to temperature gradient can predict performance in the field. All the SWEAT test provides is the relative performance of a metal system under conditions of extreme Joule heating which, as explained above, we are not interested in. For example, in their original paper, Root and Turner indicate that "much of the random variability found in the previous tests can be removed". The loss of variability is evidence that the proper statistically based failure sites are not being exercised and that the failure is being limited to the temperature gradient failure mode which, though consistent, is irrelevant.

## BEM

Closely related to the SWEAT test is the "Breakdown Energy of the Metal" or BEM test introduced at the same time as SWEAT by Hoang and Crook.(4,17,18) In this similar test, extreme accelerations are also used, but compounding the error, is a test structure which purposely creates huge temperature gradients by placing the sample conductor in intimate thermal contact with the underlying silicon substrate via carefully placed holes in the oxide. The measured quantity is the total energy dissipated, obtained by integrating the power over the test time until failure. Obviously, this test suffers from all of problems outlined above for the SWEAT test, and, similarly, initial apparent success and the ability to get data very quickly has made it seductive.

## Why?

The question is naturally asked, why has so much time and effort been spent on these tests if they are so obviously flawed. The answer lies in that some of the early data was misinterpreted in such a way as to make the test look remarkably successful, yet the results could have been predicted easily. In both early papers, experiments were performed which purported to demonstrate a correlation with conventional EM testing. Under the circumstances, there was a correlation, in that those failing the SWEAT and BEM test earlier would also fail standard test earlier, but the correlation was not absolute, nor did it scale temporally. The spread in the lifetimes was always smaller for the quick tests than for the conventional test results.

The reason for the correlations, however, did not require such a test to discover, nor did they relate to important quantities related to reliability. In some cases, samples which were more resistive failed earlier and those with thicker passivation failed later. The more resistive samples were either thinner, or possessed material with a higher specific resistivity, but this difference could be measured with an ohmmeter. Those with thicker passivation would have failed later in the BEM test because of better heat sinking provided by the passivation layer.

The only area where these tests might provide a unique advantage is in discovering material with poor quality step coverage. The extreme current density would produce Joule heating at the thin regions and, consequently, short lifetimes which should provide an alarm for poor step coverage, (19) but the data should only be used after considerable experimentation and no attempt should be made to extrapolate the data obtained into lifetimes under operating conditions. Unfortunately, recent tests of this type, complicated by the presence of passivation and its ability to transfer significant heat from the area of poor coverage, have been found to be only partially successful. (20)

The bottom line is that the SWEAT and the closely related BEM test, because of the extreme thermal gradients produced, do not represent realistic failure modes and therefore cannot be used to predict lifetime in the field. Perhaps, in modified form, they might be able to provide information on poor quality step coverage, but this has yet to be demonstrated.

## TRACE

The next clever acronym we will address is the "Temperature and Resistance Analysis to Characterize Electromigration" or TRACE technique developed by Jim Schwarz. (1,21-24) Being more benign than the previous tests, TRACE may hold more promise as a candidate for a true "quick test". In this technique the resistance of a sample is carefully monitored as the temperature is ramped at a calibrated rate while a high current density is passed. The current density is carefully chosen so not to promote temperature gradient failure obviating the problems of the SWEAT and BEM tests.

As the temperature is increased the sample resistance initially increases linearly until some point where it deviates positively from linearity. The deviation from linearity is assumed to reflect the electromigration-induced damage. This experiment is reminiscent of the early experiments of Baluffi for measuring the vacancy concentration as a function of temperature. (25) After suitable manipulation, the activation energy for the resistance increase can, in principle, be extracted from a single sample.

The results of this experiment were interesting. A surprising development was that the activation energy for damage varied considerably from sample to sample, but that if the average of a group were calculated it agreed with the activation energy calculated from standard failure time experiments, which has had interesting consequences when incorporated into failure models. This was later confirmed by a different technique. (5) Of additional interest is the observation that the effect of passivation was found to increase the activation energy for diffusion in a way expected from mechanical effects, also experimentally confirmed. (26)

The reason this experiment seemed to work well may be that in contrast to the other fast tests, temperature gradient failure was not induced, and presumably the same mechanisms responsible for film failure were also responsible for the "damage" reflected by the observed increases in resistance. However, it must be pointed out that a correspondence on sample by sample basis of lifetime to TRACE results was not achieved. (23) The technique appears to have the ability to predict the activation energy of failure, which may be useful, since conventional methods to obtain this quantity are extremely time consuming. However, attempts to find a way to "decelerate" the results to use conditions has been unsuccessful.

It is interesting that this method has not received the (unpublished) attention that the others have. Perhaps this is because the test is not quite as quick, taking several hours rather than seconds to produce results. Although not fatally plagued by thermal gradient failure, it has not been shown to be a suitable replacement for life testing and can only be considered a useful, albeit limited, adjunct to the conventional accelerated life test.

## AC Bridge

This technique was developed by Lloyd and Koch in 1987 (5) to assess (erroneous) unpublished criticism that the TRACE technique was merely measuring an increase in temperature and not anything to do with electromigration. A test was developed where temperature changes would not affect the data and any observed effect must be due to electromigration. In this test an AC Whetstone Bridge was constructed where two legs of the bridge are adjacent sections of the same stripe connected to a common ground. A low current, low frequency AC signal is applied and a lock-in amplifier used to measure the bridge voltage to high accuracy. A stress current is applied to one half of the test structure and the bridge is balanced. The other side of the test sample remains unpowered. The bridge voltage then represents the difference in the resistance between the unstressed and the stressed samples. Since the temperature history of the two legs are identical and any variations in oven temperature are equally experienced by both sides, any changes in the bridge voltage are necessarily the effect of electromigration. The current density is limited to where the Joule heating is only on the order of a few degrees.

The use of the AC bridge and the lock-in amplifier allows very small changes in the resistance to be observed, often on the order of parts per million, so that data can be obtained under conditions much closer to use conditions. In actual use, the method suffers from a few sources of error, but the effect is usually small.

The data obtained confirmed that there were indeed resistance changes consistent with the TRACE technique and a variation in the activation energy for resistance increase from stripe to stripe was also seen, as well as a difference in the activation energy between Al and Al/Cu.

Interestingly, when the stress current was turned off, the resistance slowly decayed. (5,27) The decay rate was too slow to be accounted for by cooling or the outdiffusion of vacancies. The decay could not be fitted to a simple exponential decay, so it was impossible to obtain an activation energy for the decay process. This decay was later confirmed by other techniques. (28)

Although this test was very sensitive and showed promise, a causal link of the resistance increase with individual failure times was never established, although some reasonable correlations were found. The drawback with this test is that it isn't really a "quick" test in terms that it may take several hours to obtain relevant data. The technique has been used as a research tool, and the results of experiments have been incorporated into reliability models, particularly the result of the variation in activation energies from stripe to stripe. One bothersome feature, however, was that the rate of resistance increase was proportional to the current density, in disagreement with the failure times. This alone may cast suspicion on its use as a reliability monitor.

## 1/f Noise

1/f noise has been an interesting subject in the literature of physics for many years. The noise in metal films is actually a fluctuation in the resistance which possesses a 1/f distribution. The source of the resistance fluctuations has been attributed to a process, or alternately a series of processes which possess a distribution of activation energies. (29)

It has been demonstrated that a major source of 1/f noise in thin metal films of Al and Au is associated with the grain boundary and is presumably related to the grain boundary diffusion process. (30,31) Since electromigration is also grain boundary controlled, it is reasonable to expect that 1/f noise and electromigration should in some way be related. Indeed there is some indication that this may be so, at least to the correspondence of activation energies, but the unambiguous link between 1/f noise and electromigration failure has not been made, contrary to the assertions of some of the literature. The problem is that much of the work done in this field has been flawed by a misunderstanding of the electromigration failure process and how it relates to noise measurements, which have resulted in naive models with no sound physical basis.

The earliest attempts at relating 1/f noise to electromigration failure were reported by Chen et al. in 1985 (2), but the results suffered from sensitivity problems and were inconclusive. Shortly after, more sensitive measurements did indeed show that 1/f noise in the resistance of thin films was affected by electromigration damage (30). An important result of these earlier studies, and what has been the subject of extensive study over the past few years was the observation that the noise character changed near the end of life from classical equilibrium 1/f to a noise whose spectrum more closely approximated $1/f^2$. The $1/f^2$ noise became prevalent when the sample resistance began to experience measurable increases. These observations have led to reports that the noise spectrum of metal films intrinsically consists of two distinct portions, corresponding to 1/f and $1/f^2$. The 1/f noise is excited at low current density and temperature and the $1/f^2$ noise is excited at higher accelerations. (32) The noise power was naively fit to an Arrhenius relation and since the low temperature, low current density, equilibrium 1/f noise was seen not to be thermally activated, it was concluded that there was no relationship to electromigration. The more highly accelerated $1/f^2$ noise, however, which produced acceptable Arrhenius plots, was claimed to be electromigration-related. Models were proposed for the source of the noise, which if accepted, would demand a rewriting of diffusion theory. (33) Since a $1/f^2$ spectrum is expected where a drift in the resistance is superimposed upon the noise, .it is suggested here that it is easier to attribute $1/f^2$ noise to this which is within the context of presently accepted physics. A recent paper has appeared which has indeed shown that $1/f^2$ noise is accompanied by a resistance drift, hopefully putting this controversy behind us. (34) However, it may be that a $1/f^2$ noise spectrum may be a sensitive drift indicator, although a sophisticated voltmeter might do just as well.

This is not to say that 1/f noise measurements are not without merit. The equilibrium noise appears to be from grain boundary diffusion, so the technique, in principle, can be used as a "microscope" to probe what is taking place in the grain boundaries themselves. Unfortunately, the effort over the past few years has been exclusively devoted to demonstrating that 1/f noise can be used as a reliability tool and little effort has been expended in understanding exactly what is making the noise.

## BOTTOM LINE

The bottom line to this discussion is clear. There is no magic test to date that can replace the traditional electromigration life test for predicting the reliability of integrated circuit conductors. There are several tests which can be of use in determining some parameters required for reliability monitoring, such as 1/f or AC resistometric methods, or can detect misprocessing such as thin metal of poor step coverage as would a modified SWEAT test. In none of these tests can the results be extrapolated to use conditions in any way but qualitatively.

It takes time to hatch an egg. You can accelerate the process by gently applying heat, as any hen knows. But if you heat it up too much......... you get breakfast.

REFERENCES

1) R. Pasco and J.A. Schwarz, Proc. 21st Ann. Reliab. Phys. Symp., IEEE, 10 (1983)

2) T.M. Chen, T.P. Djeu and R.D. Moore, Proc. 23rd Ann. Reliab. Phys. Symp., IEEE, 87 (1985)

3) B.J. Root and T. Turner, Proc. 23rd Ann. Reliab. Phys. Symp., IEEE, 100 (1985)

4) C.C. Hong and D.L. Crook, Proc. 23rd Ann. Reliab. Phys. Symp., IEEE, 108 (1985)

5) J.R. Lloyd and R.H. Koch, Proc. 25th Ann. Reliab. Phys. Symp., IEEE, 161 (1987)

6) J.M. Towner, A.G. Dirks and T. Tien, Proc. 24th Ann. Reliab. Phys. Symp. 7 (1986)

7) T.B. Massalksi (Editor) *Binary Alloy Phase Diagrams*, American Society for Metals (1986)

8) D.R. Frear, J.R. Michael, C. Kim, A.D. Romig, Jr., and J.W. Morris, Jr., SPIE Vol. 1596 *Metallization: Performance and Reliability Issues*, 72 (1991)

9) J.H. Rose, J.R. Lloyd, A. Shepela and N. Riel, Proc. Annual Meeting of Electron Microscopy Society of America (EMSA) (1991)

10) F.M. d'Heurle and P.S. Ho, in TYhin Films, Interdiffusion and Reactions, J.M.Poate, K.N. Tu and J.W. Mayer (editors), Electrochemical Society (1975) and refrences therein.

11) J.R. Lloyd SPIE Vol. 1596 *Metallization: Performance and Reliability Issues*, 106 (1991) and references therein.

12) M. Shatzkes and J.R. Lloyd, J. Appl. Phys., 59, 3890 (1986)

13) J.J. Clement and J.R. Lloyd, J. Appl. Phys., 71, 1729 (1992)

14) Madison Avenue

15) N.G. Bui, V.H. Pham, S. Chen and J.T. Yue, 1990 Wafer Level Reliability Workshop Re port, p 133 (Unpublished)

16) C.R. Crowell, C.C. Shih and V. Tyree, Proc. 28th Ann. Reliab. Phys. Symp. IEEE, 37 (1990)

17) In preparing this review, it was discovered that although there has been considerable discussion of the SWEAT and BEM tests at many workshops and conferences, and even a JEDEC committee has been investigating the techniques, precious little has appeared in the technical journals. Perhaps this is due to the problems outlined in the text.

18) L. Bacci, C. Caprile and G. DeSanti, Proc. V-MIC Conf., 463 (1987)

19) A. Ditali and R. Cross, Solid State Technology, March 1991, p 41

20) L. Elliott, S. White and N. Teasdale. Digital Equipment Corp. (unpublished) (1991)

21) R.W. Pasco, L.E. Felton and J.A. Schwarz, MRS Proc. Vol. 25, 581 (1984)

22) J.A. Schwarz and L.E. Felton, Solid State Electronics, 28, 669 (1985)

23) L.E. Felton, J.A. Schwarz and J.R. Lloyd, Thin Solid Films, 155, 209 (1987)

24) A.J. Patrinos and J.A. Schwarz, Thin Solid Films, 196, 47 (1991)

25) R.O. Simmons and R.W. Baluffi, Phys. Rev., 129, 1553 (1963)

26) J.R. Lloyd and P.M. Smith, J. Vac. Sci. Technol. A1, 455 (1983)

27) J.R. Lloyd and R.H. Koch, Appl. Phys. Lett., 52, 914 (1988)

28) J.A. Maiz and I. Segura, Proc. 26th Ann. Int'l. Reliab. Phys. Symp., 209 (1988)

29) P. Dutta and P.M. Horn, Rev. Mod. Phys. 53, 497 (1981)

30) R.H. Koch, J.R. Lloyd and J. Cronin, Phys. Rev. Lett., 55, 2487 (1985)

31) K.P. Rodbell, P.J. Ficalora and R.H. Koch, Appl. Phys. Lett., 50, 1415 (1987) 36)

32) J.G. Cottle and T.M. Chen, Proc. V-MIC Conf., CH-2488-5/87, 449 (1987)

33) W. Yang and Z. Celik-Butler, Solid-State Electronics, 34, 911 (1991)

34) D.M. Liou, J. Gong and C.C. Chen, Jap. J. Appl. Phys., 29, 1283 (1990)

# PLANARIZED COPPER INTERCONNECTS BY SELECTIVE ELECTROLESS PLATING

M.H. KIANG, J. TAO, W. NAMGOONG, C. HU, M. LIEBERMAN AND N.W. CHEUNG
Department of Electrical Engineering and Computer Sciences
University of California, Berkeley, CA 94720

H.-K. KANG and S.S. WONG
Center for Integrated Systems
Stanford University, Stanford, CA 94305

## Abstract

Electroless deposition of planarized copper in $SiO_2$ trenches has been carried out using Pd/Si plasma immersion ion implantation (PIII) or $Pd_2Si$ deposition to form a seed layer at the bottom of trenches. Electrical resistivity of the plated Cu is $\leq 2$ $\mu\Omega$-cm for both types of seeding layers. For the PIII seeding method, we found a threshold Pd dose of $2 \times 10^{14}/cm^2$ is required to initiate Cu plating, and RBS analysis confirms intermixing of Pd, Si, and $SiO_2$ improves the adhesion of the plated Cu to $SiO_2$. Electromigration tests show both void and hillock formation under accelerated current stress testing, with an activation energy of 0.8 eV for interconnect open failure. The DC and pulse-DC median-time-to-failure (MTF) of plated Cu are found to be about two orders of magnitude longer than that of Al-2%Si at 275°C, and about four orders of magnitude longer than that of Al-2%Si when extrapolated to room temperature. Pulsed DC electromigration stressing exhibits a transition from low to high frequency behavior around 900 kHz, indicating a vacancy relaxation time much shorter than that of Al. For bipolar AC stressing, the ratio $MTF_{AC} / MTF_{DC}$ of Cu is much smaller than that of Al-2%Si, indicating a different void recovery mechanism for Cu.

## Introduction

Pure copper has an electrical resistivity of $\approx 1.7$ $\mu\Omega$-cm, about 30% lower than that of Al, which is advantageous for reducing interconnect RC time delays in integrated circuits [1]. However, patterning of narrow Cu interconnects by reactive ion etching is undesirable because the etching products are usually non-volatile unless the process is performed at high substrate temperature. For multilevel interconnection technology, it is also beneficial to maintain a planar surface to reduce depth-of-focus distortions in lithography and to improve metal step-coverage. In principle, both patterning and planarization requirements can be satisfied with selective copper filling of dielectric trenches by chemical vapor deposition [2] or electroless plating [3,4]. In this paper, we report results on electroless plated Cu interconnects seeded by plasma immersion ion implantation of Pd/Si and by $Pd_2Si$ in $SiO_2$ trenches, and the electromigration behaviors of the plated copper metallization under continuous current and time-varying current stressing conditions.

## Electroless Cu Plating

Electroless Cu deposition involves the simultaneous occurrence of non-conjugated anodic oxidation of reductant and cathodic metal deposition at the surface of a catalytic substrate. It utilizes a solution containing a cupric salt, a reducing agent (e.g. formaldehyde), and a complexing or chelating agent to keep the cupric ions in solution. In addition, since the reducing power of formaldehyde increases with the alkalinity of the solution, the baths are usually operated at pH value above 11; we therefore need a chemical, usually sodium hydroxide, to provide the required alkalinity. The plating process is autocatalytic--that is, it is catalyzed by the deposited Cu surface--so that the process continues once it starts, and is characterized by the evolution of hydrogen gas from the plated surface. Using formaldehyde as the reducing agent, the main reaction can be expressed as [5]:

$$Cu^{2+} + 2HCHO + 4OH^- \rightarrow Cu + H_2 + 2H_2O + 2HCOO^-$$

It has been demonstrated [6,7] that the overall process can be resolved into two electrochemical half reactions, which take place at two Cu electrodes. The reactions in two such half cells are :

Anode:   $2HCHO + 4OH^- \rightarrow 2HCOO^- + 2H_2O + H_2 + 2e^-$

Cathode: $Cu^{2+} + 2e^- \rightarrow Cu$

While on Pd and Pt surfaces the oxidation takes place without any gas being produced [8,9]:

Anode: $HCHO + 3OH^- \rightarrow HCOO^- + 2H_2O + 2e^-$

The electroless copper plating bath composition used in this study is given in Table 1. Cupric sulfate was the cupric salt used, and formaldehyde was the reductant. Ethylenediaminetetraacetic acid (EDTA) was the complexing agent for cupric ion, whereas potassium cyanide was used to stabilize the solution by chelating cuprous ions and to adsorb upon other freshly formed minute particles which may accidentally be present in the bath. Otherwise, the disproportionate particles would become catalytically active, and thus lead to general bath decomposition with formation of copper powder. Lastly, Gafac was the wetting agent added to aid the control of a proper plating rate to improve the quality of electroless copper deposits.

TABLE 1   Composition of Electroless  Cu Plating Solution

| | |
|---|---|
| $CuSO_4 \cdot 5H_2O$ | 7.5 gm/l |
| EDTA | 25 gm/l |
| NaOH | adjust to pH=13 |
| HCHO (37%) | 10 ml/l |
| KCN Solution ( 1gm NaOH + 1gm KCN to  1 litre of DI water) | 10 ml/l |
| Gafac (3%) | 75 ml/l |

Results of Cu plating on various metal substrates are summarized in Figure 1 where correlations of the catalytic properties to the  number of valence electrons and electronegativity are observed. In this figure, Cu is defined to be plated if a copper thickness of at least 1000Å is obtained after 30 minutes in the plating solution at 65°C. The elements that plated Cu belong to Group VIII and Group IB of the periodic table. One possible explanation is that metals with s2d6, s2d7, s2d8, and s1d10 electron configurations will take electrons from the formaldehyde to fill their outer shells to achieve more stable electronic configuration. Figure 2 illustrates the resulting negative charges at the seeding site attract positive cupric ions, and reduce them to copper at the catalytic surface. Metals with s2d10 and s2d5  configurations or with less than five electrons in their d-shell  cannot oxidize formaldehyde because they are more likely to lose valence electrons to attain stability. Thus, plating of copper is prohibited. Figure 1 also shows a threshold electronegativity value of 1.8 which sets the criterion for copper plating.

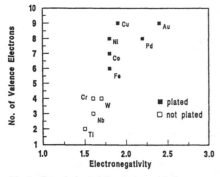

Fig.1 : Correlation of Cu plating with the number of valence electrons and the electronegativity for various metal substrates

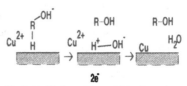

Fig.2: Oxidation of formaldehyde and reduction of cupric ion on Pd and Pt surfaces.

## Pd/Si Seeding With Plasma Immersion Implantation

The process sequence to fabricate planarized Cu interconnect in $SiO_2$ trenches is shown in Figure 3. Pd was chosen as the catalytic element in our selective plating studies because it eliminates the conventional $PdCl_2$ treatment for activating the substrate [10].By selective plating of copper into oxide trenches and/or via holes, no extra masking step is required and a completely planarized surface can be achieved . Previous studies have demonstrated that by implanting W into $SiO_2$ trenches as a seeding layer, trenches can be planarized by selective tungsten chemical vapor deposition [11,12]. We have also used implantation of Pd into $SiO_2$ trenches to seed electroless plating of Cu [13,14]. However, adhesion of the plated Cu to $SiO_2$ surfaces implanted with pure Pd is poor when the Cu acquires a thickness larger than 0.5 μm.

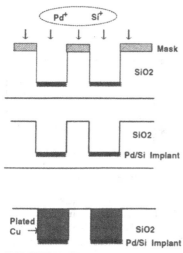

Fig. 3: Process Sequence to form planarized Cu interconnect with Pd/Si PIII

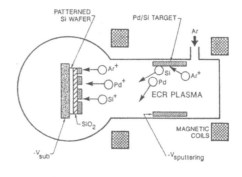

Fig.4: Schematic of PIII reactor to perform Pd/Si implantation

190

Figure 4 shows the schematic of the PIII reactor we used to produce the Pd/Si seeding layer[15]. A sputtering target consisting of Pd and Si was introduced into the Ar plasma that was regulated at a pressure of 1-10 mTorr. We used an rf (8.4 MHz) power supply to bias the target negatively 200-260 V with respect to the chamber walls; Pd and Si atoms were sputtered off from the target, with a small percent of them being ionized by the Ar plasma. A pulsed negative bias of 25 kV was applied to the wafer holder situated in the downstream plasma to facilitate both the implantation and ion mixing processes. The bias pulses had a frequency of 100 Hz and a pulse duration of 2 μsec.

The process sequences for Pd/Si seeding with PIII are summarized in Figure 5. $SiO_2$ films (1-μm thick) were grown on Si wafers by wet oxidation at 1100°C for 2.5 hours. We used photoresist to pattern the oxide, and then made the oxide trenches to a depth of approximately 5500Å by reactive ion etching (RIE) in a $CF_4$:$CHF_3$:He plasma ambient. Using the patterning photoresist on the oxide substrate as the implantation mask, we performed Pd/Si PIII to form the catalytic seed layer. To investigate the effect of Pd depth profile on Cu film adherence, two different process flows were employed during PIII : (1) One-step deposition. The rf target bias and the pulsed wafer bias were applied simultaneously, i.e., Pd and Si deposition and implantation will occur at the same time. Ar ion bombardment will facilitate ion-mixing. (2) Two-step deposition. The target was biased while no high voltage pulse was applied to the wafer. Sputtered Pd and Si atoms were deposited onto the wafer. The rf power to the Pd/Si target was turned off after the deposition had proceeded for a certain period. By applying negative high voltage pulses to the wafer, Ar ion bombardment will facilitate ion-mixing of the deposited Pd/Si film with the $SiO_2$ substrate. After PIII, the photoresist mask was then removed by acetone. Then the samples were rinsed in deionized water and subsequently immersed in an electroless Cu plating solution which was kept at a temperature of about 60°C. Samples used for electromigration stressing experiments were electroless-plated selectively with Ni (≈1000Å) on top of the Cu interconnect surface for passivation.

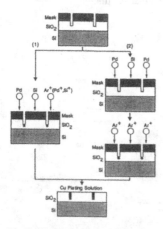

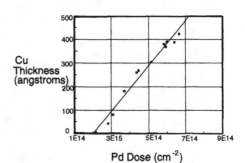

Fig.5: (1) One-step deposition: Simultaneous Pd/Si deposition and ion-mixing. (2) Two-step deposition: Pd/Si film deposition followed by Ar⁺ ion-mixing.

Fig. 6: Plated Cu thickness as a function of Pd dose. Threshold Pd PIII dose is about 2-3 ×10¹⁴ /cm². The plating conditions were for 5 minutes at 70°C.

## (A) Threshold Pd dose to initiate Cu plating

In Figure 6, we show the plated Cu thickness versus PIII Pd dose for samples that have been plated for 5 minutes at 70 °C. A threshold Pd dose is observed between 2-3 $\times 10^{14}$/cm$^2$. The threshold Pd dose for Cu plating is found to be insensitive to the co-implanted Si dose. The linear dependence of the deposited Cu thickness on the Pd dose implies that electroless Cu nucleates at separate Pd sites and developed into islands that are hundreds of angstroms in diameter before grain impingement to form a continuous layer.

This mechanism is verified by scanning electron microscopy (SEM) (Figure 7) of some plated trenches: sporadic Cu grains were forming on the sidewall of the trench, while its bottom was completely covered by continuous Cu. For oxide trenches having slightly tapered sidewalls the areal density of Pd dose on the sidewall is proportional to sinθ. With a maximum Pd dose at the bottom of the trench (θ = 90° ) greatly exceeds the threshold dose, the sidewall dose may acquire a value around the threshold dose if θ is larger than zero. Figure 8 shows an 1 μm wide trench with completely planarized Cu filling by avoiding the sidewall nucleation.

To avoid sidewall nucleation, care must be taken in anisotropic etching of oxide trenches: more vertical slope can keep the Pd dose below the threshold nucleation dose. With a plasma gas pressure at 1 mTorr, the mean free paths between gas molecules collisions is about 4 cm, which is much larger than the sheath thickness between the substrate and the plasma. Therefore, we can assume near-normal incidence of the impinging ions. The plot of Figure 9 presents the calculated maximum allowed tapering angle $\theta_M$ versus Pd dose such that the sidewall Pd dose ($\propto \sin \theta$) does not exceed the threshold of 3 $\times 10^{14}$/cm$^2$.

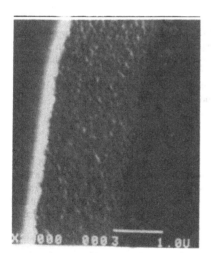

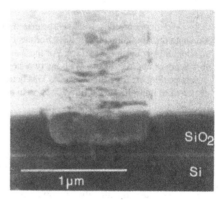

Fig.7: SEM micrograph of Cu plated in SiO$_2$ trenches at 60°C for 4 minutes. The nominal Pd dose is 5.8 $\times 10^{15}$/cm$^2$ . Cu grains are seen forming on the sidewall.

Fig. 8: SEM micrograph showing cross-section of trenches filled with Cu. The SiO$_2$ trench depth is 0.55 μm , and 1 μm wide. The trenches were treated by the two-step Pd/Si PIII process. The nominal doses for Pd and Si used were 5.8 $\times 10^{16}$/cm$^2$ and 1.2 $\times 10^{17}$/cm$^2$ , respectively.

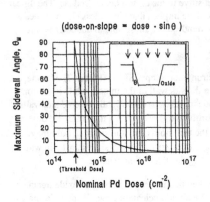

Fig.9: Calculated maximum tapering sidewall angle as a function of Pd dose.

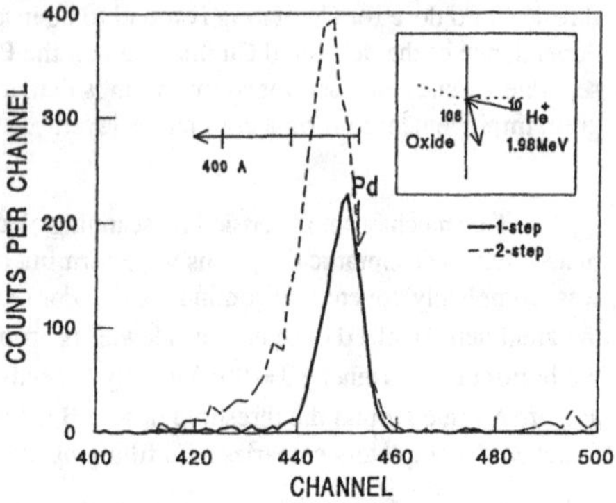

Fig. 10: Pd depth profiles measured by RBS of samples from the two PIII treatments. For the 2-step process , the Pd profile extends deeper than the case of 1-step PIII.

## (B) Implanted Pd Profiles

Pd depth profiles were characterized by RBS. The two PIII procedures shows different Pd depth distributions (Figure 10). For the 1-step process, the Pd peak concentration is located at about 80Å beneath the surface. For the 2-step process, the Pd profile extends into the substrate to a depth of approx. 400Å . With the two-step process, Cu can be plated to a larger thickness before peeling off from the substrate. This better mixing of the implanted Pd with oxide is expected to improve the Cu film adhesion to the substrate for low Pd dose samples.

## (C) Adhesion Studies

While Pd was chosen to be the seed layer material for electroless Cu plating, Si was also included in the target to improve adherence of implanted Pd with oxide substrates. We, in fact, observed better adhesion of the plated Cu films to Pd/Si implanted substrates than to substrates that were implanted with only Pd. Furthermore, we examined the adhesion property of the plated Cu films using Scotch-tape tests. Correlation of the adhesion results and RBS data shows that a higher Pd/(Pd+Si) dose ratio gave better adherence of Cu to the substrate (Table 2). Strong adhesion of plated Cu on $SiO_2$ was always obtained with a dose ratio higher than 40%. In some experiments where the second step of Ar ion-beam mixing was not used, good adhesion of the plated Cu was still obtained if the Pd dose was larger than $10^{16}/cm^2$ , whereas for samples with lower Pd doses, the Ar bombardment step was required to obtain adequate film adhesion.

TABLE 2  Adhesion Properties of Plated Cu on SiO$_2$

| Sample # | Pd Dose ($\times 10^{15}$) | Si Dose ($\times 10^{16}$) | Pd/(Pd+Si) (%) | Adhesion |
|----------|----------------------------|----------------------------|----------------|----------|
| 1 | 6.3 | 8.6 | 6.8 | No adhesion |
| 2 | 2.5 | 1.5 | 14 | No adhesion |
| 3 | 2.6 | 8.2 | 24 | No adhesion |
| 4 | 3.7 | 0.9 | 29 | Weak adhesion |
| 5 | 12.0 | 1.8 | 40 | Weak adhesion |
| 6 | 48.0 | 7.2 | 40 | Weak adhesion |
| 7 | 24.0 | 3.1 | 44 | Good adhesion |
| 8 | 63.0 | 7.2 | 47 | Good adhesion |

## Pd$_2$Si Seeding

The process sequence to form Cu interconnect using Pd$_2$Si as a seeding layer is illustrated in Figure 11 [16]. About 600Å of polysilicon was deposited onto the Si wafers covered with 5000 Å of thermal oxide. After polysilicon etching, 8000Å of SiO$_2$ was deposited, followed by oxide trenches etching. Then 2000Å Pd was evaporated and then annealed at 245°C for 15 min. in Ar ambient to react with the underlying polysilicon. The unreacted Pd was selectively removed by wet etch. The wafers were then immersed in the electroless copper bath. The plated Cu film thickness is about 3000Å. A 6000Å overcoating of SiO$_2$ layer was deposited by low temperature PECVD to protect the Cu from oxidation. For the electromigration test structures, contact holes to the bonding pad area were defined by dry etching. 1- μm thick pure aluminum was sputtered, and patterned to form the contact pads for electrical probing.

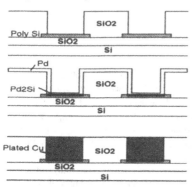

Fig.11: Process Sequence using Pd$_2$Si
as the seeding layer for Cu plating

## Electrical Properties

Electrical resistivity of the electroless-plated Cu interconnect seeded by PIII was ≤ 2 μΩ-cm, a reasonably good value compared with 1.7μΩ-cm for pure Cu. The corresponding temperature resistance coefficient (TRC) of the Cu lines is 3.6 ×10$^{-3}$/°C around room temperature, close to a reported value of 3.9×10$^{-3}$/°C for pure Cu [17]. The effective resistivity of the Cu/Pd$_2$Si stack film was measured to be 2.1 μΩ-cm, and the temperature-resistance-coefficient (TRC) of the Cu film was 3.4 ×10$^{-3}$/°C.

## Electromigration of Plated Copper Metallization

Accelerated electromigration testing of the Cu interconnects were carried out using current stressing at elevated temperatures between 200-275°C. A current source regulated by a transistor and driven by the output of a TTL gate was used to generate pulsed DC and AC currents [18]. Electromigration testing was performed on wafers placed directly on the heated stage of a probe station. The failure time was defined to be when the interconnect was completely open.

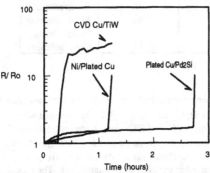

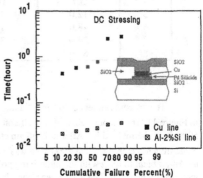

Fig.12: Change of interconnect resistance as a function of EM stressing time for (a) Plated Cu/Pd$_2$Si, (b) Ni/ Plated Cu, and (c) CVD Cu/TiW metallization. Samples (a) and (c) were stressed at a temperature of 275°C. Sample (b) was stressed at 200°C to prevent the Cu-Ni interfacial reaction [19]. All samples were stressed at a current density of $1.5 \times 10^7$ A/cm$^2$ .

Fig.13: Log-normal plots of time to failure for (a) Cu/Pd$_2$Si and (b) Al-2%Si interconnects under DC current stressing at 275 °C. The stress current density is $1.5 \times 10^7$ A/cm$^2$ .

When subjected to current stressing at temperatures from 200-275°C, both Cu interconnects seeded by PIII of Pd/Si (i.e., the Ni/Cu structure) and seeded by Pd$_2$Si (i.e., the Cu/Pd$_2$Si structure) show open failures. In Figure 12, we show the time dependence of the resistance change (normalized to initial resistance before stressing) for both structures. Both types of interconnects exhibit an initial increase of resistance, followed by an electrical open. We also show the behavior of a similar structure fabricated by CVD copper on TiW. After a 20-fold increase of resistance, the CVD Cu/TiW structure still maintains electrical continuity by shunting of the TiW underlayer. SEM was used to observe the failure sites. Cu depletion(void) and accumulation (hillock) were generally observed within the interconnects, agreeing with mass transport due to electron flow.

Because of copper's potentially better resistance to electromigration failure, it is natural to compare the Cu and Al metallization systems. Figure 13 shows log-normal plots of the lifetime distribution of Cu/Pd$_2$Si and Al-2%Si samples for $J_{DC} = 1.5 \times 10^7$ A/cm$^2$ . We can see that the median-time-to failure (MTF) of Cu is about two orders of magnitude longer than that of Al-2%Si at a stressing temperature of 275°C. The longer lifetime for Cu is probably due to the higher activation energy for atomic motion. Indeed, from the Arrhenius plot of MTF the activation energy of electroless-plated Cu is found to be about 0.8eV[16], compared with 0.4eV for Al-2%Si [18]. .

For time-varying DC current stressing waveforms, the vacancy relaxation model of Liew et al. has been used to satisfactorily explain the electromigration characteristics of Al metallization for both interconnects and contacts [16,20] . Under the special condition of a rectangular current waveform, the median-time-to-failure can be written as[20]:

$$MTF_{pulse-DC} = \frac{MTF_{DC}}{aD \left[ 1 - \frac{(1 - e^{-aD})(1 - e^{-a(1-D)})}{D(1 - e^{-a})} \right]}$$ (1)

D is the pulse DC duty factor and a = $1/(f\tau)$, where f is the pulse DC repetition frequency, and $\tau$ is the vacancy relaxation time constant. Figure 14 shows the frequency dependence of $MTF_{pulse-DC}$ under peak current density of $1.5 \times 10^7 A/cm^2$ and duty factor of 50% for both $Cu/Pd_2Si$ metallization and conventional Al-2%Si metallization. The Cu pulse DC lifetimes are still about two orders of magnitude longer than that of Al-2%Si. The predictions of Liew's Model are also shown in Figure 14 with the vacancy relaxation time as the only fitting parameter. The $\tau$ for Cu (=1.1 $\mu$s) is observed to be about 20 times smaller than that of Al-2%Si (=20 $\mu$s). In Liew's model, the vacancy relaxation time is an empirical parameter which describes the recombination of excess vacancies at sinking sites (e.g., dislocations, grain boundaries, and metal surfaces). Within the frequency range of interest for integrated-circuit operation, it is noteworth that only a single transition frequency is observed for Al or Cu metallization. More work will be required to elucidate the relationship of $\tau$ to metallization microstructures.

Figure 15 shows log-normal plots of the cumulative failure data for $Cu/Pd_2Si$ interconnects under DC and AC( i.e., bipolar) stressing conditions. A symmetrical 1MHz rectangular AC waveform with a peak current density of $1.5 \times 10^7 A/cm^2$ and duty factor of 50% was used. This small $MTF_{AC} / MTF_{DC}$ ratio for Cu ($\approx 50$) when compared with that of Al-2%Si ($\approx 2500$), strongly suggests the void recovery ability of plated Cu is inferior to that of Al.

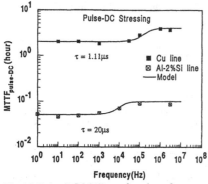
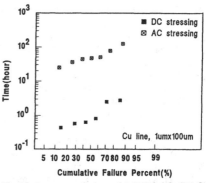

Fig.14: Pulse-DC MTF as a function of frequency for (a) Plated $Cu/Pd_2Si$ and (b) Al-2%Si interconnects. The solid lines are fitting of the vacancy relaxation model.

Fig.15: Log-normal plots of MTF for $Cu/Pd_2Si$ metallization under DC and bipolar AC stressing. The peak current densities are $1.5 \times 10^7/cm^2$ for both cases at a stressing temperature of 275°C.

## Conclusions

To resolve the difficulty in patterning Cu films for multilevel interconnects, selective copper plating of $SiO_2$ trenches has been carried out using Pd/Si plasma immersion ion implantation or $Pd_2Si$ deposition. The Pd/Si PIII process is self-aligned, with only a single masking step required to define the oxide trench, etching, seeding, and selective plating. We found a threshold PIII Pd dose of 2 to $3 \times 10^{14}/cm^2$ is required to initiate the electroless plating of Cu on $SiO_2$ surface. By careful control of the anisotropic etching of the oxide trenches and proper choice of the nominal Pd dose, 1-$\mu$m wide Cu filled lines with flat surfaces suitable for planarized multilevel metallization were successfully fabricated. Resistivities and TRC's of the plated Cu obtained by both seeding methods are very similar, close to those reported for bulk Cu.

The electromigration characteristics of electroless plated $Cu/Pd_2Si$ interconnects have been studied under constant current and time-varying current stressing. Our results show that under DC and pulse DC current stressing, the Cu lifetimes are about two orders of magnitude longer than that of Al-2%Si at 275°C , and four orders of magnitude longer than that of Al-2%Si when extrapolated to operation at room temperature. Similar to that of Al, pulse DC lifetime of Cu agrees with the vacancy relaxation model, having a vacancy relaxation time 20 times shorter than that of Al-2%Si. Under bipolar current stressing, the ratio of $MTF_{AC} / MTF_{DC}$ for Cu is only 50 versus 2500 for Al-2%Si.

## References

[1] S. P. Murarka, *Tungsten and Other Advance Metals for ULSI Applications IV*, edited by G.C. Smith and R. Blumenthal (Mater. Res. Soc. Proc., Dallas, TX, 1990) pp.179-187.
[2] A. Awaya and Y. Arita, Proceedings of VLSI Symposium, Kyoto, Japan, pp.103-104 (1989).
[3] P. L. Pai and C. H. Ting, IEEE Electron Device Lett., EDL-10, 423 (1989).
[4] C. H. Ting, M. Paunovic, P.L. Pai, and G. Chiu, J. Electrochem. Soc.,136, 462(1989).
[5] R. M. Lukes, *Plating*, 51, 1066 (1964).
[6] M. Paunovic, *Plating*, 55, 1161 (1968).
[7] S. M. El-Raghy and A. A. Abo-Salama, J. Electrochem. Soc., 126, 171 (1979).
[8] P. Buck and L. R. Griffith, J. Electrochem. Soc., 109, 1005 (1962).
[9] J. E. A. M. van den Meerakker, J. Appl. Electrochem., 11, 387 (1981).
[10] R. Sard, J. Electrochem Soc., 7, 864 (1970).
[11] D. C. Thomas and S. S. Wong, IEDM 86, 811 (1986).
[12] D.C. Thomas, N.W. Cheung, I.G. Brown, and S.S. Wong, *Tungsten and Other Advanced Metals for VLSI/ULSI Applications V*, edited by S.S. Wong and S. Furukawa (Mater. Res. Soc. Proc., San Mateo, CA, 1989) pp. 233-242.
[13] X.Y. Qian, M.H. Kiang, J. Huang, D. Carl, N.W. Cheung, M.A. Lieberman, I.G. Brown, K.M. Yu, and M.I. Current, Nuclear Instrument and Methods, B55, 888 (1991).
[14] M-H. Kiang, C.A. Pico, M.A. Lieberman, N.W. Cheung, X.Y. Qian and K.M. Yu, Materials Research Society Proceedings, Vol. 223, pp.377-383, (1991).
[15] N.W. Cheung, Nuclear Instruments and Methods, B55, 811 (1991).
[16] K-K. Kang and S.S. Wong, submitted to Electron Device Letters.

[17] CRC Handbook of Chemistry and Physics, 68th edition, F122 (1987).

[18] B.K. Liew, N.W. Cheung, and C. Hu, IEEE Transaction on Electron Devices, Vol.37, no.5, 1343 (1990).

[19] J. Li, Y. Shacham-Diamond, J.W. Mayer and E.G. Colgan, Proceedings of 8th International VLSI Multilevel Interconnection Conference , pp. 153-159(1991).

[20] Jiang Tao, K.K. Young, N.W. Cheung, and C. Hu, Proc. 30th International Reliability Physics Symposium, 338 (1992).

## Acknowledgments

The authors would like to thank members of the Plasma Materials Processing Laboratory at UC-Berkeley for technical assistance. This work is supported in part by the California State MICRO Program, Applied Materials Inc., and ISTO/SDIO administered by ONR under Contract N00014-85-K-0603.

This article also appears in Mat. Res. Soc. Symp. Proc. Vol 260

# ION IMPLANTATION TO INHIBIT CORROSION OF COPPER

P.J.Ding and W.A.Lanford
Department of Physics, State University of New York at Albany, Albany, NY12222
S.Hymes and S.P.Murarka
Center for Integrated Electronics, Rensselaer Polytechnic Institute, Troy, NY12180

## ABSTRACT

Ion implantation of boron and aluminium is used to passivate copper surfaces. Such a process modifies the copper only near its surface without affecting copper's desirable bulk properties. The present experiments show that implant doses as low as $10^{15}$ ions/cm$^2$ of either boron or aluminium can reduce the oxidization rate by one order of magnitude or more compared to non-implanted samples.

## I. INTRODUCTION

Copper is being developed as a material in ULSI metallization. However, exposed copper surfaces are potentially subject to corrosion. This paper presents results on the effects of ion implantation on the rate of oxidization of copper in air. The oxidization rate is followed by using both Rutherford backscattering spectrometry (RBS) [1] and four-point probe measurements of sheet resistivity. Crowder and Tan have previously investigated the effect of ion implantation on the oxidization of copper [2]. These authors and other references have shown that ion implantation can effectively protect copper from corrosion. Many elements have been tried as implants [3-9]; some are effective in reducing corrosion, some accelerate corrosion [8].

The present paper concentrates on B and Al as implant materials. Both elements when mixed in bulk copper are known to make relatively inert metals. Aluminium-copper mixtures have been widely used in the electronics and other industries for many years. Less well known is that copper with 0.01 percent B has been commercially available for decades (known as Boron Deoxidized Copper) as a corrosion resistant material with most of the properties of pure copper [10]. Both B and Al have been widely used in ion implantation. Hence, implementation of this process in industry would require a minimum of new tooling. Unlike most previous works, the use of relatively low implant doses is emphasized here with the goal of demonstrating that this process may be practical in industry.

## II. PROCEDURE

Copper films that were approximately one micron thick were sputter deposited on SiO$_2$/Si wafers. The purity of the copper is 99.99%. The wafers were then cut into square samples 2 cm on each side. The samples were implanted with B or Al at various energies and doses. To accelerate the oxidization of the copper, samples were then heated in air at 300°C for times up to 40 minutes.

The two principal analytic probes used to characterize the amount of oxidization of the copper were Rutherford backscattering spectrometry (RBS) and four-point probe measurements of the sheet resistivity. These measurements were made before ion implantation, and after implantation (before and after the 300°C anneals in air). We were concerned that annealing in air with its potentially variable amount of moisture might lead to non-reproducible oxidization rates. However, repeat measurements over several weeks showed good reproducibility.

## III. EXPERIMENTAL RESULTS

Figure 1 shows RBS spectra for the samples of copper, either non-implanted or implanted with B or Al at 50keV. Some of these samples had been annealed in air at 300°C for 10 or 40 minutes. These spectra show the dramatic reduction in oxidization rate as a result of the ion implantation.

Figure 1(a) shows RBS spectra from as deposited copper samples heated in air for 0, 10 and 40 minutes. Figure 1(b) shows a RUMP simulation of the 10 minute annealing data in Fig.1(a). From results in the literature [11-14], it is expected that the oxide thickness will

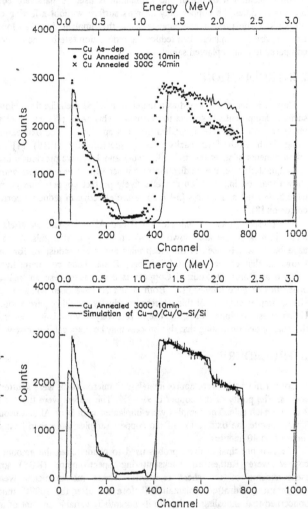

Fig.1(a). RBS spectra of non-implanted Cu films annealed at 300°C for different times.

Fig.1(b). RUMP simulation of 10 minute annealing spectrum in Figure 1(a). The simulation assumes 0.65 μm $Cu_2O$ on Cu on $SiO_2$ on Si.

increase with the square root of the oxidization time. The data in Fig.1(a) are consistent with this. The thickness of oxide X is given by $X=0.21T^{1/2}$ (μm), where annealing time T is in minutes. From the RUMP simulation, it is seen that the oxide that formed is $Cu_2O$.

Figures 1(c) and 1(d) show RBS data recorded from samples that have received the same 300°C air anneals as the samples for Figure 1(a), except that they were first implanted with 2.5E15 B/cm² and 4E15 Al/cm², respectively. Contrasted to the non-implanted samples, these samples have only a very thin surface oxide even after 40 minute annealing in air.

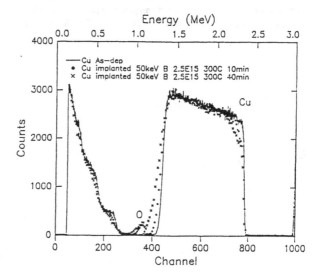

Fig.1(c). RBS spectra of Cu films implanted with B and annealed in air under same conditions as samples used in Figure 1(a).

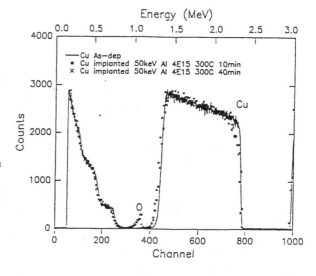

Fig. 1(d). RBS spectra of Cu films implanted with Al and annealed in air under same conditions as samples used in Figure 1(a).

Figures 2(a) and (b) show the results of a Monte Carlo calculation of the expected ion implant profiles for 50 keV B and Al in copper, respectively. These profiles were made with TRIM program [15]. These calculations indicate that the peaks in the B and Al concentrations are at approximately 100 nm and 40 nm, respectively, with the concentration decreasing near the surface. The existence of this decreased surface concentration may explain why these implanted films seem to quickly develop a thin surface oxide which then grows very slowly. Namely, the concentrations of B or Al may be too low in the very near surface region to prevent the oxidization. The data in Figure 1(c) and 1(d) show that the surface oxide is slightly thicker on the B implanted samples compared to the Al implanted samples, consistent with the results in Figure 2 which indicate that the B is depleted to a greater depth than the Al.

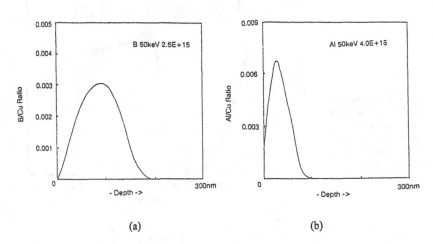

(a)                                        (b)

Figures 2(a) and 2(b). B and Al distributions in Cu films calculated by the computer program TRIM.

Four-point probe measurements of sheet resistivity have also been used on these and other samples. Results of these measurements are summarized in Figure 3 which shows the resistivity as a function of 300°C air annealing time, for both implanted and non-implanted samples.

The most prominent feature shown in these data is that the increase in resistivity with exposure time is dramatically less in the implanted samples than in the non-implanted samples. This is just another manifestation of the great decrease in the oxidization rate shown in Fig.1.

A second important point shown in Figure 3 is that the ion implantation itself increases the resistivity slightly over the non-implanted value. This is seen at the 0 time axis of these figures. As small as this is, any increase in resistivity is undesirable for electronic applications. This increase in resistivity comes from two effects. One is purely chemical. Namely, B/Cu and Al/Cu have slightly higher resistivity than pure Cu. However, considering that the expected concentration profiles (Figure 2) extend into only a small fraction of the copper film and at modest concentrations, this is expected to be a very small effect. The second contribution to the increase in resistivity with implantation is due to the damage introduced into the film during the implantation. The distribution of damage may extend well beyond the range distribution of the implanted atoms. It should be possible to remove these defects by annealing in an inert atmosphere. We have begun to explore the observed increase in resistivity with ion implantation by studying changes of resistivity with implant dose, energy, and post implant inert gas annealing. These experiments are still underway.

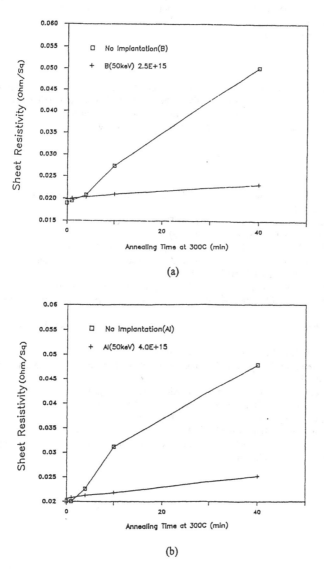

(a)

(b)

Figures 3(a) and (b). The sheet resistivities of non-implanted and implanted with B and Al copper films with annealing times.

We have also begun to determine what are the optimal ion implantation conditions in terms of dose and energy. Consistent with the results in Figure 2, generally the lower the implant energy, the more effective the implant is in reducing oxidization. Unfortunately, we have not been able to operate our implanter below 30 keV, so we do not know the optimal energy yet. The question of how large an implant dose is required is related to the implant energy and the degree of protection needed. However, implant doses as low as $10^{15}$ ions/cm$^2$ have dramatically beneficial effects.

## IV. SUMMARY

Ion implantation of B and Al is very effective in inhibiting corrosion of copper, even at low implant doses. While further work needs to be done, it should be possible to find an implantation energy, dose, and post implant (inert gas) annealing conditions that provide this corrosion protection with little or no increase in sheet resistivity above that for as deposited copper.

## References:

[1] W. -K. Chu, J. W. Mayer and M. -A. Nicolet, Backscattering Spectrometry, Academic Press, New York, 1978.
[2] B. L. Crowder and S. I. Tan, IBM Tech. Disclosure Bull. 14, 198(1971)
[3] L. G. Svendsen, Corros. Sci. 20, 63(1980)
[4] L. G. Svendsen, S. S. Eskildsen and P. Borgesen, Thin Solid Films, 110, 237(1983)
[5] O. Henriksen, T. Laursen, E. Johnson, A. Johansen, L. Sarholt-Kristensen and J. L. Whitton, Nucl. Instr. and Meth. B15, 356(1986)
[6] L. G. Svendsen and P. Borgesen, Nucl. Instr. and Meth. 191, 141(1981)
[7] P. Zhou, R. P. M.Procter, W. A. Grant and V. Ashworth, Nucl. Instr. and Meth. 209/210, 841(1983)
[8] H. M. Naguib, R. J. Kriegler, J. A. Davies and J. B. Mitchell, J. Vac. Sci. Technol. 13, 396(1976)
[9] R. J. Ratcliffe and R. A. Collins, Phys. Stat. Sol.(a) 108, 537(1988)
[10] Boron Deoxidized Copper 1170, Publication TP-58, 2nd Edition, Anaconda American Brass Company(1967)
[11] J. Bardeen, W. H. Brattain and W. Shockley, J. Chem. Phys.14, 714(1946)
[12] Y. Ebisuzaki and W. B. Sanborn, J. Chem.Educa. 62,341(1985)
[13] S. K. Roy, S. K. Bose and S. C. Sircar, Oxid. Met. 35, 1(1991)
[14] M. Uhrmacher, A. Bartos and W. Bolse, Mat. Sci. Eng. A116, 129(1989)
[15] J. F. Ziegler, J. P. Biersack and U. Littmark, The Stopping and Range of Ions in Solids, Pergamon Press, New York, 1985.

This article also appears in Mat. Res. Soc. Symp. Proc. Vol 260

# THE EFFECT OF COPPER ON THE TITANIUM-SILICON DIOXIDE REACTION AND THE IMPLICATIONS FOR SELF-ENCAPSULATING, SELF-ADHERING METALLIZATION LINES

STEPHEN W. RUSSELL, JIAN LI, JAY W. STRANE AND JAMES W. MAYER
Department of Materials Science and Engineering, Cornell University, Ithaca, NY

## ABSTRACT

Co-evaporated Cu-Ti films on thermally-oxidized Si substrates were annealed in vacuum at temperatures between 300 and 700°C. Reactions within the films and between film and substrate were monitored by Rutherford backscattering spectrometry, X-ray diffraction, transmission electron microscopy and sheet resistance. We found that, despite the competition from intermetallic compound formation, the Cu-Ti films react with $SiO_2$ beginning at 400°C, with Ti migrating to the $SiO_2$ interface to form both a silicide, $Ti_5Si_3$, and an oxide and to the free surface to form additional oxide. The reaction leaves relatively pristine Cu. Above 600°C, however, Cu begins to react with the underlying silicide. By comparison, pure Ti reacts with $SiO_2$ only above ~700°C.

## INTRODUCTION

Copper is an attractive replacements for aluminum in ULSI metallization schemes because of its low resistivity and potentially high resistance to electromigration.[1] Unfortunately, Cu doesn't adhere well to oxides such as $SiO_2$ or $Al_2O_3$. In addition, it will oxidize in air over extended periods of time, a particularly damning problem as line dimensions are reduced.[2] Thus any metallization scheme using copper will require an adhesion promoter between metal and oxide and a passivation layer surrounding each line to prevent degradation due to Cu-oxide formation.

Titanium has been used as an interfacial bonding layer in several schemes. Thin Ti and Cr layers have been used to aid in the adhesion of several metals to oxides such as glass. It is widely believed that these refractories undergo a surface reaction with both the underlying oxide and the overlying metal, forming a chemically-bound stack that enhances the overall adhesion. In addition, TiN is often used as a diffusion barrier in metal-silicon systems.[3]

With these properties in mind, we initially explored the effects of adding small amounts of Ti (up to 25% at.) to Cu films. In our initial research, we studied the reaction of these films in an ammonia ambient, based on earlier work nitriding Cu(Ti) films in $N_2$.[4] The following reactions were found to occur at 550°C:

$$Cu(Ti) + NH_3 \ ---> \ TiN + H_2 + Cu \tag{1}$$

$$Cu(Ti) + SiO_2 \ ---> \ Ti_5Si_3 + TiO_x + Cu \tag{2}$$

with (1) occuring at the free surface and (2) at the $SiO_2$ interface and at the free surface.[5] The surface Ti-oxynitride encapsulates the Cu metal, protecting it from oxidation. The underlying $Ti_5Si_3/TiO_x$ stack enhances adhesion of Cu to the $SiO_2$.

We chose to focus on reaction (2), which we found to occur upon vacuum annealing over temperatures from 400 to 600°C. Among the issues we considered were the competition with intermetallic phase formation within the Cu-Ti film. For example, Cu/Ti bilayers were found to form a series of compounds, including $Cu_3Ti$, $Cu_2Ti$ and CuTi, at 350-400°C.[6] We used sheet resistance measurements to follow the overall reaction progress, Rutherford backscattering spectrometry (RBS) to profile elemental composition versus depth, X-ray diffraction (XRD) to determine phase identification, cross-section transmission electron microscopy (XTEM) to directly observe the reaction front.

Mat. Res. Soc. Symp. Proc. Vol. 265. ©1992 Materials Research Society

EXPERIMENTAL

Samples were deposited by electron beam co-evaporation (deposition rates 1-20Å /sec) onto thermally-oxidized Si (100) wafers (oxide thickness 7300Å) at a base pressure of ~$10^{-7}$ torr. Film thicknesses were typically 2000-5000Å. Several compositions were deposited. We were primarily concerned with those in excess of 75% Cu, but others were prepared as well.

Anneals were performed in a vacuum tube furnace with a base pressure of 5-8x$10^{-8}$ torr, except for those samples used in the *in situ* crystallization study. The furnace was brought to temperature prior to sample loading; samples were removed from the furnace to achieve rapid cooling.

Sheet resistance measurements were taken using a Prometrix Versaprobe VP10 dual polarity four-point probe at a current of 50 mA. RBS was performed on a General Ionex 1.0 MV Tandetron accelerator using a 2.8 MeV He$^{2+}$ beam. For XRD, a Scintag $\Theta$-$\Theta$ Diffractometer was used in both glancing angle and conventional modes. Crystallization of $\gamma$-CuTi was studied by *in situ* XRD using a hot stage affixed to the diffractometer and kept at a pressure of ~1x$10^{-6}$ torr. TEM was performed using a JEOL 200CX at 200kV. Specimens were prepared by dimpling to 10-20 μm thickness, followed by ion milling with 4 keV Ar$^+$ until a hole had formed.

RESULTS

The composition Cu$_{76}$Ti$_{24}$ was examined in detail in this paper simply because its relatively-high Ti content facilitates determination of the overall reaction progress. It is likely, however, that any realistic application of the encapsulation process would utilize films of considerably lower Ti composition. Compositions more Cu-rich than this one exhibited essentially the same reaction and resistance behavior.

Fig. 1 depicts the normalized sheet resistance of this sample as a function of processing temperature. Samples were annealed for one hour at the temperature indicated; resistance measurements were taken at room temperature after annealing.

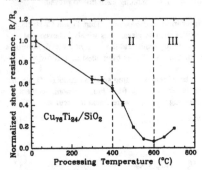

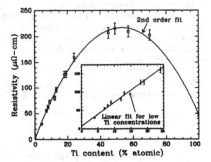

Figure 1. Normalized sheet resistance as a function of processing temperature for Cu$_{76}$Ti$_{24}$/SiO$_2$. Samples were annealed 1 hour in vacuum.

Figure 2. Electrical resistivity of thin films of Cu-Ti as a function of composition.

Three regimes of interest are observed. Up to 400°C, there is a gradual decrease in resistance to around 60% of its original value. The regime between 400 and 600°C is characterized by a sharp decrease in resistance to 10-20% of the initial value followed by a plateau. Above 600°C, the resistance inches upward again, rising by a factor of two between 600 and 700°C.

For additional information, we compared the resistivity of several films of varying compositions, with the data given in fig. 2. The fits are merely a guide to the eye. Resistivities were calculated according to $\rho = R_{sheet}*t$, where t is the thickness as determined by RBS. For reference, the bulk resistivities of Cu and Ti are 1.7 and 46$\mu\Omega$-cm, respectively.[7]

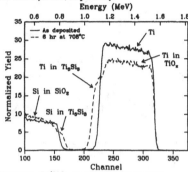

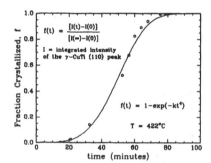

Figure 3. Ti/SiO$_2$ reaction as observed by Rutherford backscattering spectrometry (2.2 MeV He$^{2+}$). Initial Ti thickness was 3700Å.

Figure 4. Fraction of Cu$_{50}$Ti$_{50}$ crystallized into γ-CuTi as measured by *in situ* X-ray diffraction (Cu Kα radiation, λ = 1.54Å).

Two things are of note, one being that a very small addition of Ti to Cu greatly increases the resistivity. Adding 4% Ti to Cu increases the resistivity by a factor of 15. Secondly, the resistivity peaks in the middle of the composition field. Clearly, any metallization scheme utilizing Cu(Ti) lines must endeavor to remove nearly all Ti from the Cu during subsequent processing to maintain acceptable line conductivity.

With our goal being to understand the competing reactions potentially occuring in the Cu(Ti)/SiO$_2$ structure, we first studied the Ti/SiO$_2$ system (fig. 3).[8,9] Spectra of both the as-deposited and a sample annealed 6 hours at 708°C are shown.

In this case, we found the reaction to occur beginning at 700°C. The SiO$_2$ decomposes, with the Si reacting with Ti to form Ti$_5$Si$_3$. This in turn frees O, which diffuses upward to form a TiO$_x$ layer of gradually-increasing O composition and a mix of Ti-oxide phases. At x ~ 1.2, the reaction ceases.

Next, we studied the crystallization of amorphous Cu-Ti films. To do this, we deposited films of composition Cu$_{50}$Ti$_{50}$ and employed *in situ* TEM and XRD.[10,11] By measuring the integrated area of the (110) peak of γ-CuTi (tetragonal, a = 3.1Å, c = 5.3Å) as a function of time at several temperatures, we were able to monitor the crystallization kinetics. Results shown in fig. 4 show a characteristic Avrami exponent equal to 4 consistent with TEM observations of continuous nucleation and constant radial growth rates.

With Cu-Ti intermetallic phases forming at temperatures several hundred degrees lower than that of the Ti/SiO$_2$ reaction, the Cu(Ti)/SiO$_2$ reaction could exhibit fairly complicated behavior. All four samples indicated in fig. 1 were analyzed extensively by RBS and XRD to follow the composition profile and phase sequence as a function of temperature.    Up to 400°C, no change occurs in the RBS spectra of any of our samples. Apparently, no substrate reaction occurs in this temperature regime. The XRD spectra indicate sharpening of the Cu peaks, but otherwise, no additional phases are observed to form.

In fig. 5, a Cu$_{76}$Ti$_{24}$ film is shown before and after annealing at 500°C for 1 hour as observed by RBS. We found Ti to segregate to both the SiO$_2$ and free interfaces in this temperature regime, with the coincident appearance of X-ray peaks corresponding to Ti$_5$Si$_3$ and several Ti-O phases. Correspondingly, the Cu signal shows a distinct increase due to the Cu phase being depleted of Ti. We performed glancing angle (10°) X-ray diffractometry was used to identify phases. We found distinct Cu peaks in the as-deposited films for compositions under 15% Ti; between 15 and 25% Ti, the deposited film structure was highly disordered. Annealing at temperatures below 400°C leads to the gradual appearance of the Cu peaks. The Ti/SiO$_2$ reaction, with Ti$_5$Si$_3$ and Ti-O phases forming, occurs between 400 and 600°C. XRD spectra for this sample as-deposited and annealed (1 hr at 500°C) are shown in fig. 6

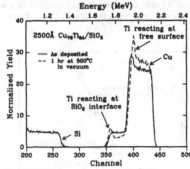

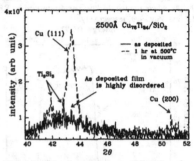

Figure 5. RBS spectra of $Cu_{76}Ti_{24}/SiO_2$ as deposited and reacted 1 hr at 500°C. Beam energy 2.8 MeV $He^{2+}$.

Figure 6. Glancing angle x-ray diffraction of $Cu_{76}Ti_{24}/SiO2$ as deposited and annealed 1 hour at 500°C.

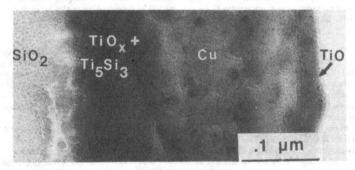

Figure 7. Bright field XTEM of $Cu_{76}Ti_{24}/SiO2$ annealed 1/2 hr at 500°C.

For further study, we employed XTEM to directly observe the reaction interface. The bright field micrograph of a sample of $Cu_{76}Ti_{24}/SiO_2$ annealed 1/2 hour at 500°C is given in fig. 7. We believe that the contrast at both the free and $SiO_2$ interfaces is due to Ti migration. Ti is known to ion mill at a somewhat slower rate than either Cu or $SiO_2$, and therefore we believe the contrast observed is due to the relatively greater thickness of these regions.

Upon annealing at temperatures higher than 600°C, the back edge of the Cu signal forms a step, indicative of Cu reacting at the bottom interface. Due to the weakness in intensity, no additional x-ray peaks are observed. However, the RBS spectra clearly show reactions occuring at the $Cu/TiO_x$ interface, with both Ti and Si diffusing upward and Cu diffusing inward through the oxide layer. Fig. 8 is an RBS spectrum of a sample annealed 1 hour at 700°C exhibiting the overall atomic motion.

DISCUSSION

Results are best understood in terms of the regimes denoted in fig. 1. Below 400°C, no reaction with

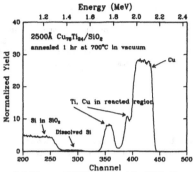

**Figure 8.** Cu$_{76}$Ti$^{24}$/SiO$_2$ annealed 1 hour at 700°C analyzed by RBS. Beam energy 2.8 MeV He$^{2+}$.

the substrate is observed by according to RBS. However, a general rise in the Cu X-ray peak intensities lead us to conclude that a lattice relaxation process is responsible for the overall decrease in resistance seen.

Above 400°C, the Ti backscattering signal shows a definite increase at both the high and low energy edges, while the Cu signal increases in intensity. This accumulation of Ti we attribute to the reaction of Ti with SiO$_2$, according to eqn. (2). As in the pure Ti/SiO$_2$ system, the liberated O diffuses away from the silicide. However, apparently oxide nucleation occurs both next to silicide and at the free surface. Thus, Ti diffuses out of the Cu in both directions. To verify the accumulation of O at the free surface, we performed 3.05 MeV $^{16}$O($\alpha,\alpha$)$^{16}$O nuclear reaction resonance on one sample and found an increase in the O concentration matching that of the Ti.

The dramatic decrease in sheet resistance in regime II we attribute to the out-diffusion of Ti from the Cu layer. As Cu is much more conductive than Ti (a factor of 40 improvement over bulk Ti; oxides and silicides are typically lower still), we can assume that all electrical conduction occurs within the Cu-rich layer. Thus the dip in overall film resistance is due to a decrease in the resistivity of the Cu layer, which is due to Ti out-diffusion to both interfaces (see fig. 2). Upon completion of the reaction in this regime, a TiO/Cu/TiO$_x$/Ti$_5$Si$_3$/SiO$_2$ stack is created. Nearly all of the Ti is removed from the Cu.

We can see that the Cu(Ti)/SiO$_2$ reaction occurs several hundred degrees lower than the that between pure Ti and SiO$_2$ (fig. 3). Clearly, the effect of Cu on this reaction is one of enhancement. To examine further this interesting phenomenon, we deposited very thin (5-150Å) layers of Cu on SiO$_2$ prior to Ti deposition in order to see if even very small amounts of Cu could somehow promote this reaction. We found (1) that this Cu dissolved into the Ti and possibly reacted at temperatures well below that of the Ti/SiO$_2$ reaction, and (2) this reaction began to occur around 600°C, below the reaction temperature of pure Ti but higher than what was found for Cu-rich co-evaporated films.

At temperatures above 600°C, the Cu begins to diffuse through the TiO$_x$ layer to react with the underlying silicide. In addition, Ti and Si diffuse through this oxide and react and/or dissolve into the overlying Cu layer, as seen in fig. 8. Simulations were attempted using RUMP,[12] and indicate a highly complex elemental profile. The overall sequence is (from the top down) TiO$_2$/Cu/Cu$_3$Ti/TiO$_x$/Ti$_5$Si$_3$/SiO$^2$, with Cu dissolved or reacted in the Ti$_5$Si$_3$ layer, Ti dissolved in the Cu layer and Si dissolved in both the Cu and Cu$_3$Ti layers. All of these effects raise the resistivity of the Cu layer, and we believe them to be responsible for the rise in sheet resistance found in regime III in fig. 1. Due to the narrow width of the Ti-Si-Cu reacted layer and its apparent lateral non-uniformity, we were unable to identify its composition. No new X-ray peaks were observed at these temperatures.

CONCLUSION

The Cu(Ti)/SiO$_2$ system offers some interesting possibilities for use in metallization lines in ULSI technology. The Ti/SiO$_2$ reaction is favored over Cu-Ti intermetallic phase formation in the intermediate (400-600°C) temperature regime. Since the Ti segregates out of the Cu layer and leaves nearly-pure Cu in its wake, the line resistance should not be unduly compromised. Competition from Cu-Ti intermetallic phases is not observed despite the fact that such phases are stable at these temperatures and compositions and have been found to form under similar conditions (bilayers, more Ti-rich coevaporations) at even lower temperatures. Processing above 600°C would need to be avoided in order to prevent Cu from reacting with previously-formed Ti-silicide.

## ACKNOWLEDGEMENTS

We would like to thank Dr. Maura S. Weathers of the Materials Science Center X-ray Facility for her assistance and advice. One of us (S.W.R.) would like to note the support of the AT&T Foundation for a Ph.D. Scholarship.

## REFERENCES

1. J.D. McBrayer, R.M. Swanson and T.W. Sigmon, J. Electrochem. Soc. 133, 1242 (1986).

2. Jian Li, J.W. Mayer and E.G. Colgan, J. Appl. Phys. 70, 2820 (1991).

3. Jian Li, Y. Shacham-Diamand, J.W. Mayer and E.G. Colgan, 8th Int'l VLSI Multilevel Interconnect Conf. (IEEE, Santa Clara, 1991).

4. K. Hoshino, H. Yagi and H. Tsuchikawa, 6th Int'l VLSI Multilevel Interconnect Conf. (IEEE, Santa Clara, 1989), 226.

5. Jian Li, E.G. Colgan, J.W. Mayer and Y. Shacham-Diamand, to be published in Appl. Phys. Lett. (1992).

6. J.L. Liotard, D. Gupta, P.A. Psaras and P.S. Ho, J. Appl. Phys. 57 (1985), 1895.

7. CRC Handbook of Chemistry and Physics, 67th ed., Boca Raton, FL, CRC Press, 1986-87.

8. S.Q. Wang and J.W. Mayer, J. Appl. Phys. 67 (1990), 2932.

9. S.W. Russell and J.W. Mayer, to be published.

10. Jian Li, S.W. Russell and J.W. Mayer, submitted to Physical Review B.

11. S.W. Russell, Jian Li and J.W. Mayer, to be published.

12. L.R. Doolittle, Ph.D. thesis, Cornell University (1987).

# STRESS-RELATED PHENOMENA IN
# CAPPED ALUMINUM-BASED METALLIZATIONS

CAROLE D. GRAAS
Semiconductor Process and Design Center, Texas Instruments Inc., P.O. Box 655012, MS 944, Dallas, Texas 75265.

## ABSTRACT

The effect of a TiW capping layer on the mechanical behavior of AlSi(1%)Cu(½%) thin films was evaluated. The aluminum films studied were deposited at temperatures ranging from 150 to 450°C, on various barrier layers (Ti, TiW, and TiN). Measurement of the stacks stress during anneal up to 450°C revealed that for films deposited at 150°C, the capping layer improved yield strength regardless of the type of barrier layer used. The uncapped stacks exhibited the traditional behavior of hillock formation during heating, and hillock/grain collapse during cooling. The capped stacks, however, showed considerably reduced, or suppressed hillock formation during heating, and active tensile stress relief mechanisms during cooling included cracking, voiding, and grain boundary grooving. These observations correlated well with XRD data characterizing the extent of tensile stress buildup in the aluminum films. For films deposited at 300°C, grain deformation during cooling after deposition presumably caused hardening which prevented cracking as well as changes in texture during anneal. This was observed regardless of the presence of a capping layer. Although the stacks studied were unpassivated, this data suggests that capping layers may enhance stress migration issues in integrated systems.

## INTRODUCTION

As performance requirements for IC metallization systems become more stringent, development efforts must focus on understanding and controlling electromigration induced damage phenomena and stress-related issues in stacked thin films. Layered aluminum-based metallizations have been studied for some time, for they offer unique hillock formation resistance and enhanced electromigration lifetime properties [1-3]. In particular, trilayer metal stacks (i.e. barrier layer/aluminum alloy/capping layer) are strong candidates for use over planarized, high-aspect ratio contacts and vias in advanced multilevel systems [4,5]. Although the enhanced performance of these metallizations is well documented, the complex mechanical behavior of these stacks is not well understood. This paper examines the effect of capping layers on the mechanical response of AlSi(1%)Cu(½%) films subject to thermally-induced stresses. In-situ wafer curvature measurements during anneal are correlated with data obtained by surface profiling, SEM and XRD analyses. A model is proposed to explain microstructural and morphological phenomena observed for a variety of barrier layer, aluminum deposition temperature, and capping layer combinations.

## EXPERIMENTAL FLOW

The films studied were deposited by DC magnetron sputtering on passivated p-type <100> silicon wafers. The underlayer passivation material was a 1 µm thick oxide film deposited by plasma-enhanced decomposition of a TEOS(TetraEthylOrthoSilicate)/$O_2$ source, with intrinsic stress values in the range -20 to -50 MPa ((-)=compressive). In the metal stacks, TiN and TiW barrier, and Ti, TiN and TiW capping materials were evaluated. TiW and Ti films were sputter-deposited sequentially with the aluminum. TiN barrier films were formed by reaction of a Ti film with $N_2$ in a furnace at 675°C, while TiN capping layers were formed by reactive sputtering of Ti in a $N_2$ ambient. TiN formation/deposition and aluminum deposition were not sequential. The

AlSi(1%)Cu(½%) films were deposited at substrate temperatures in the range 150-450°C. After deposition of the metal stacks, a 250 μm profilometry scan (stylus force 15 mg, speed 50 μm/sec) was obtained to characterize the surface roughness. The wafers were then annealed in argon and in-situ wafer curvature measurements were made throughout the thermal cycle using a Flexus F2400 apparatus. The ramp rate was 15°C/min, followed by a plateau at 450°C for ½hr, and spontaneous cooling. The surface roughness was again characterized by profilometry, taking as many scans as required to obtain representative data. For texture and local stress determination before and after annealing, X-Ray diffraction (XRD) analysis was carried out using a conventional powder diffraction technique. The radiation wavelength was CuKα1, and the machine error estimated at 2θ = 0.008° by calibration for precision lattice parameter calculation. Electron micrographs (SEM) were obtained after annealing to examine the surface morphology of the stacks.

## STRESS EVOLUTION DURING THERMAL CYCLING

Because of its practicality and excellent repeatability, the wafer curvature measurement technique is now widely accepted as a powerful method for characterizing changes in thin film stress levels during thermal cycling. The method has been described by several authors, notably D.S. Gardner et al. [6]. By measuring the displacement of a laser beam reflected off of the surface of the wafer during temperature cycling, one can calculate the change in wafer radius of curvature, and therefore the change in film stress. The film stress σ is related to the wafer curvature R by the expression:

$$\sigma = E t_{si}^2 \ / \ 6(1-v)R t_f \qquad (1)$$

where E is the Young's modulus of the Si-wafer, $t_{si}$ is the Si-wafer thickness, and v is the Poisson's ratio of the Si-wafer. For this study, we used a biaxial elastic modulus value of:

$$E \ / \ (1-v) = 1.805E11 \ Pa \qquad (2)$$

appropriate for <100> Si-wafer substrates. The wafers thickness was 625 μm. If the initial wafer flatness $R_i$ is measured prior to depositing the film, and the stress in the film $R_0$ is measured before starting the temperature cycling experiment, then the radius of curvature R used in equation (1) is expressed by:

$$R = R_i R_0 \ / \ (R_i - R_0) \qquad (3)$$

which gives an absolute measurement of the film stress for each data point taken. The method is directly applicable to single films. For film stacks, if the film thickness entered in equation (1) is that of the stack, then the obtained stress data represents an average value over the entire stack. Under certain conditions, the effect of a given layer can be substracted from the average data to better characterize the behavior of the remaining films. Ideally, this should be done only when no chemical reaction takes place between the layer to be substracted and its neighbors, and this layer should behave elastically over the entire thermal cycle. For the stacks studied, these conditions were not met, even for the passivation TEOS-derived oxide film. Therefore, the stress-temperature data presented in this paper represent average values over the entire stacks, including the passivation layer.

The stress behavior of aluminum-alloy thin films is well characterized and most features of the stress-temperature signature curves have been correlated with physico-chemical phenomena directly observed by other techniques [6,7]. We found the same features on the stress-temperature curves obtained for the stacks studied. In addition, a combination of the passivation oxide layer and a TiW barrier layer produced a featureless stress-temperature curve -- except a significant compressive stress buildup during the 450°C plateau. Therefore, the

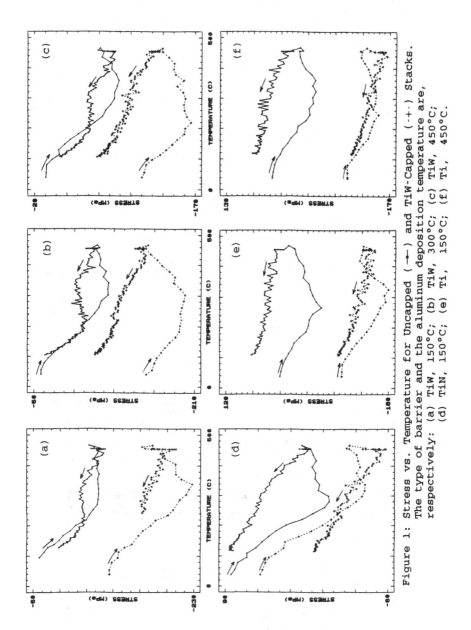

Figure 1: Stress vs. Temperature for Uncapped (+) and TiW-Capped (-+-) Stacks. The type of barrier and the aluminum deposition temperature are, respectively: (a) TiW, 150°C; (b) TiW, 300°C; (c) TiW, 450°C; (d) TiN, 150°C; (e) Ti, 150°C; (f) Ti, 450°C.

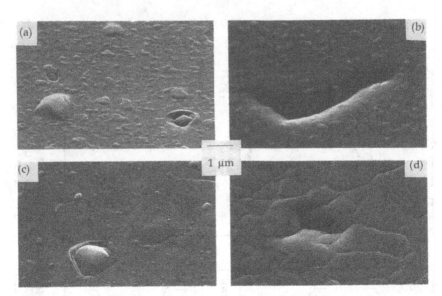

Figure 2: Samples Surface after Annealing: (a) TiN/Al(150°C);
(b) Ti/Al(150°C)/TiW cap; (c) Ti/Al(450°C); (d) Ti/Al(450°C)/TiW cap

Figure 3: Aluminum Films Stress from
XRD Spectra. (TiW/Al/TiW)

Figure 4: Aluminum Films Texture from
XRD Spectra. (TiW/Al/TiW)

Table I: Stress Relief Features Observed after Anneal

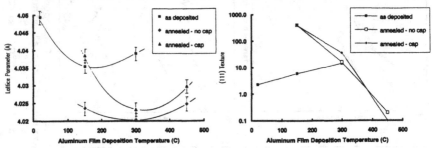

| STACK | | | COMPRESSIVE STRESS RELIEF | TENSILE STRESS RELIEF |
|---|---|---|---|---|
| BARRIER | Al TEMP.(C) | CAP | | |
| TiW | 150 | | Hillocks (3200 A) | Hillock/Grain collapse |
| | 150 | TiW | | Cracks |
| | 300 | | Hillocks (1700 A) | Hillock/Grain collapse |
| | 300 | TiW | Small hillocks (200 A) | |
| | 450 | | | |
| | 450 | TiW | | Voids |
| TiN | 150 | | Hillocks (2500 A) | Hillock/grain collapse |
| | 150 | TiW | | Cracks |
| | 150 | TiN | Small hillocks (500 A) | Cracks |
| | 150 | Ti | | |
| Ti | 150 | | Hillocks | Hillock/Grain collapse |
| | 150 | TiW | | Cracks |
| | 450 | | Hillocks (grain boundaries) | Hillock/Grain collapse |
| | 450 | TiW | | Grain boundary grooving, Cracks |

features observed on the curves obtained for the stacks represent a signature of the aluminum film behavior or its interactions with its barrier and capping layers. Subtle effects due to the addition of a capping layer can then be studied from the stress-temperature curves obtained for the complete stacks.

Figure 1 illustrates the effect of a TiW capping layer on the mechanical behavior of the aluminum during anneal. The TiW deposition process produced compressive films, therefore the capped systems are initially under a stronger compressive stress than their uncapped counterparts. Overall, the stacks exhibit an increasingly compressive state during the initial stage of the heating cycle. This is due to the thermal expansion coefficient mismatch between the aluminum film and the other layers, particularly the silicon substrate. The slopes of the curves are similar regardless of the presence of a cap. At some point between 200 and 400°C, a yield point is reached, and the compressive stress starts to be relieved. The presence of a cap does affect the yield strength, increasing it at low aluminum deposition temperatures (Fig.1(a), (d) & (e)) indenpendently of the type of barrier layer used, and decreasing it at high aluminum deposition temperatures (Fig.1(c) & (f)). During cooling, the stacks become more tensile, due again to the thermal expansion coefficient mismatch between the aluminum and the silicon substrate. The capping layers signature in this part of the anneal cycle is to reduce the plateau found centered around 300°C for stacks containing a TiW barrier layer. A similar increase in yield point during heating was observed for stacks were the TiW capping layer was replaced with either a TiN or a Ti capping layer.

## MICROSTRUCTURAL ANALYSIS

Hillock formation and grain growth are two forms of local mechanisms for compressive stress relief in thin films. Tensile stresses, on the other hand, can be relieved by grain/hillock collapse, grain rotation, or crack/hole formation [8-10]. We have observed most of these phenomena in our uncapped and capped stacks as summarized in Table I. Note that some stacks did not show clear deformation features. Films deposited on TiW at 450°C were very rough but did not exhibit significant differences before and after anneal. The electron micrographs shown in Figure 2 illustrate those phenomena for various types of stacks and aluminum deposition temperature. The uncapped aluminum films exhibit the traditional behavior of hillock formation upon heating, and hillock/grain collapse upon cooling. The capped material, however, shows very limited hillock growth upon heating, and a pronounced tendency to relieve tensile stresses upon cooling by crack/void formation, grain rotation, or grain boundary grooving.

Stress levels in the aluminum film itself can be evaluated by comparing the films lattice parameter with that of an unstrained reference powder sample -- typically 4.0494 Å (Figure 3). The smaller the lattice parameter, the larger the residual tensile stress in the film. The deposition temperature affects these residual stress levels because of the effect of post-deposition cooling. At room temperature, we observed no residual stress, at 150°C the tensile stress reached a maximum, and above this temperature, the films yielded upon cooling, relieving part of this residual stress. This is consistent with phenomena reported by other authors [9-11] who observed extensive dislocation motion on {111} planes along <110> directions -- the primary slip systems in FCC materials. We also observed this type of deformation by TEM analysis in our high-temperature films. Upon annealing, all films build up additional tensile stress as seen on the stress-temperature curves. However, the presence of a cap enhances tensile stress relief, especially for films deposited at 150°C or 450°. These are also the samples for which we observed the most extensive crack/void formation.

The films texture was estimated from the XRD spectra by comparing the peaks relative intensity to that of a powder sample. The data presented in Figure 4 indicates that the main factor driving the texture evolution of the aluminum films upon anneal is the deposition temperature rather than the presence of a cap. Figures 3 and 4 refer to the TiW/AlSiCu/TiW stack, but we found the same trends for all combinations of barrier and caps studied.

## DISCUSSION AND CONCLUSIONS

During annealing, a large number of microstructural phenomena take place. For films deposited at 150°C, high initial residual stress levels allow to reach relatively high annealing temperatures (>300°C) before onset of compressive stress relief mechanisms occurs. This onset occurs at higher temperatures for capped samples, possibly because of differences in grain growth kinetics. This is mostly visible for films deposited at low temperature 150°C, which are dislocation-free at the beginning of the anneal cycle. Only uncapped material shows signs of local compressive stress relief by forming hillocks, while compressive stress levels are reduced regardless of the presence of a cap. This suggests that capping layers may enhance compressive stress relief mechanisms of a more global nature. Upon cooling, uncapped films can relieve some of the tensile stress levels by hillock/grain collapse, while capped films, which have no hillocks, crack or form voids. For films deposited on a TiW barrier layer, the observed "plateau" in tensile stress buildup during cooling may be related to the hillock/grain collapsing phenomenon, as suggested by previous studies on single aluminum films [6], since it is not observed for the corresponding capped stacks. However, stress-temperature curves during cooling for capped and uncapped samples deposited on other types of barrier layers do not show this effect, so it may be also related to chemical interaction phenomena. Grain rotation seems to take place in both types of stacks, since all films deposited at 150°C show significant texture improvements. For films deposited at 300°C, post-deposition cooling has caused mechanical yielding, and therefore hardening. The texture is "locked-in" during annealing, and grain rotation, as well as crack formation, are limited. For higher deposition temperature films (450°C), larger initial grains reduce the yield strength. The films recrystallize at relatively low annealing temperatures (240°C), loose their texture, and relieve tensile stresses by the same mechanisms seen for films deposited at 150°C.

These conclusions suggest that stress-migration phenomena may be an issue for capped films used in integrated systems. Optimization and control of these mechanisms must include such film stack parameters as deposition temperature and composition. Future work will focus on these issues, as well as the study of passivated films.

## ACKNOWLEDGEMENTS

Hung-Yu Liu from TI Central Research Labs provided excellent X-Ray Diffraction analysis support. The Semiconductor Process Laboratory Operations staff helped process the samples and perform the SEM analyses.

## REFERENCES

1. B.W. Shen, T. Bonifield and J.W. McPherson, Proc. IEEE VMIC, 114 (1985).
2. H.H. Hoang and J.M. McDavid, Solid State Technology 30, 121 (1987).
3. J. Nulty, G. Spadini and D. Pramanik, Proc. IEEE VMIC, 453 (1988).
4. T. Fujii, K. Okuyama, S. Moribe, Y. Torii, H. Katto and T. Agatsuma, Proc. IEEE VMIC, 477 (1989).
5. J.J. Estabil, H.S. Rathore and E.N. Levine, Proc. IEEE VMIC, 242 (1991).
6. D.S. Gardner and P.A. Flinn, IEEE Trans. Electron Devices 35 (12), 2160 (1988).
7. D.S. Gardner, H.P. Longworth and P.A. Flinn, Proc. IEEE VMIC, 243 (1990).
8. U. Smith, N. Kristensen, F. Ericson and J-A. Schweitz, J. Vac. Sci. Technol. A 9 (4), 2527 (1991).
9. M. Murakami, J. Vac. Sci. Technol. A 9 (4), 2469 (1991).
10. R. Venkatraman, J.C. Bravman, W.D. Nix, P.W. Davies, P.A. Flinn and D.B. Fraser, J. Electron. Mater. 19, 1231 (1990).
11. R. Venkatraman, S. Chen and J.C. Bravman, J. Vac. Sci. Technol. A 9 (4), 2536 (1991).

# PART IV

# Oxide and Device Reliability

# HARDNESS AND MODULUS STUDIES ON DIELECTRIC THIN FILMS

Chien Chiang,   Gabi Neubauer[*], Anne Sauter Mack, Ken Yoshioka,
George Cuan, Paul A. Flinn and David B. Fraser

Components Research and [*]Materials Technology, Intel Corporation,
3065 Bowers Ave., Santa Clara, CA 95052

## ABSTRACT

We report hardness and Young's modulus measurements on various
dielectric thin films. Hardness and modulus information was derived
from indentation experiments with a Berkovich triangular-based
diamond indenter in an ultra micro-indentation instrument (UMIS).
We studied the effect of moisture content and phosphorous doping on
hardness and Young's modulus of low temperature Chemical Vapor
Deposition (CVD) Si-oxides and found that dehydration and
densification tend to harden samples, whereas  increased P-doping
results in a lower hardness. Hardness values of silicon nitride,
silicon oxynitride, sputtered oxide, spin-on-glass and APCVD Si-
oxides are compared. We also discuss how deposition conditions and
chemical compositions correlate to dielectric properties such as
stress as well as moisture uptake, thermal expansion coefficients
and hardness and modulus values. Using these results, thermal
stresses in encapsulated Al lines have been calculated and the
calculated stress in Al is higher when encapsulated with dielectric
films with higher moduli.

## INTRODUCTION

Dielectric thin films are used as insulating materials between
conducting layers in integrated circuits. A reliable film not only
needs to have good electrical properties, such as high breakdown
voltage, low leakage current and a low dielectric constant but also
needs to have good mechanical properties,  such as stability to
environment and a low stress level. When the dimensions of ICs get
smaller, requirements on interconnect systems become much more
stringent. Thermal stress induced metal voids form one of the
critical reliability issues to be addressed. The dielectric film
properties and their compatibility to metallization need to be
carefully characterized. We have studied hardness and Young's
modulus on a series of dielectric thin films using a commercially
available instrument. We have also studied the effect of water in
dielectric films on film stress and thermal expansion coefficients.
The stress of Al encapsulated by these dielectric films has been
calculated. This led to a better understanding of interconnect
system reliability.

## EXPERIMENTAL SETUP

Hardness values and elastic moduli, which we present here,
were determined from indentation experiments, performed with an

Ultra Micro-Indentation System (UMIS 2000). The UMIS 2000 is a commercial indentation instrument and is employed in our laboratory for measuring mechanical properties of thin films used in VLSI fabrication. Force and depth resolutions are 10 mN and 10 nm, respectively. Maximum forces and depths of 200 mN and 2000 nm, respectively, can be reached. The contact force, present upon start of an indentation experiment, is 0.1 mN. Depth and force sensitivities, which can be obtained with this machine, are less than those of the well known Nanoindenter. The UMIS indenter, however, has several advantages for our applications, such as easy handling, a stiff design to account for vibrational isolation, and operation without significant delays in achieving thermal stability.

The indentation instrument is a "force driven static measuring" system: It is "force driven" in the sense that the indenter is driven into the surface until a resistance equal to a desired maximum force is met. The loading and indentation of the sample is carried out so that the load is increased in increments at a rate linearly proportional to the square root of the force, until the maximum desired force is reached, to simulate a constant strain rate. At each increment, an electronic feedback loop ensures that

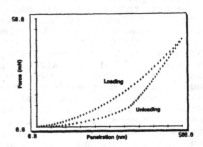

Figure 1. Indentation curve of a thermal oxide

penetrations as well as loads are measured under conditions of force equilibrium. In this sense then, the instrument is "static measuring". When the maximum force is reached, the increments are reversed and the data upon unloading are acquired. As an example, Figure 1 shows such loading and unloading data, averaged over 5 single indentations, as measured in an indentation experiment on thermal oxide with a maximum load of 2 mN. The indenter used in our system, is a Berkovich triangular base diamond pyramid indenter, face angle 65.3°. Special software is provided to account for non-ideal tip shapes at shallow indentations.

Principle

The two fundamental parameters obtainable from the indentation experiments are hardness and elastic moduli of materials. The analysis is based on several assumptions: First, it is assumed that on loading, the material deforms both elastically and plastically and on unloading relieves the elastic component only. Further assumptions include that sufficient plastic deformation is produced to keep the shear stress at all points in the stress field at or below the yield stress, that the total indentation depth $h_t$ is simply the sum of plastic depth $h_p$ and elastic deformation $h_e$, and that the hardness H of a material is connected to the contact area of the indentation in the following way:

$$H = F / A. \tag{1}$$

Here, F indicates the load and A the contact area, respectively. For an indenter of known geometry, the contact area can be expressed in terms of the plastic depth $h_p$. For the Berkovich indenter, it is:

$$A = 24.5 \, h_p^2 \tag{2}$$

The plastic depth is obtained by extrapolating the initial portion of the unloading curve, which represents the elastic recovery of the material upon unloading to zero load [1]. By this procedure, using equations (1) and (2), the hardness of a material can be obtained from a single indentation with a well defined indenter.

Finally, it is assumed that the elastic deformation resulting from depression of material surrounding the indentation is similar to that which would be produced by a friction-less circular punch with a projected area equal to that of the indentation. Then the elastic modulus of the material E can be obtained in the form of $E/(1-v^2)$ with $v$ being Poisson's ratio, according to Sneddon's solution [2]

$$dh/dF = (1/h_p) \ (\pi/24.5)^{1/2} \ (1/E_r) \tag{3}$$

where

$$1/E_r = ((1-v^2)/E) + ((1-v^2)/E_o). \tag{4}$$

F and h are measured force and depth, respectively, dh/dF is the elastic recovery rate and $E_o$ the elastic modulus of the indenter. This small additive contribution of the indenter to the measured elastic recovery rate dh/dF is the machine compliance. In the UMIS system it is obtained by indenting standard materials with well known properties. The machine compliance is then automatically accounted for and subtracted when displaying loading/unloading curves.

By means of Sneddon's approach the elastic modulus of a material can then be obtained, similarly to the hardness, from a single indentation experiment.

RESULT AND DISCUSSION

Figure 1 is the indentation curve of a thermally grown oxide film. The loading curve is a combination of elastic and plastic deformation. The first portion of the unloading curve represents the elastic recovery of the material and is used to determine the recovery rate dh/dF. Young's modulus and hardness are then derived, according to the equations (1) through (4). For the thermally grown oxide, we find a hardness of 12 GPa and a Young's modulus, corrected with Possion's ratio, of 127 GPa. The Young's moduli measured and discussed in this paper are $E/(1-v^2)$ values.

## Dielectrics

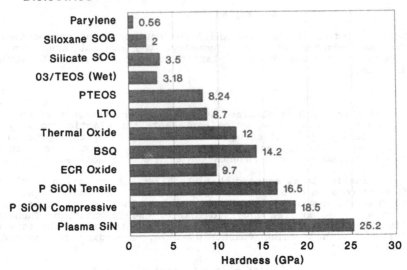

Figure 2. Hardnesses of various dielectric thin films.

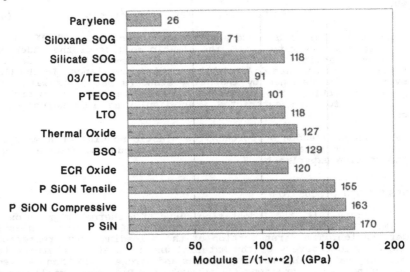

Figure 3. Moduli of various dielectric thin films

We have studied thin films of low temperature oxide (LTO) with different amounts of phosphorous doping, as well as thin films of plasma nitride and plasma oxynitride, biased sputtered and CVD oxide, ozone/TEOS oxide and spin on glasses (SOG). The hardness and Young's modulus measurement results are shown in Figure 2 and 3 respectively.    The data are compared to thermal oxide as a reference since it is a very stable and reproducible film.    The different dielectric thin films can be separated into the following groups.

## 1.The LTO system

The low temperature oxides (LTO) were deposited in a LPCVD (Low Pressure Chemical Vapor Deposition) reactor using silane, oxygen and phosphine. One of the major concerns for the low temperature oxide is the water in the film.  We have studied the stress as a function of temperature in a home made stress gauge evolved from a older system  described in reference [3]. The system is capable of measuring stress from -40 C to 1000 C in a controlled ambient. Figure 4 shows a hysteresis  in a stress vs. temperature (up to 450 C) curve which is  due to water evolution during heating. If we continue to heat the sample, the film densifies around 700 C, as shown in Figure 4. Without densification, the water which had been driven out of the film could be reabsorbed, as shown in Figure 5. In this case, the film stress is driven more compressive.

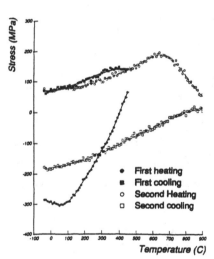

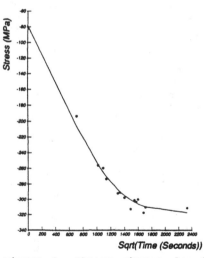

Figure 4. Stress behavior of 0%P LTO.

Figure 5. Stress change due to water absorption by LTO at room temperature.

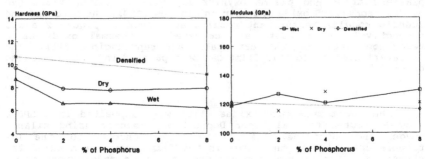

Figure 6. Hardness of P-LTO.    Figure 7. Modulus of P-LTO.

Figure 6 shows the hardness of several sets of LTO samples with different amounts of phosphorous doping. One set of samples was measured with the film saturated with moisture, the other set was baked at 450 C for one hour to drive out the water without changing the film structure and the third set of the samples was densified at 900 C for one hour. The measurement showed that hardness decreased with increased phosphorous dopant concentration. We believe that phosphorous serves as a network modifier which makes the film more plastic. Figure 6 also indicates that hardness is increased when the film is drier. This can be explained by the changing of hydroxyl groups which are dangling end-groups, to silicon-to-oxygen bonds. Figure 7 showed that the modulus seems to be independent of the water content or phosphorous dopant.

The slope of the cooling curve in the stress vs. temp cycle is due to the thermal expansion mismatch between oxide film and silicon substrate and can be used to calculate the thermal expansion coefficient. To study the dependence of thermal expansion coefficients with water content, we did the following experiments: first, we correlated the amount of water in the film quantified by Moisture Evolution Analyzer (MEA) [4] to the size of the hysteresis in stress temperature cycle as shown in Figure 8. The size of the hysteresis is defined as the difference in stress at room temperature before and after heating to 450 C. Roughly estimated, for a film containing 1 weight percent of water, the stress hysteresis is 120 MPa. We then cycled the LTO sample to various temperatures. For each cycle, the size of the hysteresis can be measured which can be converted to water content and the slope of the cooling curve can be measured which can be converted to thermal expansion coefficient and thus the relation between moisture and thermal expansion can be derived. Illustrated in Figure 9, the thermal expansion of coefficient can be greater than that of silicon depending upon the amount of water in the film. We also found that the thermal expansion coefficient of P-LTO varied with the doping level of phosphorous as shown in Figure 10.

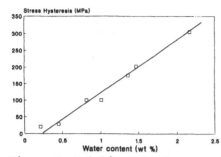

Figure 8. Relation of water content in the film and the size of stress hysteresis.

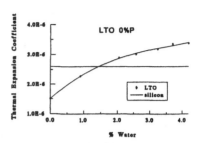

Figure 9. Relation of thermal expansion coefficient of 0%P LTO and water content in the film.

## 2.Nitride and oxynitride

As shown in Figure 2 and 3, plasma deposited nitride [5] was found to be harder than plasma oxynitride, followed by thermal oxide. The modulus in Figure 3 showed a similar trend. Nitride had the highest modulus, and oxynitride the next highest, followed by thermal oxide. The reported density of nitride is 3.1 which may attribute to the high values of hardness and modulus.

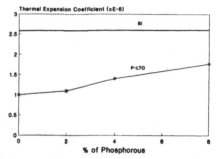

Figure 10. Thermal expansion coefficients of dry P-LTOs.

## 3.Ozone/TEOS oxide

Plasma TEOS and Ozone/TEOS oxides are gaining popularity for interlayer dielectric applications [6] due to their good stepcoverage. The stepcoverage of ozone/TEOS films is better than that of Plasma TEOS films, however, the water level in ozone/TEOS film has been a major concern. Figure 11 shows the water content of an oxide film deposited in an APCVD (Atmospheric Pressure Chemical Vapor Deposition) reactor using ozone and TEOS with various ozone concentrations. The water content, as measured immediately after deposition, indicated much lower water content for films deposited with higher ozone concentrations. However, the water content drifted up with time which is similar to the case of LTO. Again, we observed that film hardness correlated well with water content as shown in Figure 12. Films containing more water were softer. The plasma TEOS is generally drier and harder due to the ion bombardment during deposition.

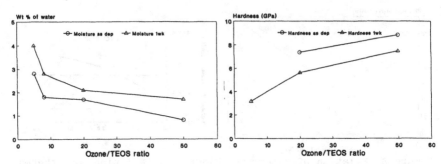

Figure 11. Water content vs. ozone/TEOS ratio.

Figure 12. Hardness of ozone/TEOS oxide.

Note that the moduli in Figure 3 are comparable for the plasma TEOS film and the Ozone TEOS film although the hardness values for these two films were quite different. This further verifies that the modulus is primarily determined by the bonding.

4.Polymers

There are two types of spin-on-glasses (SOGs) which are commonly used [7]: silicate and siloxane. Silicate contains no organic side groups, whereas siloxane has organic side groups. In Figure 2, both SOGs showed lower hardness values than thermal oxide with the hardness of silicate SOGs being slightly higher than that of siloxane SOGs. Both SOGs have been cured to 400 C and left in room ambient, thus picking up moisture. The silicon to carbon ratio which indicates the amount of organic side groups can vary from 1:1 to a few time to one. The organic side group is believed to further increase the plastic nature. However, the Young's moduli of these two SOGs were quite different. The silicate glass which consists primarily of silicon-oxygen bonds showed a value much closer to that of thermal oxide, compared to siloxane glass which consists of both silicon-oxygen and silicon-carbon bonds.

Parylene [8] is another type of polymer deposited by vapor phase polymerization. The hardness value for parylene is the lowest among all the dielectric films studied. The modulus for parylene is much lower than that of the thermal oxide due to the organic composition.

5.Other Dielectric Films.

Biased ECR oxide [9] and BSQ are two oxide films deposited with energetic ion bombardment ($^-$100 V). The ion energy improved the film quality. The water content in the ECR film can be as low as .05 wt % and in BSQ is essentially undetectable. The hardness of ECR oxide, shown in Figure 2, was similar to that of the densified LTO film while the hardness of BSQ seems to be higher than that of thermal oxide. We believe that with energy bombardment, some degree of implantation and compaction may occur and thus increasing the density of the film. However, shown in Figure 3, the moduli of ECR oxide, thermal oxide and BSQ were relatively close.

IMPLICATION

One of the motivations to study the mechanical properties of the dielectric films was to understand the impact on the reliability of the interconnect system especially for the stress induced voids problem [10]. We studied the implications of three dielectric films, oxide, nitride and parylene using the mechanical parameters measured.

Calculation Procedures

Calculations of thermal stresses in encapsulated Al lines on (100) silicon, with various passivations and aspect ratios were performed using the ANSYS finite element program [11] on an IBM RS/6000 520 workstation. Figure 13 shows the geometry used in the calculations. The width of the line is w, the height h, and the thickness of the passivation over the line is t. For the purposes of these calculations, t=h. Aspect ratios of w/h = 0.5, 1.0 and 4.0 were considered.

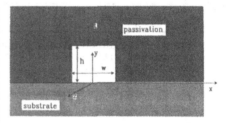

Figure 13. Geometry of passivated Al line.

Plane strain 4-noded elements were used, and only the right half of the line was modeled due to symmetry. In the model, the substrate dimensions are five times those of the line, and the substrate is fixed in space at its lower left corner. The x=0 plane has the symmetry boundary condition applied, and all other surfaces are free. Perfect bonding between all dissimilar materials is assumed. Thermal stresses are calculated on cooling from a stress-free state at 400 C to room temperature. The substrate behaves purely elastically, and the Al line behaves elastic-perfectly plastically. Three passivations are considered: SiN, $SiO_2$, and parylene. The former two are purely elastic in these calculations, while the parylene is elastic-plastic. The temperature-dependent materials properties used for the silicon, Al, SiN, and $SiO_2$ are reported elsewhere [12]. For the parylene, a Young's modulus of 25 GPa, a thermal expansion coefficient of $2.8 \times 10^{-6}$/C and a yield strength of 174 MPa were used. The SiN and $SiO_2$ are much stiffer than the parylene, having Young's moduli of 160 and 120 GPa, respectively. The film properties used in the calculation is listed in Table 1.

|  | E (GPa) | $\alpha$ ($10^{-6}$) | $\sigma_y$ (MPa) |
|---|---|---|---|
| Al | 71.5 | 23.6 | 185 |
| Si | 130.2 | 2.6 | -- |
| SiN | 160 | 3.0 | -- |
| SiO$_2$ | 120 | 0.55 | -- |
| parylene | 25 | 2.8 | 174 |

Table 1. Properties of the films used in the calculation.

| Passivation | $\sigma_x$ | $\sigma_y$ | $\sigma_z$ | $\sigma_{hydrostatic}$ |
|---|---|---|---|---|
| SiO$_2$ | 610 | 620 | 760 | 660 |
| SiN | 660 | 640 | 800 | 700 |
| parylene (elastic) | 225 | 215 | 375 | 270 |
| parylene (plastic) | 180 | 170 | 340 | 230 |

Table 2. Calculated stress components in Al lines (w/h=1) encapsulated with different dielectric thin films.

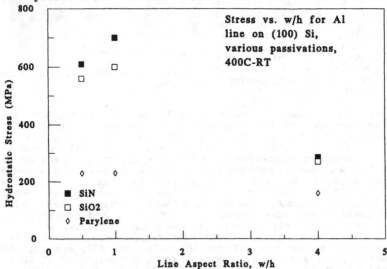

Figure 14. Thermal stresses on aluminum lines encapsulated by passivations with different hardness and moduli.

Results

The calculated stress components in Al (w/h=1) are listed in table 2. Figure 14 shows the hydrostatic stress as a function of aspect ratio for the three different passivations. For the nitride and oxide passivations, the square cross-section line has the highest stress for all components. The stresses in the w/h=4 lines is less than half that in the w/h=1 lines, and the stress in the w/h=0.5 lines is also smaller than in the square lines. This is due to the nature of the substrate and passivation constraints on the lines. The thermal expansion difference between the line and the substrate is the largest contributor to the thermal strain in the line, while the passivation acts as a mechanical constraint. For the square cross section, both of these effects conspire to create the largest stress. In the wider lines, the passivation constraint is reduced near the center of the line, allowing stresses to relax. For the narrower lines, there is proportionally less material directly attached to the substrate and proportionally more passivation constraint on the sides of the line. Thus, the stresses relax in the x and z directions, but do not relax in the y direction, compared to w/h=1 lines. The ultimate effect is a lower hydrostatic stress.

For the lines passivated with parylene, the stresses are much smaller than for the other passivations. Note also that, while 400C is a typical deposition temperature for the nitride and oxide films, parylene would be deposited at close to room temperature. Therefore the actual thermal stresses induced in lines passivated with parylene would be even lower than those calculated here. A deposition temperature of 400C was assumed for the parylene for these calculations for comparison purposes. Because the parylene is so compliant compared to oxide and nitride, and even compared to aluminum, there is very little mechanical constraint provided by this passivation. For this reason, there is much less variation in the stress with aspect ratio than for the other passivations.

CONCLUSION

Hardness and modulus of various dielectric thin films have been studied. We have noticed that water in the film lowered the hardness, changed the film stress and affected the thermal expansion coefficient. Doping with phosphorous lowers the hardness as well as the thermal expansion coefficient. Silicon nitride based materials were harder than thermally grown oxide and polymers such as SOG and parylene were even softer materials. We have also noticed that the modulus on dielectric thin films was not affected by water absorption. The modulus was primarily affected by the bonding. We have also calculated the thermal stress for Al lines encapsulated by different dielectric thin films. The simulation concluded that the stress in Al was higher when encapsulated with dielectric materials with higher moduli. It suggested that materials like SOG/parylene might help the stress induced void problem.

ACKNOWLEDGEMENTS

The authors would like to thank Prof. R.C. Budhani from UCLA for providing plasma nitride samples, J. Sisson from Wakins-Johnson for ozone/TEOS samples, R. Olson from Novatran, S. Dabral and J. McDonald from RPI for providing parylene samples. We also like to thank N. Cox, R. Hsu and M. Tang at Intel for useful discussions.

REFERENCES

1. M. F. Doerner, W. D. Nix, J. Mat. Sci. 1, 601 (1986).
2. I. N. Sneddon, Intern. J. Eng. Sci. 3, 47 (1965).
3. P. A. Flinn, D.S. Gardner and W.D. Nix, IEEE Trans., March 1987, Vol. ED-34, pp 689-699.
4. J. Neal Cox, G. Shergill, M. Rose and J. K. Chu. Proceedings of VLSI Mutilevel Interconnection Conference, 1990, pp. 419-424.
5. R.C. Budhani, R.F. Bunshah and P. A. Flinn, Appl. Phys. Lett., Vol. 52, No. 4, 1988, pp. 284-286.
6. D.N.K. Wang, S. Somekh and D. Mayden, 1st ULSI symposium, ECS, May, 1987, pp 712-722.
7. C. Chiang and D. B. Fraser, Proceedings of VLSI Multilevel Interconnection Conference, June, 1989. pp. 397-401
8. N.Majad, S. Dabral and J. F. McDonald, J. of Electronic Materials, Vol.18, No. 2, 1989
9. C. Chiang and D. B. Fraser, 2nd ULSI symposium, ECS meeting, 1989
10. P.A. Flinn and C. Chiang, J. Appl.Phys. 67(6), 1990, pp. 2927-2931
11. ANSYS Engineering Analysis System, Swanson Analysis Systems, Houston, PA (1990).
12. Anne Sauter and W.D. Nix, Mat. Res. Soc. Symp. Pro. Vol. 188, pp.15-20, 1990 MRS Spring Meeting.

# THE ROLE OF "ANTENNA" STRUCTURE
# ON THIN OXIDE DAMAGE
# FROM PLASMA INDUCED WAFER CHARGING

Sychyi Fang, and James P. McVittie
Center for Integrated Systems, Stanford University, Stanford, CA 94305-4070

## ABSTRACT

A new, physically-based model has been developed to successfully explain the roles of device structure and plasma nonuniformity on charge damage. The model includes an equivalent circuit for the charging of MOS antenna structures exposed to a nonuniform plasma, and the use of SPICE, a circuit simulator, to correlate plasma measurement to breakdown measurements. The model is applied to analyze thin oxide damage in an $O_2$ magnetron plasma asher. The simulation results show good agreement with experimental damage data of "antenna" capacitors using ramp voltage breakdown measurements.

## 1   INTRODUCTION

Plasma discharges are widely used in VLSI manufacturing to lower process temperature and to obtain directional ion bombardments. However, during plasma processing thin oxides under a gate can be severely degraded by wafer surface charging [1] [2] [3]. Moreover, the gate oxides are becoming more vulnerable to the surface charging as they become thinner to meet device performance needs. Since the gate oxide degradation is a direct concern of VLSI yield and reliability, understanding plasma induced oxide damage is essential.

Recently, it has been reported that this damage depends on the uniformities of plasma especially in high density plasma like magnetron plasma [4]. In addition, the role of device structure on this damage has also been reported [5] [6] [7]. However, the mechanisms of this charge build-up and the role of device structure on subsequent damage are not well understood.

In this paper, a physically-based model, which successfully explains the role of plasma nonuniformity on charge damage [8], is applied to analyze the role of device structure on thin oxide damage in a magnetized $O_2$ plasma. In particular, antenna capacitor structures [7] have been used, where large area poly-Si pads ($A_f$) serve as "antennas" to increase thin oxide damage during plasma processing.

## 2   A MODEL FOR PLASMA SURFACE CHARGING

A wafer placed into a rf discharge is subjected to charge particle bombardment by both positive ions and electrons. Figure 1(a) shows the case of an uniform plasma, which we define as one in which the local ion and electron conduction currents to the wafer surface balance each other everywhere across the wafer surface over the rf cycle. In this case, the mean sheath

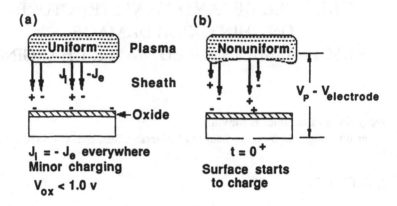

**(a)**  **(b)**

Uniform Plasma  Nonuniform

Sheath

Oxide  $V_P - V_{electrode}$

$J_i = - J_e$ everywhere  $t = 0^+$
Minor charging  Surface starts
$V_{ox} < 1.0$ v  to charge

Figure 1: A schematic diagram showing wafer charging for (a) oxide exposed to a uniform plasma, (b) oxide exposed to a nonuniform plasma at initial state.

field at the surface is typically in the few KV/cm range. Since oxide damage requires MV/cm fields, charging should not be a problem for a uniform plasma.

In a nonuniform plasma, the ion and electron currents do not locally have to balance over the rf cycle, at least initially. As schematically shown in Fig. 1(b), a nonuniform plasma can have "low" points where the sheath is thinner and the plasma potential, $V_P(x,t)$ is lower. It is reasonable to expect that at these "low" points the oscillating sheath reaches the near electrode region first and allows significant electron current to flow before it becomes significant at other points. Thus the mean electron current is higher at these points. This effect will be enhanced by the exponential dependence of electron current on the barrier height or sheath voltage, $V_{sh}(x,t)$. The ion current, which is dependent on the plasma density and electron temperature, should be more uniform than the electron current and will dominate in the "thick" regions of the sheath.

If the surface is an oxide, this imbalance causes the surface to charge until either the surface potential becomes high enough to bring the local sheath conduction currents into balance or the tunneling current (F-N) balances the difference in the local particle currents. It is this tunneling current that leads to oxide degradation.

Figure 2(b) shows the equivalent circuit for the charging of antenna capacitors. For an antenna capacitor test structure (Fig. 2(a)) with fully exposed electrodes, the injected plasma current to the gate is given by $I_{plasma} = A_f \times (J_i + J_e + J_{disp})$, where $J_i$, $J_e$, and $J_{disp}$ are the current density for the electron, the ion and the displacement currents, respectively. This current flows to the substrate as shown in Fig. 2 by three parallel paths: the two oxide capacitors, and an equivalent diode for the F-N oxide current. Using the measured discharge parameters, and the plasma currents ($J_i$, $J_e$, $J_{disp}$) to the wafer [8], SPICE simulations were used to determine the F-N current for antenna capacitors.

## 3  EXPERIMENTAL

The $O_2$ plasma was generated in two reactor configurations. The first consisted of a parallel plate design with the wafer on the grounded electrode. In the second case, a permanent

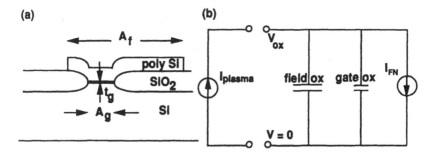

Figure 2: (a) An antenna structure of MOS capacitors, and (b) an equivalent circuit for wafer charging.

magnet was added under the grounded electrode to distort the discharge as shown in Fig. 3. Both were operating at 13.6 MHz, 100 mTorr, and 200W. To characterize the plasma, a tuned Langmuir probe was used for both the nonmagnetron [9] and the magnetron plasmas [10] to measure discharge parameters such as plasma potential and ion density etc.

Polysilicon gate MOS antenna capacitors on 5 ohms-cm <100> n-type Si wafers were used in this study. The thin gate oxide was grown in dry $O_2$ at 850°C to a thickness of ($t_g$) ranged from 6 nm to 12 nm. The antenna area ($A_f$) was varied from 0.064 to 0.4 mm$^2$ with the gate oxide areas ($A_g$) of $20 \times 20$ $\mu m^2$. Ramp voltage breakdown testing was used to characterize oxide damage after $O_2$ plasma exposure and the % failure is percentage of early breakdowns with accumulating voltage ramp.

## 4   RESULTS AND DISCUSSION

Whereas the nonmagnetron plasma was visibly uniformly intense across the wafer, the magnetron plasma was visibly most intense near the N and S poles and least intense at the center. The measured distribution of plasma potential, $V_P$, is shown in Fig. 4. Thus, the nonmagnetron discharge is relatively uniform, while the magnetron plasma is nonuniform in the N-S direction. Regarding plasma induced damage, wafers exposed to the nonmagnetron $O_2$ discharge show only intrinsic breakdown. With the magnetron discharge exposure, Fig. 5 shows a wafer map of antenna capacitors with significant early breakdown. The damage is nonuniform in the N-S direction and is most serious at the center, with some additional damage at the wafer edge close to the N and S poles. Clearly our experimental results show that plasma nonuniformity correlates well with the oxide damage. In addition, antenna capacitors at the wafer center region are used to study the role of device structure on charge damage. Table 1 shows that the oxide failure % at the wafer center increases with increasing antenna size $A_f$, and decreasing gate oxide thickness $t_g$.

For the plasma current sources considered, the simulated surface charging reaches steady states within 1 ms. The simulation results show good agreement with experimental damage data

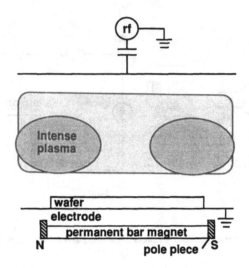

Figure 3: A schematic diagram showing a magnetron plasma asher used in this study.

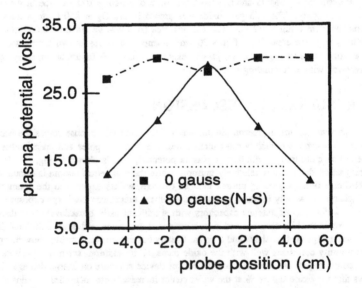

Figure 4: The spatial dependence of the timed averaged plasma potential, $V_P$, for an $O_2$ plasma measured by a tuned Langmuir probe.

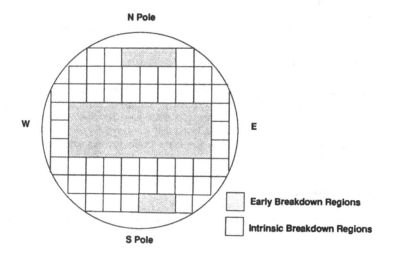

Figure 5: A wafer map of antenna capacitors after a magnetron $O_2$ plasma exposure.

| antenna pad area, $A_f$(mm²) | 0.04 | 0.2 | 0.4 | 0.2 | 0.2 |
|---|---|---|---|---|---|
| gate ox thickness, $t_g$(nm) | 12 | 12 | 12 | 7 | 12 |
| % Failure(plasma) | 2 | 29 | 85 | 72 | 29 |
| $I_{FN}$ ($\mu$A)(simulated) | 0.24 | 0.94 | 1.8 | 1.65 | 0.94 |

Table 1: The role of device structure on charge damage at wafer center.

of antenna capacitors across the wafer [8]. Furthermore, Table 1 shows the correlation between the simulated oxide current and measured oxide failure percentage. The model clearly captures the role of collection area ($A_f$) and oxide thickness ($t_g$) on this damage. With increasing $A_f$, the collected current increases, hence causing more oxide degradation. A thinner gate oxide has a smaller turn-on voltage which results in larger current imbalance, hence more damage.

# 5  CONCLUSIONS

A plasma damage model which relates plasma nonuniformity to oxide gate charging was applied to analyze the role of antenna structure parameters on oxide breakdown integrity during plasma processing. The simulation results show good agreement with experimental damage data of antenna capacitors. It was found that the increase in damage is simply an increase in the collected plasma current ($J_i$-$J_e$). For a locally exposed antenna capacitor, the increase scales with antenna area.

## Acknowledgements

The authors would like to thank Dr. I.-W. Wu and Prof. K.C. Saraswat for their helpful discussions. This research was supported by SRC and DARPA.

## References

[1] I.-W. Wu, M. Koyanagi, S. Holland, T.Y. Huang, J.C. Mikkelsen, Jr., R.H. Bruce, and A. Chiang, J. Electrochem. Soc., 136, 1638 (1989).

[2] W. M. Green, J. B. Kruger, and G. Kooi, J. Vac. Sci. Technol. B, 9(2), 366 (1991).

[3] C.T. Gabriel, J. Vac. Sci. Technol. B, 9(2), 370 (1991).

[4] S. Samukawa, Ext. Abstr. 38th meeting, the Japan Soc. of Appl. Phys. , 499 (1991).

[5] K. Tsunokuni, et al., Solid State Devices and Materials, Tokyo, 195 (1987).

[6] K. Hashimoto, D. Matsunaga, and M. Kanazawa, Proc. Symp. Dry Process, 93 (1991).

[7] S. Fang, A.M. McCarthy, and J.P. McVittie, Proc. 3rd International Symp. on ULSI, Electrochem. Soc., 91(11), (473) 1991.

[8] S. Fang, and J.P. McVittie, to be published in IEEE Electron Device Lett, June, 1992.

[9] A. P. Paranjpe, J. P. McVittie, and S. A. Self, J. Appl. Phys., 67, 6718 (1990).

[10] J. G. Cook, S. R. Das, and T. A. Quance, J. Appl. Phys., 68, 1636 (1990).

# RELIABILITY OF SUBMICRON MOSFET'S WITH DEPOSITED GATE OXIDES UNDER F-N INJECTION AND HOT-CARRIER STRESS

MING-YIN HAO*, JACK C. LEE*, IH-CHIN CHEN**, and CLARENCE W. TENG**
* Dept. of Electrical and Computer Engineering, University of Texas at Austin, Austin, Tx 78712
** Semiconductor Process and Design Center, Texas Instruments Incorporated, Dallas, Tx 75265

## ABSTRACT

In this paper, submicron nMOSFET's with thin LPCVD gate oxides deposited using chemical reaction of dichlorosilane and nitrous oxide at 800 °C were evaluated. Fowler-Nordheim injection was performed to compare the resistance to electron trapping and interface-state generation of these LPCVD oxides to that of the thermally grown oxides at 850 °C. It was found that with an 850 °C 20-minute post-deposition nitrogen anneal, the LPCVD oxides exhibit comparable immunity to electron trapping as the thermal oxides. Transistor characteristics such as $\Delta g_m/g_m$, $\Delta I_D/I_D$ revealed to be similar for both oxides, but there was an order of magnitude higher $\Delta V_t$ detected for the LPCVD oxide films (with/without anneal) than the thermal oxides. The mechanism responsible for the hot-carrier induced device degradation was also characterized in this study. It was found that the major source of hot-carrier damage was electron trapping for the thermal oxides and the $N_2$-annealed LPCVD oxides; while for the as-deposited LPCVD oxides, both electron trapping and interface-state generation contributed to device degradation.

## I. INTRODUCTION

In ULSI/VLSI applications such as FPGA (field-programmable gate array) and SOI (silicon-on-insulator) structures, conformity of the oxide films is a crucial requirement. It is well known that low-pressure chemical-vapor-deposited (LPCVD) oxides yield good conformity and step coverage suitable for these applications [1]. The lower process temperatures of LPCVD oxides may result in better dopant profile control, which makes the deposited oxides even more attractive. It has also been reported that the defect density of the LPCVD oxides deposited involving the reaction of $SiH_4$ and $O_2$ could be lower than that of the thermally grown $SiO_2$ since most of the oxide defects are incorporated from the substrate [2].

Recently, nitridation of $SiO_2$ or oxidation of Si in $N_2O$ ambient to grow $SiO_2$ [3] has been used to improve the characteristics of ultrathin dielectrics - e.g. reduced interface-state generation under electrical stressing and better diffusion barrier to impurities. The improved integrity is believed to be due to the nitrogen incorporation at the $Si/SiO_2$ interface. However, significant electron trapping occurs in the nitrided oxides because of the inclusion of H in the dielectric during the nitridation process in $NH_3$ ambient. On the other hand, $N_2O$ oxidation is a simple one-step process to incorporate nitrogen in the dielectrics. Unfortunately, the amount of nitrogen incorporation, dielectric thickness, and oxidation temperature are interrelated and cannot be easily optimized independently. Furthermore, because of the slower oxidation rate, to grow oxide with $t_{ox} > 120$ Å would require very high thermal budget. In this study, we investigate the possibility of utilizing LPCVD process involving dichlorosilane (DCS) and nitrous oxide ($N_2O$) at 800 °C for thin gate dielectric applications. Maximizing advantages of LPCVD process and nitrogen incorporation from $N_2O$ is the primary purpose of this study. Both MOS capacitors and

Mat. Res. Soc. Symp. Proc. Vol. 265. ©1992 Materials Research Society

MOSFET's were fabricated using control thermal oxides and these new LPCVD oxides, and subject to Fowler-Nordheim and hot-carrier stressing. It was found that these LPCVD oxides exhibit comparable characteristics in terms of electron trapping, interface-state generation and transconductance degradation under Fowler-Nordheim and hot-carrier stressing; but they suffered from threshold voltage instability problem after hot-carrier stress. Device degradation mechanisms for these dielectrics have been investigated and will be discussed in detail.

## II. DEVICE FABRICATION

Standard nMOSFET's were fabricated on (100) p-type silicon wafers with a resistivity of 10-15 $\Omega$-cm. After LOCOS isolation, thermal oxide ($\sim$ 100 Å) was grown at 850 °C (in pure $O_2$ or $O_2$+5% HCl) on the control wafers; while for other wafers, LPCVD oxides of approximately the same thickness were deposited involving reaction of dichlorosilane and nitrous oxide gases at 800 °C. The flow rates were 100 sccm for dichlorosilane and 250 sccm for nitrous oxide such that stoichiometric silicon-dioxides were formed. The process pressure was 300 mTorr and the resulting deposition rate was ~2.3 Å/min. Some of the samples with LPCVD oxides underwent an anneal in nitrogen at 850 °C for 20 mins immediately following the gate oxide deposition. *In-situ* phosphorus doped polysilicon was deposited and patterned, then followed by a phosphorus implant (Dose = 4 x 10$^{13}$ cm$^{-2}$) to form the LDD regions. After implementing 0.15 $\mu$m thick oxide spacers, a source/drain implant (As, 3 x 10$^{15}$ cm$^{-2}$) and a thermal annealing at 900 °C for 20 mins in nitrogen were performed. Aluminum/TiW stacked layers were evaporated afterwards, and a final sintering was done in a $H_2$ ambient at 450 °C for 20 mins.

## III. RESULTS AND DISCUSSION

To characterize the charge trapping immunity of the dielectrics, constant current stress was performed within the F-N tunneling regime (stress current density = 30 mA/cm$^2$). The measurements were done on transistors with dimensions of W/L = 25/25 (in $\mu$m), and the effective oxide thickness is 102 Å for control oxides and 105 Å for LPCVD oxides (with/without $N_2$ anneal). Fig. 1 shows the gate voltage change ($\Delta V_g$) necessary to maintain a constant current as a function of stress time. A positive gate voltage shift ($\Delta V_g$) indicates that electrons are trapped as well as electron trap generation are taking place in the oxide films [4]. The control oxide shown in Fig. 1 has higher resistance to electron trapping than the un-annealed LPCVD oxides. The increase in electron traps in the un-annealed LPCVD oxides could be due to the hydrogen incorporation from the DCS gas while nitrogen from $N_2O$ is insufficient in suppressing the high trapping rate. However, an additional post-deposition nitrogen anneal did suppress the electron trapping in the LPCVD oxide films. As can be seen in Fig. 1, the resistance to charge trapping of the nitrogen annealed LPCVD oxide can be as good as the thermally grown

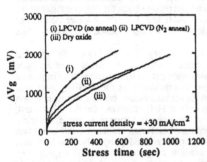

Fig.1 Charge trapping characteristics under F-N tunneling stress.

control oxides. Also, an optimization of DCS and $N_2O$ ratio might increase the nitrogen incorporation and thereby improve the trapping characteristics of the as-deposited LPCVD oxide.

The interface-state generation was characterized by measuring low frequency capacitance-voltage (LFCV) curves before and after a constant current stress ($J = 10$ $mA/cm^2$ for 10 sec). The advantage of constant current stressing technique is that it ensures a constant electric field near the charge-injection electrode, thereby making the measured interface-state density independent of the amount of charge trapping in the dielectric [5]. Compare the CV curves in Fig. 2a and Fig.2b, one can conclude that the thermal oxide has higher immunity to interface-state generation than both types of LPCVD oxides. Moreover, the post-deposition anneal performed on LPCVD oxides was not effective in terms of reducing the generation of interface states. The results suggest that perhaps a short $O_2$ anneal is needed in order to improve the interfacial quality of LPCVD oxides.

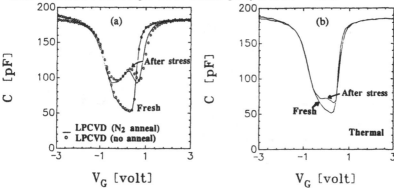

Fig. 2 Low Frequency C-V curves for (a) LPCVD oxides (with/without $N_2$ anneal) (b) thermal oxides before and after F-N stress (stress condition: 10 $mA/cm^2$ for 10 sec). The distortion of the stressed LFCV curves is an indication of the interface-state generation.

Hot-carrier stress at peak $I_{sub}$ conditions ($V_D=7V$; $V_G=V_D/3$ -- $V_D/2$) was performed on nMOSFET's with $W_{eff} = 5 \mu m$ and $L_{eff} = 0.8 \mu m$ while drive currents ($I_D$), transconductances ($g_m$), and threshold voltages ($V_t$) were monitored to evaluate the effects of hot-carriers on device reliability for both thermally grown oxides and LPCVD oxides.

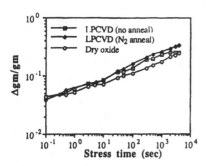

Fig.3 Transconductance degradation under hot-carrier stress.

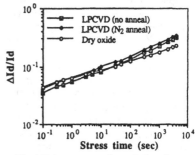

Fig.4 Drive current degradation due to hot-carrier stress. The drive current data were taken at $V_g = 5V$, $V_D = 0.1$ V with grounded source and substrate.

Fig. 3 and Fig. 4 show $|\Delta g_m/g_m|$ and $|\Delta I_D/I_D|$ as functions of hot-carrier stress time. The transconductance of interest was its peak value, while the drive current shown in Fig. 4 was extracted at $V_G = 5V$. For both thermally grown oxides and LPCVD oxides, no significant difference can be observed from these behaviors. Furthermore, the nitrogen post-deposition anneal for LPCVD oxides didn't have much effect on the changes in $g_m$ or $I_D$. However, according to the $\Delta V_t$'s shown in Fig. 5, almost one order of magnitude difference can be detected between the thermal oxide and the LPCVD oxides. This suggests that the threshold voltages of the transistors with thermally grown gate oxide were much more stable under the injection of hot carriers than those with the LPCVD oxides. No appreciable difference in the threshold voltage change can be observed between the $N_2$-annealed LPCVD oxides and the as-deposited films. Thus, this additional heat treatment provided no improvement to the as-deposited LPCVD oxides under the stress of hot carriers at peak $I_{sub}$ conditions.

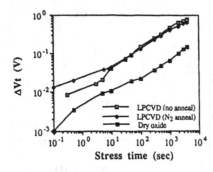

Fig.5 Threshold voltage shifts after hot-carrier stress.

As mentioned previously, electron trapping and/or interface states can be formed in the gate dielectric by hot-carrier stressing. An attempt was made to separate the contribution of these two effects to the damage of transistor characteristics in this study. The devices were first stressed at peak $I_{sub}$ conditions where both electrons and holes could be injected into the gate oxides. If the shift of the threshold voltages under such stresses was caused by the formation of electron trapping, a subsequent hole injection should be able to neutralize those trapped electrons and result in a decrease in $V_t$. However, the generated interface states could not be eliminated by the injected holes. Therefore, hole injection would not be able to reduce $V_t$ back to its initial value if part of the threshold voltage change was originated from the generation of interface states.

The experiment was carried out on transistors with W/L = 1.0/1.0 (in $\mu$m) and area $6.25 \times 10^{-6}$ cm$^2$. In the first step of the measurement, the drain voltage was adjusted to 6V for LPCVD oxides and 7V for thermal oxides ($O_2$+5%HCl samples, $T_{ox}$ = 101 Å) in order to see a similar $\Delta V_t$ value within reasonable amount of time. The gate voltages applied during these stresses were the voltages where a peak $I_{sub}$ was detected. The subsequent hot-hole injection was performed at $V_G = 1$ V with $V_D$ kept the same (i.e. $V_D$=6 V for LPCVD oxides; 7 V for thermal oxides). The threshold voltage was extracted to be the gate voltage corresponding to $I_D = 0.1$W/L $\mu$A. Fig. 6(a)(b)(c) show the threshold voltage shift due to hot-carrier and hot-hole injections as a function of stress time for different gate oxides. The abrupt drop in the $\Delta V_t$ plots was the transition point between two stress modes. As shown in these figures, a sharp decrease in the threshold voltage can be observed after the injection of hot holes. For transistors with the un-annealed LPCVD gate oxides, a 45% recovery of the threshold voltage can be achieved by hot-hole injection (see Fig. 6(a)). This suggested that 45% of the $V_t$ shift under hot-carrier injection was caused by the formation of electron trapping in these as-deposited oxide films. However, for $N_2$-annealed LPCVD oxides, the injection of hot holes gradually reduced the threshold voltage to its initial value, indicating that the dominant mechanism of $V_t$ shift under hot-carrier stress is the generation of electron trapping (see Fig. 6(b)). Similar behavior was also observed for transistors with thermally grown gate oxides. Injected hot holes shift the threshold voltage to that of a fresh device rapidly (see Fig. 6(c)). Therefore, electron trapping is responsible for the increase in $V_t$ for the thermal oxides under the stress of hot carriers.

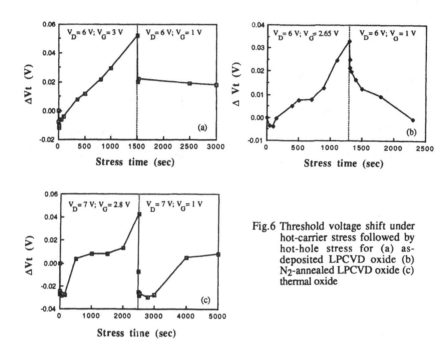

Fig.6 Threshold voltage shift under hot-carrier stress followed by hot-hole stress for (a) as-deposited LPCVD oxide (b) $N_2$-annealed LPCVD oxide (c) thermal oxide

## IV. CONCLUSIONS

Thin LPCVD oxides formed from dichlorosilane and nitrous oxide mixture at 800 °C were investigated as an alternative gate dielectric for MOSFET's. The as-deposited LPCVD oxide films exhibited more electron trapping compared to the thermal oxides grown at 850 °C. A post-deposition anneal at 850 °C in nitrogen for 20 mins improved the film quality and yielded comparable electron trapping level as the thermal oxide studied. However, the LPCVD oxides couldn't achieve the high resistance to interface-state generation as the thermal oxides, even with an additional 850 °C nitrogen anneal. Transistor characteristics such as $\Delta g_m/g_m$ , $\Delta I_D/I_D$ and $\Delta V_t$ were evaluated under the hot-carrier stress. No significant difference was detected in $\Delta g_m/g_m$ and $\Delta I_D/I_D$ among the various oxide films. But the thermal oxide demonstrated a $\Delta V_t$ almost one order of magnitude lower than those of the LPCVD oxides. The post-deposition anneal did not cause noticeable improvement for LPCVD oxides in terms of hot-carrier immunity. The mechanisms responsible for the hot-carrier induced device degradation were also identified by performing a hot-carrier injection at peak $I_{sub}$ condition followed by a hot-hole injection. This experiment revealed that the formation of electron trapping dominated the device degradation for the thermal oxides. While for the un-annealed LPCVD gate oxides, both interface-state generation and electron trapping are involved. However, apply a post-deposition nitrogen anneal to the LPCVD oxides could alter their characteristics and result in similar response to hot carriers as the thermal oxides. Optimization of LPCVD process involving adjustment of DCS and $N_2O$ ratio in CVD process and $O_2$ anneal after deposition might improve the reliability characteristics of LPCVD oxides.

[REFERENCE]

1. S. M. Sze, VLSI Technology, 2nd ed. (McGRAW -Hill publishers, New York, 1988), p. 253.
2. J. Lee, I-C Chen, and C. Hu, IEEE Electron Device Letters, 7, 506 (1986).
3. H. Hwang, W. Ting, B. Maiti, D. L. Kwong, and J. Lee, Appl. Phys. Lett. 57, 1010 (1990).
4. Y. Nissan-Cohen, J. Shippar, and D. Frohman-Bentchkowsky, J. Appl. Phys., 57, 2830 (1985).
5. I.C. Chen, S. E. Holland, and C. Hu, IEEE Trans. Electron Devices, 32, 413 (1985).

# Effects of X-ray Irradiation on MOSFET Characteristics and DRAM Leakage Phenomena

SRINANDAN R. KASI and STEVEN H. VOLDMAN
IBM Corporation
1000 River Road, Essex Junction, Vermont 05452

## ABSTRACT

The influence of 0-1000 mJ/cm$^2$ 1-2 keV x-ray radiation on the MOSFET characteristics of submicron CMOS devices is investigated. Fully-integrated LOCOS- and STI-defined MOSFET structures, with abrupt- and LDD-junction geometries, are used to evaluate irradiation effects on different device design features for future (post-256 Mb DRAM) logic and DRAM technology considerations. The effects of x-ray irradiation on DRAM trench storage-node leakage are also investigated and compared to high energy ion implant-related leakage behavior.

## INTRODUCTION

As dynamic random access memory (DRAM) cell density increases, the process technology demands placed by ultra large scale integration (ULSI) of devices will require the application of radiation-intensive processing techniques such as x-ray lithography, and high-density plasma and reactive ion etching. These potentially damaging process environments, together with device design features, will critically determine device performance and reliability. Therefore, the compatibility of such novel processing techniques with device requirements needs early assessment. The impact of x-ray radiation on fully-processed metal oxide semiconductor field effect transistor (MOSFET) devices has been studied extensively and forms a useful starting point for further work [1]. However, a major portion of the work has focused on the effects of x-ray fluence and post-exposure annealing on the MOSFET operating characteristics of thick-oxide, LSI-era devices. Device-reliability issues (e.g., DRAM leakage phenomena) and the combined effects of device design and irradiation have not been completely investigated.

Recent MOSFET reliability work has begun to address some of the latter issues. The effects of x-ray irradiation on the MOSFET gate-induced drain leakage phenomena of 0.25-$\mu$m (T$_{ox}$=7 nm) CMOS devices has been studied and reported to be completely annealable by a standard forming gas process [2]. The impact of irradiation on MOSFET short-channel effects has also been examined [3, 4]. In this work, we assess the effects of x-ray radiation on the electrical characteristics of LOCOS-defined, graded-junction MOSFETs [5, 6] and on device design features using shallow-trench isolated (STI), abrupt and lightly doped drain (LDD)-junction structures [7]. Measurements are also made of the effect of radiation on DRAM-trench leakage phenomena.

### LOCOS-Defined Graded-Junction N- and P-Channel MOSFETs

X-ray exposure experiments to study changes in MOSFET behavior were done on abrupt-junction, LOCOS-isolated, submicron N- and P-MOSFET structures (T$_{ox}$=15 nm) fabricated on 5 in. dia. 0.008 $\Omega$cm heavily p$^+$-doped Si wafers, with a

2.5-$\mu$m p-epitaxial layer. The p-channel MOSFETs are placed in a 1.4-$\mu$m deep retrograde n-well, while devices are fully integrated in the IBM 4-Mb substrate-plate trench (SPT) DRAM technology [8]. Details of device fabrication and electrical characterization methods are discussed elsewhere [5, 6]. Dose-dependent x-ray irradiation was done using a 1.2 keV synchrotron light source, for a dose of 0-1000 mJ/cm2 (0-125 Mrad-SiO₂) [9].

Standard MOSFET measurements were made on control and irradiated submicron N- and P-MOSFET devices. NFET $I_d$-$V_d$ characteristics show an ≃10% reduction in drain current ($I_d$) after exposure. The observed reduction in $I_d$ occurs because of electron trapping at radiation-generated, acceptor-type interface states. Such electron trapping must dominate another radiation-related mechanism that increases channel current ($V_T$ lowering due to trapped positive charge), for an overall reduction in $I_d$. P-MOSFET $I_d$-$V_d$ characteristics show more significant changes after exposure. These changes in the $I_d$-$V_d$ characteristics of LOCOS-isolated P-MOSFETs following irradiation are shown in Figure 1. Data from devices with width-to-length (W/L) dimensions of 20/20 $\mu$m are shown for different x-ray exposures and for two different n-well configurations: Figure 1A (no implant) and Figure 1B (4X10$^{13}$ cm$^{-2}$ retrograde n-well implant dose).

Two observations can be made about data from such measurements: First, x-ray exposure reduces the drain current in the saturation regime for all samples. Examples of this behavior are shown in the IV characteristics in Figure 1. The

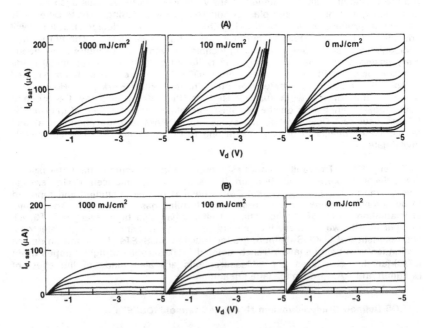

Figure 1. LOCOS-Isolated P-MOSFET $I_d$ - $V_d$ plots for (A) zero, and (B) 4 x 10$^{13}$ cm$^{-2}$ retrograde n-well implant doses. Data are shown for each exposure condition for gate biases of 0 V to 5 V.

reduction results from an increase in the threshold voltage ($V_T$) of a PFET device caused by radiation-induced positive trapped charges in the gate oxide layer. Irradiation also prompted a reduction in the overall transconductance ($g_m$) of the device. The second observation concerns a radiation-induced anomalous leakage-current component that increases exponentially at $V_d > 3.5$ V. Figure 1A shows that leakage is independent of channel conduction and increases with the x-ray dose. This current which can occur because of channel punchthrough, junction breakdown, and/or generation, depends on the W/L dimensions of the device; it increases with L, which suggests that channel punchthrough does not cause the anomalous current. To verify that junction breakdown may have contributed to the anomalous current, measurements were carried out at different temperatures in the range 25-85 °C. The onset of leakage shifted to lower drain bias with increasing temperature; behavior that is opposite of what would be expected for junction avalanche breakdown. The above observations suggest that leakage current is related to the collection of thermally generated carriers. This component is affected by the retrograde n-well implant dose. Increasing the implant dose to $4 \times 10^{13}$ cm$^{-2}$ (Figure 1B) eliminates this leakage current. These observations suggest that radiation-induced defects cause the leakage current. Radiation-induced traps and trapped-charge in the isolation layer can enhance the thermal-generation of carriers in the depletion region of the underlying substrate. When such generation occurs within diffusion length of the junction, an anomalous current contribution is measured.

## STI-Defined Abrupt and LDD-Junction N- and P-MOSFETs

The combined effects of x-ray exposure and MOSFET design features was examined using STI, and LDD-junction structures in submicron MOSFETs. The fabrication process for the STI-defined N- and P-MOSFETs ($T_{ox} = 12$ nm) has been provided elsewhere [7, 10].

N-MOSFETs show only minor changes with irradiation. Due to positive charge trapping in the gate oxide layer, the channel $V_T$ is lowered by 0.15 volts after a 100 mJ/cm$^2$ exposure. The change in N-channel saturation drain current ($I_{d,sat}$) as a function of effective channel length ($L_{eff}$) upon x-ray exposure of LDD and abrupt drain-geometry MOSFETs is shown in Figure 2A. $I_{d,sat}$'s are comparable for control and irradiated N-MOSFETs when $L_{eff} > 0.6$ μm for both drain structures. For $L_{eff} < 0.6$ μm, only the LDD structures show an increase (20%) in $I_{d,sat}$ over their control counterparts. Drain currents in the linear regime, however, show an opposite trend (not plotted); values are comparable for control and irradiated samples at $L_{eff} < 0.3$ μm, while at longer $L_{eff}$ there is a 5-15% reduction for irradiated samples. Overall, device transconductance values are unchanged with exposure for LDD structures and show a $<5\%$ decrease for abrupt geometries.

Irradiated P-MOSFET structures show more significant changes than the exposed N-MOSFETs. There is a 0.7 V increase in the $V_T$'s (for $L_{eff} > 0.5$ μm) following a 200 mJ/cm$^2$ x-ray exposure. Irradiation reduces PFET $I_{d,sat}$'s by nearly 50% at different $L_{eff}$ for both drain structures. In the linear regime, drain currents are degraded $\simeq 50\%$ for both structures (Figure 2B). Overall, saturation transconductance ($g_{m,sat}$) values are reduced 25-30% for the LDD, and 15-20% for the abrupt-junction geometries, where the $L_{eff} = 0.25$-$0.5$μm. There is enhanced degradation of the $g_{m,sat}$ at $L_{eff} < 0.3$ μm for the LDD structures. Radiation-induced trapped charge and interface states located in the gate oxide region overlapping the LDD extensions cause both enhanced channel-shortening and mobility degradation for LDD-based P-FETs.

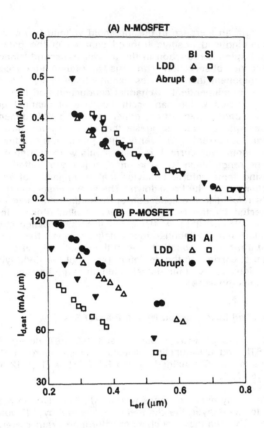

**Figure 2.** Plot of $I_{d,sat}$ vs. $L_{eff}$ for (A) N-MOSFET and (B) P-MOSFET structures, with LDD and abrupt-junction geometries. Data is shown for the devices, before x-ray irradiation (BI), and after 1000 mJ/cm$^2$ x-ray irradiation (AI).

## LOCOS-Defined DRAM Trench Leakage

The 4-Mb SPT cell DRAM trench (storage-node)-leakage characteristics for the samples used in this study (constant, high-energy implanted, retrograde n-well samples with doses in the range 0-4X10$^{13}$ cm$^{-2}$) have been reported [5, 6]. It was shown that trench leakage increases with a retrograde n-well implant dose. In this work, x-ray radiation induced modulation of trench leakage is examined for structures bordered by either LOCOS-isolation or a p$^+$-implant region. The changes in leakage behavior at 85 °C for 0-1000 mJ/cm$^2$ irradiation of LOCOS-isolation bounded trenches are shown in Figure 3. A 100X increase in leakage is observed for the maximum dose. Leakage currents generated by radiation-induced trapped charge and interface states in the LOCOS-isolation layer (bordering both the trench and the p+ drain implant edges) and trench dielectric layer contribute to the overall increase in trench leakage. However, isolating their individual contributions is not possible with this structure.

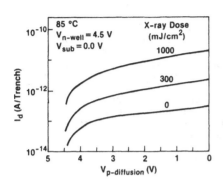
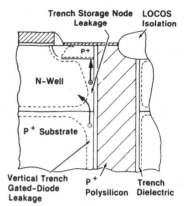

**Figure 3.** X-ray radiation-induced changes in the SPT-cell DRAM trench leakage I-V characteristics. LOCOS-Isolation bounded trenches (3.2K array) were used in the 0-1000 mJ/cm$^2$ x-ray exposure measurement. A schematic of the device used in the measurement is shown.

It is possible to study separately the leakage from radiation-induced defects in the trench dielectric layer and at the trench-substrate interface using the p+ implant-bounded 0.4-$\mu$m trench array structure. Leakage characteristics from such an array for the cases of no retrograde n-well doping (bottom) and $4 \times 10^{13}$ cm$^{-2}$ retrograde implant dose (top) are shown in Figure 4. Leakage currents increase 1-10X for both samples upon 0-1000 mJ/cm$^2$ irradiation. However, samples with no retrograde well implant show a larger differential increase in leakage. When measurements were made for all the samples with implant doses in the range 0-4$\times 10^{13}$ cm$^{-2}$, it was observed that leakage current depended on both the x-ray fluence and implant-dose. Leakage from irradiated, low implant-dose samples was comparable to non-irradiated, high implant-dose samples. This suggests that x-ray irradiation and high-energy ion implantation result in enhanced storage node leakage caused by similar physical mechanisms. Both generate bulk traps and trapped charges in the different dielectric layers and interface states that promote leakage.

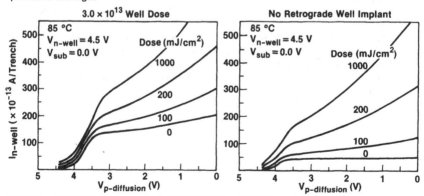

**Figure 4.** X-ray radiation-induced changes in the SPT-cell DRAM trench leakage I-V characteristics. P-diffusion bounded trenches (3.2K array) were used in the 0-1000 mJ/cm$^2$ x-ray exposure measurement. Data is shown for different retrograde n-well doses.

## CONCLUSIONS

Submicron dimension CMOS structures have been irradiated by x-rays and changes in device properties investigated. Specifically, the effects of drain design features and CMOS well-doping levels on the radiation-induced changes in MOSFET characteristics are studied. Charge trapping in the isolation and transfer-gate oxide layers and the formation of interface traps represent the primary changes introduced by exposure. Positive-trapped charge and acceptor-type interface traps cause PFETs to be more significantly affected by exposure than NFETs. Drain geometries, such as the LDD structure, result in enhanced device degradation; short-channel effects set in earlier for these structures over the abrupt drain geometry. Trench leakage increases because of radiation-generated trapped charges, bulk oxide traps, and interface states in the dielectric layers of the storage node and isolation structure.

### Acknowledgements

The authors gratefully acknowledge L. C. Hsia of IBM, East Fishkill, NY, for his assistance with the x-ray exposures.

### REFERENCES

1. T. P. Ma and P. V. Dressendorfer, Ionizing Radiation Effects in MOS Devices and Circuits, (Wiley-Interscience Publication, New York, 1989).
2. A. Acovic, L. C. Hsia, C. C. -H. Hsu, J. M. Aitken, and A. Balasinki, Proc. Third Workshop on Radiation-Induced and/or Process-Related Electrically Active Defects in Semiconductor-Insulator Systems (Microelectronics Center of North Carolina, 1991), p. 147.
3. C. L. Wilson and J. L. Blue, IEEE Trnas. Nucl. Sci., 29, 1676 (1982).
4. P. K. Bhattacharya and A. Reisman, Proc. Second Workshop on Radiation-Induced and/or Process-Related Electrically Active Defects in Semiconductor-Insulator Systems (Microelectronics Center of North Carolina, 1989), p. 262.
5. P. E. Cottrell, S. Warley, S. Voldman, W. Leipold, and C. Long, IEDM Technical Digest, 1988, 584.
6. S. Voldman and C. Long, Proc. Internat. Conf. Microelectronic Test Structures, 1990, 237.
7. D. Kenney et al., Symp. VLSI Technology, Digest Tech. Papers, 1988, 25.
8. N. C. C. Lu et al., IEEE Jour. Solid State Circuits, SC-21, 627, (1986).
9. IBM Beamline U6, Brookhaven National Laboratory, Long Is., NY.
10. P. Bakeman et al., Proc. 1990 Symp. on VLSI Technology, Honolulu, Hawaii, 1990, 11.

# POSITIVE CHARGE IN SILICON DIOXIDE
## DUE TO HIGH ELECTRIC FIELD FN INJECTION

X. Gao*, S. Yee** and H. Mollenkopf*
*Material Characterization Laboratory, SEH America, Inc., 4111
NE 112th Ave., Vancouver, WA 98682
**Department of Electrical Engineering, FT-10, University of
Washington, Seattle, WA 98195

## ABSTRACT

Electric fields with different intensities and polarities in
the range corresponding to the Fowler-Nordheim (FN) tunneling
injections were applied on the MOS structures to examine the
generation behaviors of the positive oxide charges. The
concentration changes of the field-induced positive oxide
charges with the injected electron numbers were measured and a
linear relation between the saturated density of the positive
oxide charge and the average intensity of the oxide electric
field was obtained. The experimental results show that the
amount of the field-induced positive oxide charge only depends
on the intensity but not on the polarity of the oxide electric
field. There was a threshold value of the average oxide
electric field below which no positive charge could be
generated; and this implies that the generation mechanism of
the positive oxide charge due to high electric field can be
different from that due to low electric field.

## INTRODUCTION

One problem related to the stability and reliability of a
MOS device is the accumulation or creation of positive charges
in $SiO_2$. Positive charges can be introduced into $SiO_2$ by
electric field or various radiations such as photon, electron
and ions. With the decrease of the device dimension due to the
increase of the device integration, the effects of positive
oxide charges created by electric fields on the device
performance have become more important.[1,2]

Although extensive work has been done on the issues
related to the field-induced positive oxide charges, their
behaviors and generation mechanisms still remain unclear. In
order to understand the nature of the field-induced positive
oxide charges, their dependences on both the polarity and the
magnitude of the oxide electric field are examined in this
paper. We restrict ourselves only in the high electric field
case, i.e., the electric fields which correspond to the Fowler-
Nordheim (FN) tunneling injection.

## EXPERIMENTAL

The devices used in our experiments were aluminum gate MOS
capacitors fabricated on (100) boron-doped silicon substrates
of resistivity 10 $\Omega$-cm. Silicon dioxide was thermally grown to
a thickness of 63 nm at 920 °C in dry oxygen gas with 3% TCA. A
15 minutes in situ annealing at 920 °C in nitrogen gas was
followed. Aluminum gates with area of 0.32mm$^2$ and backside

aluminum contacts were deposited with vacuum evaporation technique. The post-metallization annealing was performed at 400 °C for 30 minutes in forming gas (nitrogen with 6% hydrogen). The ionic contamination was in the $10^{10}$ cm$^{-2}$ range or lower as it was determined by the triangle voltage sweep method at 150 °C.

The constant voltages corresponding to electric field intensities from 6.2 to 10.2 MV/cm with both positive and negative polarities were applied on the gates of the MOS devices. In order to avoid the current limitation from the generation of the minority carrier in the bulk silicon during positive-bias FN injection, continuous illumination with an ordinary light of proper intensity was used. The intensity of the illumination was determined by I-V measurements which confirmed the FN tunneling current rather than the minority carrier generation current. Because illumination could give additional energy to the electrons injected into the conductance band of $SiO_2$, illumination with same condition was also applied during negative-bias FN injections in order to compare the effect of the bias polarity. C-V measurements were performed at a frequency of 400 KHz. The density of the oxide charge was determined by automatically measuring the shift in the capacitance curve at a point corresponding to the mid-point of the silicon bandgap, because this method can reduce the effect of the field-induced interface traps when it is compared with the flatband shift measurement.[3] The injection was periodically stopped and the C-V shift was measured within a few seconds. All the measurements were automatically controlled by computer.

## RESULTS AND DISCUSSION

The average electric field E in the oxide layer is defined by

$$E = (Vg - \Phi_{Si} - \Phi_{MS}) / d_{ox} \tag{1}$$

where Vg is the applied gate voltage, $\Phi_{Si}$ is the surface potential of silicon, $\Phi_{MS}$ is the work function difference of the electrodes, and $d_{ox}$ is the oxide thickness.

The density change of the positive oxide charge $\delta N_+$ is given by

$$\delta N_+ = -\delta V_{mp} \epsilon_{ox} / d_{ox} q \tag{2}$$

where $\delta V_{mp}$ is the shift of the voltage in the C-V curve at a point corresponding to the mid-point of the silicon bandgap, $\epsilon_{ox}$ is the dielectric constant of $SiO_2$, and q is the elementary charge.

The concentration dependences of the positive oxide charges on the injected electron number for two different oxide electric fields of 8.73 and 7.25 MV/cm are given in Fig. 1. One important feature showed in Fig. 1 is that the generation of the positive oxide charge only depends on the average intensity but not on the polarity of the oxide electric fields. In addition, the positive oxide charges increase with the increase

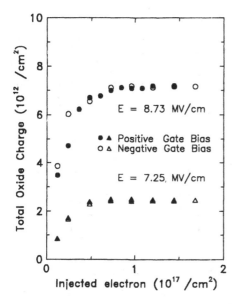

Fig. 1. Total oxide charge vs. the injected electron under positive and negative gate biases with light illumination for two different average oxide electric fields.

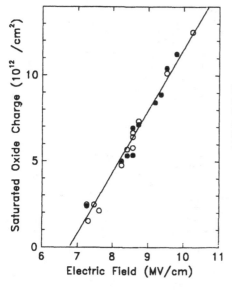

Fig. 2. The saturated oxide charge vs. average oxide electric field under positive and negative gate biases with light illumination.

● Positive bias
○ Negative bias

of the injected electrons. The generations of the positive oxide charges are saturated after the injections with large amount of electrons into the oxides.

The saturated concentration of the positive oxide charge vs. the average intensity of the oxide electric field is given in Fig. 2. A linear relation between the saturated concentration of the positive oxide charge and the average intensity of the oxide electric field is obtained. It is shown again that the saturated density of the positive oxide charge only depends on the average intensity but not on the polarity of the oxide electric field.

The concentration changes of the saturated positive oxide charges with the average intensity of the oxide electric field for both positive and negative polarities were also investigated by Fischetti.[4] Like in our experiment, continuous illumination was used during positive gate bias FN injection in his experiments to avoid the current limitation due to the electron generation in the bulk silicon. His results showed the similar linear relations between the saturated density of the positive charge and the average intensity of the oxide electric field as showed in Fig. 2. In contrast to our results, his results showed a difference of the saturated positive oxide charge density between the positive and negative gate biases. In his experiments, the constant current FN injections rather than constant voltage FN injections were used; and their average oxide electric field was determined from the gate voltage measured at the moment the injection was halted. Because the gate voltage increases during constant current FN injection as reported previously,[5,6] the average intensity of the oxide electric field at the moment the injection was halted is larger than those during the FN injection. Therefore, the average intensity of the oxide electric field obtained from the finally measured gate voltage was a overestimated value. If the increase rates of the gate voltages during positive bias injections were different from during negative bias injections, the overestimations of the average oxide electric fields will be different. Hence, there was a difference of the saturated positive oxide charge densities between the positive and the negative gate biases. In our experiment, the constant voltage FN injections were used. Because the average oxide electric fields were not changed during the constant voltage FN injections, more accurate results were obtained in our experiment.

Nissan-Cohen et al[7] also reported the dependence of the positive charge density on the average intensity of the oxide electric field. In their experiments, n+ regions were used to provide the needed electrons to the reverse layers of the silicon surface during the positive polarity FN injections. No bias polarity dependence of positive oxide charge density was observed in their experiments, which is consistent with our above mentioned results. However, their results showed logarithmic dependence of positive charge density on the average intensity of the oxide electric field but not linear dependence as observed by us and others.[4,8] By examining their experiments, we find that their positive charge densities were measured after electron injections of $10^{-3}$ C/cm$^2$ which corresponds to an electron number less than $10^{16}$ /cm$^2$. Their experiment results also suggested that the generations of the positive charges were not saturated after an electron injection less than $10^{-3}$ C/cm$^2$ in most of their measurements. However, the experimental results in Fig. 1 and reported by others[4,6] also show that the concentrations of the positive oxide charges do not reached to saturation after an electron injection less than $10^{16}$ /cm$^2$. Therefore, the logarithmic relations they obtained were not the relations between the saturated density of the positive oxide charges and the average intensity of the oxide electric fields.

It can be seen from Fig. 2 that there is a threshold value of the average oxide electric field below which no positive oxide charge can be generated. However, it has been reported in the literatures[4,9] that positive oxide charge could also be induced by applying low electric field in the range of avalanche injections. We have no explanation for this discrepancy, other than to suggest that the generation mechanism of positive oxide charge in the low electric field range is different from that due to the high electric field reported in this paper.

The generations of positive oxide charges during low field avalanche injection and high field tunneling injection were compared by Fischetti.[4] He concluded that both low and high electric fields generated the same type of positive charge. Examining his results, we can find that the linear fitting of his data which includes the low electric field values indicates larger deviation than those in Fig. 2 and reference [8] which only include the high electric field data. If his low electric field data are excluded, better linear fitting with smaller deviation can be obtained.

The existence of the threshold electric field was confirmed previously by several authors.[4,8,10] Although the threshold values of the electric fields varied from experiments to experiments, an electric field larger than 6 MV/cm seems to be necessary to generate this kind of high field-induced positive oxide charge. The differences of the measured threshold value of the oxide electric field among different experiments is maybe from the differences of the oxide growth conditions and the sample history. An accurate determination of the threshold electric field seems to be necessary to understand the generation mechanism of the high field-induced positive oxide charge.

## CONCLUSIONS

The generation of the positive oxide charge from FN injection due to high electric field only depends on the intensity but not on the polarity of the oxide electric field. There is a linear relation between the saturated density of the positive oxide charge and the average intensity of the oxide electric field. A threshold value of the average oxide electric field was observed, below which no positive oxide charge could be generated. This threshold value implies that the generation mechanism of the positive oxide charge at high electric field is different from that at low electric field.

## ACKNOWLEDGMENT

The authors would like to thank J. Matlock and D. Gupta for the critical reading of this manuscript. This work was supported by SEH America, Inc.

## REFERENCES

[1] Z. A. Weinberg, in *The Physics and Technology of Amorphous SiO₂*, edited by R. A. B. Devine (Plenum Press, New York, 1988), p. 427.

[2] L. P. Trombetta, F.J. Feigl, and R.J. Zeto, J. Appl. Phys. 69, 2512 (1991).

[3] Z. A. Weinberg and T. N. Nguyen, J. Appl. Phys. 61, 1947 (1987).

[4] M. V. Fischetti, J. Appl. Phys. 57, 2860 (1985).

[5] I. C. Chin, S. E. Holland, and C. Hu, IEEE Trans. Electron Devices 32, 413 (1985).

[6] S. Holland and C.Hu, J.Electrochem. Soc. 133, 1705 (1986).

[7] Y. Nissan-Cohen, J. Shappir, and D. Frohman-Bentchkowsky, J. Appl. Phys. 57, 2830 (1985).

[8] M. Shatzkes and M. Av-Ron, J. Appl. Phys. 47, 3192 (1976).

[9] J. M. Aitken and D.R. young, IEEE Trans. Nucl. Sci. **NS-24**, 2128 (1977)

[10] Z. A. Weinberg, W. C. Johnson, and M. A. Lampert, J. Appl. Phys. 47, 248 (1976).

# ELECTRICAL AND RELIABILITY CHARACTERISTICS OF SILICON-RICH OXIDE FOR NON-VOLATILE MEMORY APPLICATIONS

BIKAS MAITI AND JACK C. LEE
Microelectronics Research Center, University of Texas, Austin, TX 78712

## ABSTRACT

Non-stoichiometric films are particularly attractive for non-volatile memory applications. In this work low-pressure chemical vapor deposition (LPCVD) using silane and nitrous oxide gases were used to deposit thin silicon-rich $SiO_2$ films. Reliability issues concerning these non-stoichiometric oxides have been studied in comparison to ultrathin conventional thermal oxides. It has been found that with an increase in the silicon content, the current injection efficiency at a given electric field increases. This has significant advantage in terms of low programming voltage applications. There is also a considerable reduction in electron trapping, extremely large increase in charge-to-breakdown and negligible interface state generation in comparison to ultrathin thermal oxides. These characteristics of non-stoichiometric oxide films can be explained by a modified conduction mechanism which in turn is due to dispersed crystallites in the oxide film.

## INTRODUCTION

EEPROM cell structures currently use very thin tunneling oxides grown thermally from silicon substrate or oxides grown from polysilicon [1]. Polyoxides use the rough polysilicon surface as an efficient electron injector for non-volatile memory applications. These polyoxides, however, suffer from higher rate of electron trapping which limits the number of write/erase cycles [2]. On the other hand, for even lower programming voltage applications further scaling of already thin tunneling oxides would be required. The reliability of ultrathin tunneling oxides is severely degraded in the form of early breakdowns when the thickness of these oxides are aggressively scaled down. This restricts the scaling of thin oxides for low programming voltage applications. APCVD silicon-rich oxides (or non-stoichiometric oxides) have also been proposed for enhanced electron injection [3]. The dielectric thickness used was relatively thick (~ 200Å). In this study we have investigated the electrical and reliability characteristics and the suitability of LPCVD thin silicon-rich oxides (<100Å) as alternative to ultrathin tunneling oxides. LPCVD process should provide additional advantages in terms of uniformity and conformal step coverage, while reducing dielectric thickness would further decrease the programming voltage.

Mat. Res. Soc. Symp. Proc. Vol. 265. ©1992 Materials Research Society

## EXPERIMENTAL

MOS capacitors were fabricated using standard etch-back of field oxide and n+ poly gate technology. After a standard RCA clean, the silicon-rich oxides were deposited at 800°C using low-pressure chemical vapor deposition technique on 10-15 Ω-cm p(100) Si wafers for various $N_2O/SiH_4$ gas flow ratios (50, 80, 100) controlled by varying the silane flow while keeping $N_2O$ flow constant at 200 sccm. The deposition rate decreases with increasing $N_2O/SiH_4$ ratio as shown in Figure 1. The amount of silicon in the oxide film was increased by decreasing the

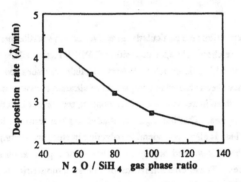

Figure 1.     Deposition rate of LPCVD oxide at 800°C for various $N_2O/SiH_4$ mole ratios by varying $SiH_4$ flow ($N_2O$ flow = 200 sccm).

ratio of the reactant gases, $R = N_2O / SiH_4$. The deposited oxide was then annealed for 30 minutes at the same temperature in $N_2$ ambient under low pressure in the same deposition furnace. The thickness variation across a 4" wafer is less than ±3%. Ultrathin thermal oxides were also grown at 850°C and 900°C for the purpose of comparison with these deposited non-stoichiometric oxides. The effective oxide thickness in all the cases were between 62~73Å as measured by high frequency capacitance-voltage method and by assuming the dielectric constant to be 3.9 for the values of R considered in our study.

## RESULTS AND DISCUSSIONS

Figure 2 shows the current density (J) versus electric field (E) characteristics of oxides deposited for various $N_2O / SiH_4$ ratios (R) and ultrathin thermal oxide (~72Å). Enhanced electron injection was observed for both voltage polarities. This is probably due to the presence of silicon clusters and various silicon oxide phases in these films [4]. A. J. Learn, et. al. [5] have suggested that the oxide deposited for R=100 is stoichiometric from the refractive index of the oxide films, but the J vs E data presented here indicate that some amount of excess atomic silicon still exists in the oxide as manifested by the relative low-field current conduction as compared to the Fowler-Nordheim tunneling in stoichiometric thermal oxide. The as-deposited phase of the oxide films is amorphous for the values of R used in the study [6]. Furthermore, with the oxygen concentrations used here (>40%), the diameter of silicon microcrystals in the

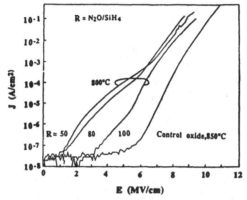

Figure 2. Current density (J) vs electric field (E) characteristics of LPCVD oxide for various $N_2O/SiH_4$ ratios. Current injection is higher at lower electric fields which can advantageous for programming applications.

dielectric is probably less than 10Å [4]. But subsequent annealing (800°C, 30 min) could lead to the formation of small crystallites [6]. Current conduction in the Si-rich oxide layer is

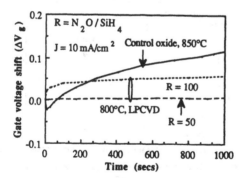

Figure 3. Gate voltage shift ($\Delta Vg$) of MOS capacitors with time for a constant current stress. Non-stoichiometric oxides have lower trapping due to electron tunneling.

believed to be due to direct tunneling from one Si region (oxygen-poor region) to an adjacent one [7]. This enhanced current conduction at low voltages could be used for lower programming voltages in non-volatile memory applications.

A constant current density ($J = 10$ mA/cm$^2$) stress was performed to study the charge trapping. The change in the gate voltage was monitored as shown in Figure 3. It is seen that the electron trapping is significantly lower for Si-rich oxides with trapping rate decreasing as R

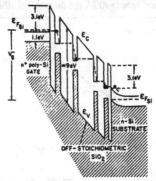

**Figure 4.** Schematic energy band representation of the conduction mechanism involving electron tunneling between tiny Si islands in the Si-rich SiO$_2$ layer [10].

is decreased (i.e., more silicon rich). In our study the amount of excess atomic silicon is about 1~5% for the values of R used in depositing oxide films [9]. Thus, these films are comprised of a Si-phase and a stoichiometric SiO$_2$-phase. Electrons tunnel between microscopic silicon

| Oxide Growth Process | $Q_{BD}$ (C/cm$^2$) |
| --- | --- |
| | $J_{stress} = 50$mA/cm$^2$ |
| SiH$_4$ & N$_2$O CVD, 800°C R=N$_2$O/SiH$_4$ | |
| R = 50 | > 500 |
| R = 80 | > 500 |
| R = 100 | 300 |
| Thermal oxide | |
| 850°C | 55 |
| 900°C | 60 |

**Table I.** Comparison of charge-to-breakdown ($Q_{BD}$) under constant current density (50 mA/cm2) stress for various oxide processes. N$_2$O-SiH$_4$ CVD oxides show a vast improvement in $Q_{BD}$ as compared to ultrathin thermal oxide (~65Å).

islands and hence, are more difficult to be captured (Figure 4). Moreover, this modified conduction model also suggests that the trapped carriers can be easily detrapped by means of tunneling across these tiny Si islands under appropriate electric fields [8]. For a given current density, the electric field across the oxide is lower for the Si-rich oxides and hence, electron trap generation rate is also lower. This can be beneficial for use in EEPROMs where reduced charge trapping can minimize threshold voltage shift and increase the ability for extended write/erase cycling. The other effect of reduced charge trapping is the dramatic increase in charge-to-breakdown ($Q_{BD}$). These oxides show very high charge-to-breakdown ($Q_{BD}$) as compared to thin thermal oxides as seen in Table I. For R = 50 or 80, $Q_{BD}$'s > 500 C/cm$^2$ at stress current density of 50 mA/cm$^2$.

Interface state generation under constant current stressing (J = 1 mA/cm$^2$) was measured using high-low frequency capacitance-voltage method. Figure 5 and 6 show quasi-static C-V curves for both Si-rich oxide and thermal oxides respectively. The initial interface state density for silicon rich oxides is about $5 \times 10^{10}$/eV-cm$^2$. Lower interface state generation is believed to

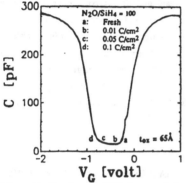

Figure 5. Quasi-static C-V characteristics of N$_2$O-SiH$_4$ CVD oxide MOS capacitor after Fowler-Nordheim constant current stress (J=1 mA/cm$^2$) show negligible interface state generation.

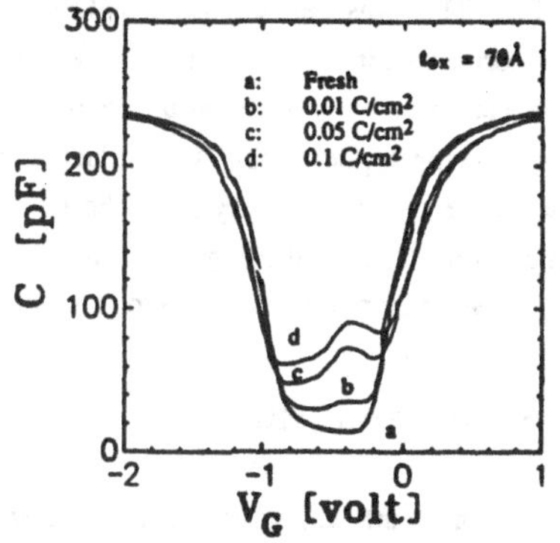

Figure 6. Quasi-static C-V characteristics of thermal oxide MOS capacitor after Fowler-Nordheim constant current stress (J=1 mA/cm$^2$) show substantial interface state generation.

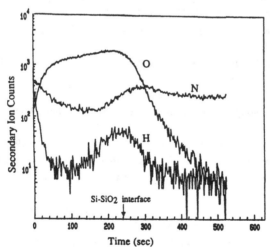

Figure 7. SIMS profile for 70Å LPCVD oxide deposited at 800°C using SiH$_4$ and N$_2$O.

be due to lower electric field across the bulk of the oxide for the same stress current density. SIMS analysis indicate some amount of nitrogen pile-up at the $Si/SiO_2$ interface as shown in Figure 7 which could also be responsible for the lower interface state generation [11-12].

## CONCLUSIONS

Electrical and reliability characteristics of Si-rich non-stoichiometric oxide have been studied. The films were deposited by LPCVD using silane and nitrous oxide. Excess silicon can be incorporated in a controllable manner by adjusting the ratio of the gas phase reactants. This excess atomic silicon increases the conductivity of these films which could be utilized for lower programming applications. The conduction mechanism in these oxides is modified and leads to lower charge trapping, very high charge-to-breakdown ($Q_{BD}$) and negligible interface state generation. The results suggested that these highly reliable Si-rich oxides may be used as alternative to ultrathin thermally grown tunneling oxides in non-volatile memory applications.

## ACKNOWLEDGEMENT

This work was partially supported by SRC/SEMATECH under contract # 88-MC-505.

## REFERENCES

1.  S. K. Lai, V. K. Dham and D. Guterman, IEDM Tech. Digest, p. 580-583, 1986.
2.  H. A. R. Wegner, IEDM Tech. Digest, p. 480, 1984.
3.  D. J. DiMaria, K. M. DeMeyer, C. M. Serrano, and D. W. Dong, J. Apl. Phys., vol. 52, p. 4825, 1981.
4.  M. Hamasaki, T. Adachi, S. Wakayama, and M. Kikuchi, J. Appl. Phys. 49, p 3987, 1978.
5.  A. J. Learn and R. B. Jackson, J. Electrochem. Soc., p. 2975, 1985.
6.  A. V. Dvurechensky, F. L. Edelman and I. A. Ryazantsev, Thin Solid Films, vol. 91, p. L55, 1982.
7.  D. J. DiMaria and D. W. Dong, J. Appl. Phys., vol. 51(5), p. 2722, 1980.
8.  D. J. DiMaria, D. W. Dong and F. L. Pesavento, J. Appl. Phys., vol. 55, p. 3000, 1984.
9.  D. Dong, E. A. Irene and D. R, Young, J. Electrochem. Soc., vol. 125, no.5, p. 819, 1978.
10. D. J. DiMaria, T. N. Theis, J. R. Kirtley, F. L. Pasavento, D. W. Dong, and S. D. Brorson, J. Appl. Phys., 57(4), p. 1214, 1985.
11  S. K. Lai, J. Lee and V. K. Dham, IEDM Tech. Digest, p. 190, 1983
12. H. Hwang, W. Ting, B. Maiti, D-L Kwong and J. Lee, Appl. Phys. Lett. 57(10), p. 1010,1990.

# HIGHLY RELIABLE STACKED THERMAL/LPCVD OXIDES FOR ULTRATHIN GATE DIELECTRIC APPLICATIONS

BIKAS MAITI AND JACK C. LEE
Microelectronics Research Center, University of Texas, Austin, TX 78712

## ABSTRACT

One of the major problems associated with thin dielectrics is defects in the form of micropores or voids in ultrathin $SiO_2$ when oxides are thermally grown from the silicon substrate. This study describes the synthesis of thin multilayer stacked oxides by the sequential growth of thermal oxide layer and the deposition of a layer of LPCVD oxide followed by a densification/reoxidation step (3-step process) to reduce the density of these micropores. The electrical characteristics of these stacked oxides were compared with those of conventional thermal oxide and reoxidized-deposited oxide (2-step process). The stacked oxides formed by the 3-step process exhibit a drastic reduction of the defect density. This improvement is believed to be due to the misalignment and filling up of the micropores in the thermal oxide layer by the deposited oxide layer. Variation of the thickness of the deposited oxide shows that LPCVD oxide of about 30Å in the 3-step process is sufficient in reducing defect density and improving the breakdown characteristics of the stacked oxide. Using this simple stacked oxide process, the yield of ultrathin oxides can be improved significantly.

## INTRODUCTION

VLSI/ULSI MOS devices require highly reliable thin gate oxides. The major challenge in achieving this is in growing ultrathin oxides with low defect density and interface states. The defect density in thin gate oxides has been attributed to the presence of voids or micropores in the conventional thermal oxides [1]. High resolution transmission electron microscopy (HRTEM) studies have shown that these micropores are about 10-25Å in diameter and separated by about 100Å [2]. P. K. Roy, et.al. has shown that by the use of a stacked oxide layer the defect density of the dielectric can be substantially improved [3]. The stacked oxide layer is formed by a 3-step process i.e., by a sequential growth of thermal oxide layer and the deposition of a layer of LPCVD (low-pressure chemical vapor deposition) oxide followed by a densification/reoxidation step. The use of a deposited oxide layer on top of the thermally grown $SiO_2$ layer helps in filling and misalignment of the micropores in the bottom layer. Thus, the interconnecting pores and weak spots are reduced and thereby, the yield is increased.

In this paper, we have compared this 3-step process with a 2-step process (a deposited layer followed by thermal oxide growth) and the conventional thermal oxide (1-step process). The effects of thickness of each layer and process sequence have also been studied.

## EXPERIMENTAL

Polysilicon gate MOS capacitors were fabricated on p-type Si(100) 12-20 $\Omega$-cm wafers. After a standard field-oxide etch-back isolation process followed by a standard RCA clean of the wafers, the 3-step stacked oxide was fabricated as follows. A thermal oxide was first grown in a hot-wall atmospheric furnace in $O_2$, TCA and Ar at 900°C. This was followed by a low temperature (450°C) oxide (LTO) deposition in an LPCVD furnace. The deposition rate (~3 Å/min) was controlled by adjusting the flow rate of silane and oxygen to 1.5 sccm and 45 sccm, respectively. The final step in the stacked oxide process was a short $O_2$ reoxidation/ $N_2$ densification anneal at 900°C. Conventional thermal oxide was also grown at 900°C. The 2-step process consisted of growing a layer of thermal oxide after the deposition of LPCVD oxide. The total oxide thickness in all the cases were about 120-130Å. A brief description of the various process splits is given in Table I. Standard polysilicon gate technology was used to complete the MOS capacitor fabrication process.

Table I. Fabrication process for MOS capacitors showing thickness and temperature of processing of each oxide layer.

| Oxide processing | Temperature (°C) | Thickness (Å) |
|---|---|---|
| **1-step process**<br>Thermal | 900 | 120 |
| **2-step process**<br>CVD / Reoxidation<br>CVD / Thermal | 450 / 900<br>450 / 900 | 116 / 16<br>60 / 68 |
| **3-step process**<br>Thermal / CVD / Reoxidation<br>Thermal / CVD / Reoxidation | 900 / 450 / 900<br>900 / 450 / 900 | 55 / 60 / 10<br>72 / 30 / 20 |

## RESULTS AND DISCUSSIONS

Charge trapping in these oxide structures was investigated by using a constant Fowler-Nordheim current stress of 10 mA/cm². The change in gate voltage with time to maintain a constant current was measured for various cases. The charge trapping characteristics of the 2-step process and the control oxide do not show any significant difference (Figure 1). But the stacked oxides exhibit lower charge trapping as compared to the control oxide (Figure 2). This

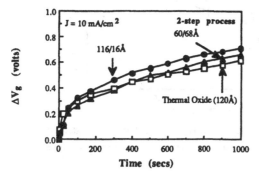

Figure 1. Charge trapping characteristics measured as gate voltage shift for a 1-step and a 2-step process. Stress current density, J = 10 mA / cm$^2$.

reduction could be related to the smoother oxide/silicon interface formed in the reoxidation step for the stacked oxides [4]. Moreover, the additional reoxidation step might reduce traps in the bulk of the oxide.

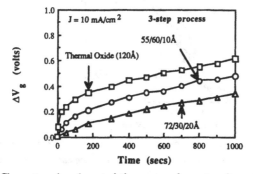

Figure 2. Charge trapping characteristics measured as gate voltage shift for a 1-step and a 3-step process. Stress current density, J = 10 mA / cm$^2$. Stacked oxides show much lower trapping rates compared to control.

Interface state density, $D_{it}$, was calculated using high-low frequency capacitance-voltage (C-V) method. The stress current density was 1 mA/cm$^2$. The initial interface state is found to be lower for stacked oxides in comparison to control oxide and the two step process. The interfacial oxide grown in the reoxidation step of the 3-step process appears to reduce the stress at the Si/SiO$_2$ interface as observed by X-ray micro-diffraction [4]. The initial interface

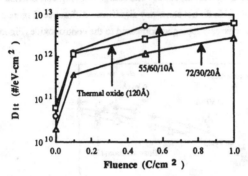

Figure 3. Interface state density (Dit) for thermal and 3-step stacked oxides for various fluence levels. Longer reoxidation of the stacked layer seems to improve the initial interface state density.

state density of $6 \times 10^{10}$/eV-cm$^2$ for control oxide reduces to $2 \times 10^{10}$/eV-cm$^2$ for stacked oxide with a 20 Å interfacial oxide grown in the reoxidation step. However, the rate of interface state generation in the stacked oxide is comparable to that of conventional thermal oxide.

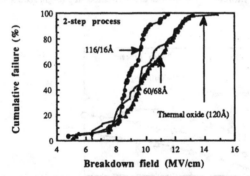

Figure 4. Cumulative failure (%) for thermal oxides and reoxidized deposited oxides (2-step process) vs electric field. Reoxidation does not appear to reduce early breakdowns since it does not compensate for the micropores and voids in the oxide.

The major improvement by using the stacked oxide structure is observed in the improved yield or reduced defect density (Figure 4 and 5). Voltage ramp stresses were performed on capacitors (area=$1 \times 10^{-3}$ cm$^2$) by applying a negative voltage ramp and by measuring the voltage at destructive breakdown. Destructive breakdown is defined as the point in the ramped I-V measurement at which a concurrent decrease in device-voltage and increase in device-current occurs. Fifty samples were measured for each cumulative breakdown characteristics. In the 2-step process the cumulative breakdown failures do not show marked improvement over the conventional thermal oxides. Growing a thicker thermal oxide layer

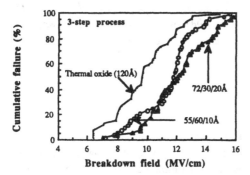

Figure 5. Cumulative failure (%) for thermal oxides and stacked oxides (3-step process) vs electric field. Deposition of a thin layer of LPCVD oxide is effective in reducing defect density by filling and misaligning the micropores in the bottom oxide layer.

below the deposited layer only makes the distribution tend towards that of control oxide. This suggests that this 2-step process does not seem to compensate or fill up the micropores in the initial layer of oxide. But the growing/depositing/growing sequence in the 3-step stacked oxide process does seem to compensate for the defects in the first layer of oxide. The 3-step process not only shows lower percentage of breakdown at lower electric fields but also a dramatic increase in yield at higher electric fields. This strongly supports the contention that the deposited layer is very effective in filling up or misaligning the micropores and thereby, effectively reducing the weak spots in the oxide layer.

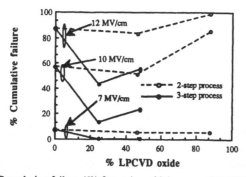

Figure 6. Cumulative failure (%) for various thicknesses of LPCVD oxides for 2-step and 3-step process plotted for various breakdown fields. This plot shows that a 30Å LPCVD oxide layer in a 3-step process is sufficient in reducing the defect density considerably.

The effects of varying the thickness of each layer was also studied. In Figure 6 percentage cumulative failure has been plotted vs percentage of LPCVD oxide for various breakdown field criteria for all the processes. It shows that a 30Å LPCVD oxide in a 3-step process is sufficient to improve the yield substantially although it is not very apparent whether a thicker deposited layer would increase the defect density.

## CONCLUSIONS

This study describes the synthesis of thin stacked $SiO_2$ structures. The stacked dielectric films is formed by growing, depositing and growing $SiO_2$ layers by thermal oxidation, LPCVD oxide deposition and densification/reoxidation, respectively. This kind of stacked oxide structure was compared with conventional thermal oxide and a 2-step process ( i.e., thermal oxide grown after a layer of oxide was deposited). The stacked oxide structure formed by the 3-step process do not suffer from high density of electron traps. The interface state density reduces after a thin layer of interfacial oxide was grown in the reoxidation step. The drastic lowering of the defect density in the 3-step process is believed to be due to filling up or misalignment of micropores by the deposited layer on the thermal oxide layer. But in the 2-step process the defects in one layer does not seem to be compensated by the other layer. Moreover, a 30Å deposited oxide layer is sufficient in improving the breakdown characteristics. This study suggests that the 3-step process could be used to improve the yield of ultrathin dielectrics for future generation MOS devices.

## ACKNOWLEDGEMENT

This work was partially supported by SRC/SEMATECH under contract # 88-MC-505.

## REFERENCES

1.  E. A. Irene, Semiconductor International, p92, 1985.
2.  J. M. Gibson and D. W. Dong, Journal of Electrochem. Soc., vol. 127, no. 12, p. 2722, December 1980.
3.  P. K. Roy, R. H. Doklan, E. P. Martin, S. F. Shive, and A. K. Sinha, IEDM Tech. Dig., p. 714, 1988.
4.  P. K. Roy and A. K. Sinha, AT&T Technical Journal, p. 155, November/December 1988.

# A NEW POLYMERIC MATERIAL FOR ESD AND CORROSION PROTECTION

J. P. Franey and R. S. Freund   AT&T Bell Laboratories 600 Mountain Ave. Murray Hill, NJ 07974

R. Sias, and B. Beamer Baxter Industrial Div. 27200 N. Tourney Rd. Valencia, CA 91355

## ABSTRACT

A new material for permanent electrostatic discharge (ESD) and corrosion protection has been developed. This material, called Reactive Polymer, consists of a blend of polymer resin (eg. polyethylene, polypropylene, ABS, PVC) with stable solid-state additives. It can be formed into bags, tubing, sheets, bubble-pack, foam, trays, IC carriers, tape-and-reel, etc. ESD and corrosion protection is provided for components, circuit packs, assembled systems, etc. Protection from corrosion or tarnish by atmospheric $H_2S$, COS, and HCl is provided for many forms of silver, copper, and copper alloys such as the stock metals, electrical connectors, bus bars for electric power transmission, and electroplated metals.

## 1    Introduction

Because ESD can have a devastating effect on the quality and reliability of manufactured electronics, it has become of concern throughout the industry. Electronic devices often include circuitry designed to minimize their sensitivity to ESD, but such protection is usually not sufficient. The standard practice, therefore, is to be sure that during manufacture, shipment, and storage, sensitive components and assemblies contact only static dissipative materials. In the coming years, as device dimensions shrink, the importance of ESD control can only increase.

The principle measure of an ESD dissipative material is a surface resistivity in the range of $10^5$ - $10^{11}$ ohms/square. Various methods have been used to obtain materials in this range, but present materials are known to suffer from a number of disadvantages.

Carbon-loaded polymers are often used as static dissipative materials. Certain carbon blacks are highly conductive, so dispersing them in a polymer at a high enough concentration can provide conductivity in the appropriate range. Unfortunately, this conduction mechanism is characterized by a percolation threshold [1,2,3,4]. At concentrations below this threshold there are no continuous conducting paths through the material. For concentrations above this threshold, there are many random conducting paths and the resistivity drops to very low values. Manufacture of carbon-loaded polymeric materials with resistivity in the ESD range is difficult because the resistivity is a sensitive function of the carbon black concentration, and probably of the processing conditions as well. An additional disadvantage of carbon-loaded polymers is that the carbon particles tend to shed (Typically a count of 20,000 to 40,000 particles/ft$^3$ that are $>0.5$ microns). Thus, carbon particulate loaded polymers are normally unacceptable for use in clean rooms .

A commonly used approach to ESD dissipative materials is to coat an organic chemical on a plastic part. These organics are usually amines or amides, and when atmospheric water is adsorbed the surface becomes conductive in the ESD range. When coated on polyethylene, the resulting product is sometimes known as "pink poly ". Organics of this class are also coated on IC shipping tubes by dipping them in the liquid. This approach to ESD control has three major drawbacks. One is that the organic compounds are volatile, so after a period of several months to several years, much of the coating evaporates and the material becomes insulating [5]. Unfortunately, there is no visible indication that ESD protection has been lost. A related problem is that the volatilized organic may deposit on the part being protected, thus coating it with an organic film which may be corrosive and insulating. The third drawback is that these materials do not function in low humidity, since they rely on adsorbed water. Thus in dry climates, and certainly in dry packages, where the relative humidity can drop below 5%, they become insulating.

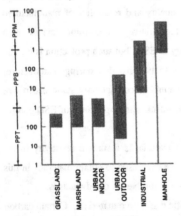

Figure 1.Hydrogen Sulfide Atmospheric Concentrations.

Corrosion of critical conductor surfaces by trace atmospheric gases is another reliability concern besides ESD. Although the rate of corrosion may be slow, the effect is cumulative [6]. Thus parts in storage or in shipment are particularly vulnerable. This problem is most severe for copper and silver which react with hydrogen sulfide ($H_2S$), carbonyl sulfide (COS), and hydrogen chloride (HCl). These trace gases are widespread [7]. Consider hydrogen sulfide: its ambient concentrations (Fig. 1) are generally in the range of 0.1 part per billion (ppb) to 10 ppb, although in certain industrial areas, in manholes, and in areas near oil fields, pulp mills, or sewage treatment plants its concentrations can exceed 100 ppb. Laboratory measurements [8] of $H_2S$ corrosion of copper, silver, and bronze (Fig. 2), show

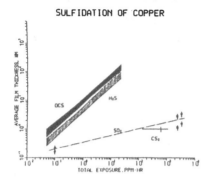

Figure 2. Sulfur Gas Corrosion Of Copper.

that corrosion product film thickness is proportional to the $H_2S$ concentration times the duration of exposure. One year of exposure of copper to 1 ppb of $H_2S$ can grow a sulfide film on copper about 50 nm thick. The traditional method of protecting electronic components from corrosion, other than gold plating, involves depositing a protective coating called a volatile corrosion inhibitor (VCI). A volatile organic is soaked into a piece of paper and then sealed in a container with the part to be protected. The resulting thin insulating film which forms on the part excludes water from the surface. Two limitations of this method, as for pink poly, are that the organic material can evaporate without warning over a period of time, and that it coats the part with an insulating film, often requiring an extra cleaning step before use.

## II.  Reactive Polymers

The material described in this paper provides both ESD and corrosion protection, and has properties which surpass those of the available alternatives. It has been manufactured with stable, controllable surface resistivity in the range $10^3$ to $10^{12}$ Ohms/square. It performs independently of the humidity. It does not contain volatile additives, so it does not contaminate the parts it is used to protect. It consists of a polymer matrix such as polyethylene, polypropylene, ABS, or PVC, containing copper particles as additives to react with corrosive atmospheric gases and thus prevent their penetration through the polymer. Additional non-volatile solid state additives, in synergy with the copper particles, provide excellent ESD properties. The material is called Reactive Polymer because of the reactivity of copper particles with corrosive atmospheric gases. Slight variations of conventional polymer processing techniques have been used to successfully produce Reactive Polymer in various forms.

The AT&T Bell Labs ESD Group, Murray Hill, NJ, the E.I. DuPont Nemours ESD Group, Wilmington, DE, and the Baxter Healthcare Industrial Div. ESD Group, Valencia, CA, have performed the following measurements in agreement with each other.

The surface resistivity (Fig. 3) was measured according to ASTM D257 "Standard Test

Method for Measuring Surface Resistivity". Multiple instrumentation was used to cross check any possible inconsistencies among commonly used measurement apparatus that might preclude the proper measurement of the characteristics of the newly developed Reactive Polymer Technology.

The Instrumentation used was:

1) Keithley #617 Electrometer with Keithley #6105 Concentric Ring Adaptors. Measurements were taken with 30 Volts applied to the samples.

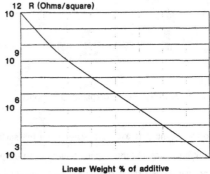

2) ASTM D 257 Parallel Bars and a Keithley #617 Electrometer with 100 Volts applied.

3) Voyager #SRM 110 Handheld meter.

Figure 3.  Linear Weight % of Additive

4) Monroe #626 Handheld meter.

5) Trek #150 Handheld meter.

All instruments were factory calibrated within the previous 12 months. Surface resistivity variations on material were < ±4x10$^1$ Ohms/Sq on manufactured materials. The smooth variation of resistivity with concentration of additive (Figure 3) makes it possible to manufacture material with the desired resistivity within one order of magnitude. Moreover, since none of the Reactive Polymer constituents are volatile, the resistivity is expected to be stable for greater than a decade.

Tests of surface resistivity as a function of relative humidity have been made at RH of 50%, 12.5%, and 5%. The instrumentation and method were the same as above. Results show no measurable change from 50% to 12.5% RH. When the samples were taken down to 5% RH a decrease of 0.3% was statistically calculated. The decrease can be attributed to the contribution of the conductivity of the surface water to the inherent resistivity of the polymer. No decrease from this value is expected to occur at humidities of < 5%.

Samples were subjected to MIL B-81705C Elevated Temperature Test (160° F for 12 days), and 24hour water wash test as per Federal Test Method 101-4046. None of the samples exhibited any measurable degradation by these tests.

Static decay time was measured according to MIL B-81705 C using an ETS 406C analyzer. Since the static decay time is maximized at low humidities all samples were tested at 5% RH. All samples tested had static decay times less than the detectable time of the test instrumentation, (<0.01 seconds) with a 20 kv charge which was increased from the 5KV standard for more sensitivity. The maximum time allowed is 2 seconds.

Particle shedding was measured using a Helmke Tumble Tester. Particulates > 0.5 micron in diameter were measured. An average of 1.6 particles per cubic foot was counted (Fig. 4). The AT&T specification for class 10 approval is a maximum of 10 particles/ft.$^3$. As a reference, cleanroom approved paper sheds 2 particles/in.$^2$

Corrosion protection from the sulfur and chlorine containing gases H$_2$S, COS, and HCl results as their diffusion through the Reactive Polymer is stopped by reaction with additives to form stable compounds. The effectiveness of the Reactive Polymer for corrosion has been tested by sealing test coupons of copper inside a bag of Reactive Polymer and placing it inside a chamber filled

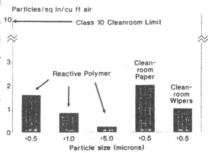

Figure 4. Comparison of Particle Shedding.

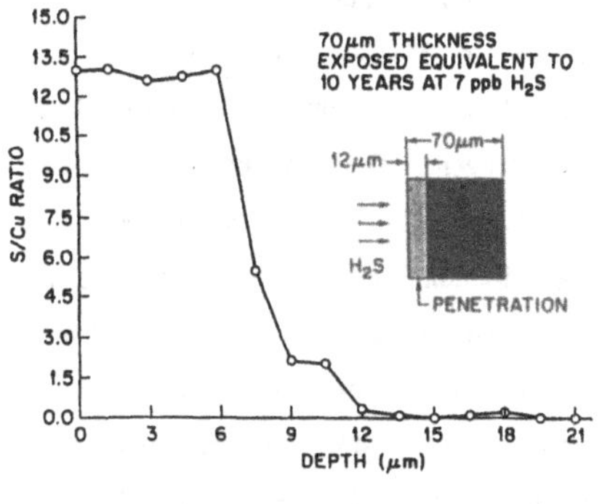

with a measured concentration of H$_2$S. After completion of exposure, some bags have been cut in cross section and examined with a scanning electron microscope with energy dispersive x-ray analysis to determine the location of sulfur. The result (Fig. 5) shows that H$_2$S has penetrated only part of the way through the polymer. Controls run with polyethylene without the additives show H$_2$S beginning to penetrate 1000 times sooner.

Figure 5. Penetration of Sulfur into Reactive Polymer.

ESD characteristics were tested after corrosive gas exposure and were unaffected.

Additional tests were performed using "Clam Shell" film holders, much like the ASTM cup device for measuring water weight loss. With 100% of HCl corrosive gas applied to one side of the film the concentration of permeating gas was measured on the reverse side with a Mast Oxidant Monitor. The time delay of the application of gas to one side to the measurement of the gas on the reverse side was the test criterion. Typical polyethylene time delays of 16 to 18 minutes were measured. For the same resins containing the Reactive Polymer additives time delays were 10,000 to 28,000 minutes.

A similar test for moisture permeation has been carried out by the ASTM cup method. The results in Fig. 6 show that the Reactive Polymer has a water vapor permeability less than half that of a sheet of polyethylene of the same thickness. Samples were run for 5 weeks at 23° C.

## Corrosion Induced Electrical Overstress

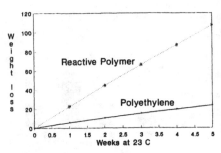

Figure 6. Water Vapor Permeability of Polyethylene and Reactive Polymer Polyethylene.

Corrosion protection is an important part of "Electrical Overstress Protection" (EOS).EOS can be caused by the high power dissipated when current flows through a high resistance junction. The following describes an experiment designed to demonstrate the relationship between corrosion and EOS.

Thirty six copper bus bars (0.125" x 0.750" x 3") were manufactured, acid dipped, packaged, and stored for 4 weeks in air containing approximately 2 parts per billion of hydrogen sulfide. 24 bars were left bare, and 12 were packaged in a 3 mil bag of Reactive Polymer. 12 bars were buffed clean with hand held power buffers and No-Ox ® coated. The individual bars were paired, bolted, and torqued to form 6 individual junctions of each protective scheme. The junctions were then tested for contact resistance using a model #398R Universal Systems 200 Ampere DC power supply to supply current across the junctions. Junction voltage drop was measured with a Fluke #8840A Digital Multimeter. Junction resistance was calculated by R= E/I and dissipated power by I × E.

Figure 7 shows the results. The unprotected bars formed poor junctions. The bars protected by the Reactive Polymer, however, provided junctions as good as the labor intensive buffed and No-Ox ® coated junction.

## III    Applications

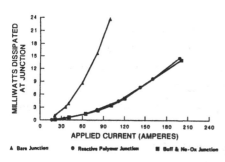

Figure 7. Comparison of Reactive Polymer Protection of Bus Bar Juntions compaired With No Protection and Conventional Cleaning and Coating.

The desirable properties of the Reactive Polymer and the absence of undesirable properties make it attractive for use in a wide variety of applications. These fall in the general categories of industrial ESD control, commercial corrosion prevention, and consumer product corrosion prevention.

For ESD control in the electronics industry, thin polyethylene Reactive Polymer can be used for packaging in the form of bags, sheet, bubble wrap, and pallet wrap.    PVC, polycarbonate, and glass filled sheet molding polymers can be used to manufacture equipment cases and process trays. Polypropylene can be used for workbench tops and vacuum formed parts. Other possible products are tote boxes, IC carrier tubes, tape and reel, and aluminum foil laminations for dry pack requirements.

Corrosion protection of copper and silver is needed by many industries. In electronics, connectors and bare printed wiring boards are subject to corrosion, as are all silver plated components. One expects that reduced corrosion will reduce the number of connector-related circuit failures and enhance solderability or reduce the demands on the solder flux. Other industries dependent on clean copper are the electric power industry which uses massive copper conductors and the electroplating industry, which needs to protect finished parts during shipment.

## ACKNOWLEDGEMENTS

We wish to thank the following for helpful discussions and work on the project: T. L. Welsher, D. Robinson Hahn, G. T. Dangelmayer, L. Murphy, D. J. Boyle, T. E. Graedel, W. Mac Farland, R.G. Renninger, and M. C. Jon of AT&T. and H. Huang, H. Gold, and D. Robson of E. I. DuPont Experimental Station, Wilmington, DE.

REFERENCES

1. S. Kirkpatrick, Percolation and conduction, *Rev. Modern Phys.*, *45*, 574, 1973.

2. E. K. Sichel, J. I. Gittleman, and P. Sheng, Tunneling conduction in carbon-polymer composites, in *Carbon Black-Polymer Composites*, E. K. Sichel, Ed., pp. 51-77, Marcel Dekker, New York, 1982.

3. K. K. Mohanty, J. M. Ottino, and H. T. Davis, Reaction and transport in disordered composite media: Introduction to percolation concepts, *Chem. Eng. Sci.*, *37*, 905, 1982.

4. B. J. Jachym, Conduction in carbon black-doped polymers, in *Carbon Black-Polymer Composites*, E. K. Sichel, Ed., pp. 103-134, Marcel Dekker, New York, 1982.

5. M. R. Havens, Understanding Pink Poly, *Electrical Overstress/Electrostatic Discharge Symposium Proceedings*, pp. 95-101, 1989.

6. J. P. Franey, G. W. Kammlott, and T. E. Graedel, *Corrosion Sci.*, *25*, 2, 1985.

7. R. P. Turco, R. C. Whitten, O. B. Toon, J. P. Pollack, and P. Hamill, *Nature 283*, 283, 1980.

8. T. E. Graedel, G. W. Kammlott, and J. P. Franey, *Science 212*, 663, 1981.

# ADVANCES IN THE THERMAL-OXIDATIVE STABILIZATION OF DIVINYL SILOXANE BIS-BENZOCYCLOBUTENE POLYMER COATINGS

T. M. Stokich, Jr., D. C. Burdeaux, C. E. Mohler, P. H. Townsend, M. G. Dibbs, R. F. Harris, M. D. Joseph, C. C. Fulks, M. F. McCulloch and R. M. Dettman
The Dow Chemical Company, Central Research & Development, Midland, MI

## ABSTRACT

This paper discusses the nonhermetic performance of polymer thin film coatings derived from 1,3-bis(2-bicyclo[4.2.0]octa-1,3,5-trien-3-ylethenyl)-1,1,3,3-tetra-methyl disiloxane (mixed isomers, CAS 117732-87-3), known also as divinyl siloxane bis-benzocyclobutene or DVS bis-BCB. The stability of the dielectric constant and the mechanical properties have been examined before and during high temperature exposures of the polymer films to air for extended periods of time at high and low humidity. Infrared absorbance spectra and dielectric constant measurements have been correlated.

Formulations of the DVS bis-BCB prepolymer with a polymeric oxidation inhibitor are predicted to yield polymer films which display less than a 10% change in the dielectric constant after 40 years in air at 85°C. These films have excellent potential for use as the dielectric coating layer in nonhermetic packaging applications for Multi Chip Module (MCM) circuits.

## INTRODUCTION

DVS bis-BCB[1,2] (monomer structure depicted in I) was designed for microelectronic applications[3-7]. It is available as Cyclotene™, a partially polymerized resin (B-staged, ~35%) dissolved in mesitylene. Polymer films are easily processed from this resin[7-9] and they display: low dielectric constant (2.65) and dielectric loss ($\leq 0.0008$)[10]; hydrophobicity[11]; resistance to common processing solvents; and planarization (>90%)[3,12].

A network element in the cured DVS bis-BCB polymer has the structure depicted in II[8]; the linking element is a tetrahydronaphthalene group. Because these structures are hydrocarbon-based, they can be susceptible to air oxidation. Polymeric oxidation inhibitors are added to reduce this susceptibility.

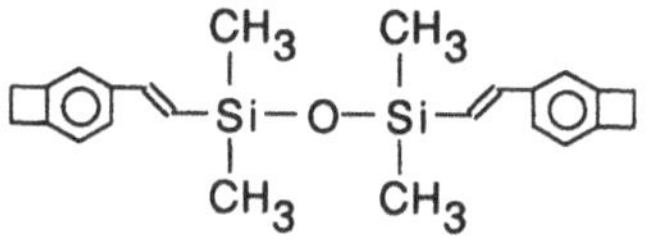

I

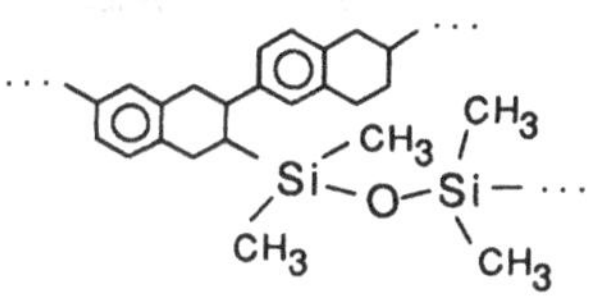

II

## MATERIALS

DVS bis-BCB prepolymer solutions in mesitylene were spin-coated onto four-inch silicon wafers to produce 5-30 μ thin films. Bare silicon wafers were used as substrates for film specimens which were monitored with FT-IR spectroscopy. Thermal oxide silicon wafers were used to fabricate capacitor test pads for measurements of dielectric constants. Formulations without the polymeric oxidation inhibitor were used as controls.

## METHODS

Films of DVS bis-BCB were exposed to a series of process and post-process assembly challenges that were designed to simulate the exposure the material could receive during MCM processing. These films were subsequently aged in air to determine the rate of change in critical properties due to oxidation.

Cure processes:
- 250°C, 1 hour (95% conversion), 1 to 6 cycles.
- 210°C, 1/2 hour (70% conversion), 1 to 5 cycles + 250°C, 1 hour, 1 cycle.

Assembly processes:
- 350°C, $N_2$, 5 min., 1 to 5 cycles.
- 200°C, dry air, 30 min., 1 cycle.
- 250°C, dry air, 30 min., 1 cycle.

Aging conditions:
- 150°C, dry air, 1000 hours.
- 85°C/85% relative humidity, air, indefinite time.
- Dry air at 85°C, 100°C, 125°C, 170°C, 180°C, 200°C, 250°C to obtain Arrhenius relationship.

Spectral measurements were made on Nicolet Model 170SX and Model 800 FT-IR spectrophotometers in transmission. Absorbance spectra of the films were computed relative to the background for a bare silicon wafer.

Effects of aging on the electrical properties of the films were determined by making capacitance and conductance measurements. With the antioxidant stabilized formulation, the dielectric constant of the cured polymer was found to be frequency independent. With no oxidation, the dielectric constant and the dielectric loss were the same for polymers with or without antioxidant.

## RESULTS AND DISCUSSION

FT-IR spectra taken during air exposures at all temperatures revealed essentially the same effects: The unoxidized polymer had an absorbance at 1500 cm$^{-1}$ (corresponding to a mode of tetrahydronaphthalene[8]) which was depleted by oxidation. From zero absorbance at the outset, absorbance bands at 1700 and 1780 cm$^{-1}$ developed concurrently during the air exposures. They are consistent with the formation of aromatic carbonyls and anhydrides, respectively. Also observed was the loss of C-H bonds. Methyl groups were much less affected and essentially none of the Si-O or Si-CH₃ bonds were affected.

Figure 1 provides an illustration of the rate of absorbance change for an air exposure at 150°C ($A_{\tilde{v}}$ denotes absorbance and b denotes sample thickness in microns). The sample specimen was a cured (250°C, 1 hour) DVS bis-BCB polymer with the antioxidant additive.

The implicit effect of oxidation on the dielectric constant, ε', was investigated by making correlations between ε' and $A_{\tilde{v}}$ (1500 and 1700 cm⁻¹) using measurements made at the same times on the same films during air exposures. Such correlations were each linear. An *oxidation time constant* (denoted as τ) was then defined as *the period of time when the change in the dielectric constant was less than 10%*. Based on the linear correlations, measurements of that time constant were made primarily from the FT-IR data. Similar linear correlations were found whether or not the stabilizer was present but there was less of a change in either ε' or $A_{\tilde{v}}$ with the stabilized coatings.

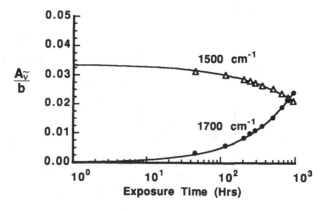

Figure 1. Oxidation rate, measured by loss of tetrahydronaphthalene (1500 cm⁻¹) and formation of carbonyls (1700 cm⁻¹) for DVS bis-BCB polymer with antioxidant. Exposure was at 150°C in air.

To describe the temperature dependence of the air oxidation process, the *oxidation time constant*, τ, was measured at temperatures from 85° to 250°C. The results are depicted in Figure 2 for specimens both with and without the antioxidant stabilizer. A point shown at 85°C is based on an *extrapolation* of FT-IR data, but all others are from interpolations made within the range of measurements. It is clear immediately that the stabilized material is more resistant to oxidation and implicitly, the dielectric constant is more stable. An Arrhenius relationship was used to obtain the depicted curve fit and "activation energies". Extrapolation of the Arrhenius curve to 85°C indicates that the antioxidant extended the *oxidation time constant* from ~2 to ~18 years. After one year at 85° and 100°C, no significant oxidation has been observed in samples having the antioxidant. However, these results were obtained on initial trials with the inhibitor and they do not indicate results that can be obtained with optimal concentrations. Therefore, a subsequent study of the concentration dependence of the oxidation inhibition was completed at 125 and 150°C. From preliminary data, a prediction based upon the use of the Arrhenius-equation indicates that the optimal concentration (one currently used for commercial formulations) will extend the oxidation time constant at 85°C to ~40 years.

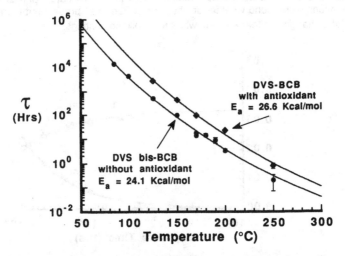

Figure 2. Temperature dependence of $\tau$ for 10 micron polymer films fabricated with and without stabilization. Oxidation time constants were determined using FT-IR data at 1500 and 1700 cm$^{-1}$.

Figure 3 illustrates measurements of $\tau$ obtained from FT-IR data for a series of the stabilizer concentrations (specific concentration values can not be disclosed in this discussion). Without antioxidant, the polymer films had *oxidation time constants* of ~100 hours in air exposures at 150°C; they improved monotonically with increasing stabilizer concentration, but reached an effective plateau where additional improvements were minimal. Addition of the stabilizer had no effect on ease of processing, the dielectric constant and loss, hydrophobicity, resistance to processing solvents or planarization.

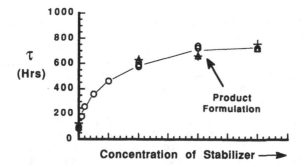

Figure 3. *Oxidation time constants* for specimens exposed to air at 150°C. Circles represent a control group that had one full cure. Diamonds are for specimens that had five soft cures and one full cure, and pluses are for samples that had six full cures cycles.

Incidental to the oxidative stability improvement, the tensile fracture stress and the strain-to-failure both *improved* to 81 MPa and ~6%, respectively. When films were exposed to air for up to 1000 hours at 150°C, neither of these properties were deteriorated.

The polymer performance under different processing and assembly conditions was a critical consideration. As shown in Figure 3, it was found that samples processed via a *single* full cure at 250°C in nitrogen for one hour had the same oxidative stability as those processed via *six* full cure cycles or those with five soft cures (210°C for 30 minutes, each) plus a full cure cycle.

Effects of four different *assembly* challenges were also evaluated. Assembly challenges were meant to represent conditions of chip attachment, TAB bonding, solder reflow and rework in MCM fabrication. Two of the challenges involved a post-cure exposure of the film to nitrogen at 350°C for either 5 or 25 minutes. For samples having the same antioxidant concentration, neither challenge resulted in measurable changes in the oxidation time constant. The other two challenges involved post-cure exposures of polymer films to air for 30 minutes at either 200° or 250°C. While the 200° challenge had no apparent effect on the oxidation time constant, the 250°C challenge in air resulted in the reduction of $\tau$. With commercial formulation concentrations of antioxidant, $\tau$ was 650 to 750 hours without this challenge but it was 450 to 500 hours with it. Care must therefore be taken when processing in air at higher temperature or for longer times.

## CONCLUSIONS

Polymer films made from DVS bis-BCB resins with a polymerized homopolymer antioxidant have been found to be oxidatively stable under 85°C operating conditions (films are predicted to have less than a 10% increase in $\varepsilon'$ over a period of 40 years). Films made from these formulations were also found to have electrical and physical properties virtually identical to those of films made with the base resin, except that the tensile fracture strength improved with the addition of the stabilizer.

Under the recommended conditions for curing DVS bis-BCB, the oxidative stability of the polymer has been unaffected by the cure process itself. MCM devices assembled under conditions consistent with this report may also perform properly for many years without expensive hermetic packaging.

## ACKNOWLEDGEMENTS

The authors gratefully acknowledge AT&T Bell Laboratories (Murray Hill, NJ) for its assistance in understanding the effects of oxidative stability and Professor Denise Denton of the University of Wisconsin for assistance in determining the degree and rate of water uptake. The authors are also grateful to CJ Carriere, EO Shaffer and TL Chritz for determining mechanical fracture properties and B Bokhart for assistance in capacitance measurements.

## REFERENCES
1. RA Kirchhoff, U.S. 4,540,763 (Sept 10, 1985).
2. AK Schrock, U.S. 4,812,588 (Mar 14, 1989).
3. DC Burdeaux, et. el., J. Electronic Materials, **19**(12), 1990, 1357.
4. RW Johnson, et. el., IEEE Trans. on CHMT, **13**, 1990, 347.
5. JJH Reche PE Garrou and JN Carr, Int. J. Hybrid Microelect., **13**, 1990, 91.
6. MJ Berry, et. el., Proc. of ECTC, Las Vegas, 1990, 746.
7. RH Heistand, et. el., Proc. International Society for Hybrid Microelectronics, Orlando, Oct. 1991, 96-100.
8. TM Stokich, WM Lee & RA Peters, Mat. Res. Soc. Proc., **227**, 1991, 103.
9. MG Dibbs, et. el., Proc. of the 6th Int. SAMPE Electronic Materials and Processes (June 23-25, 1992).
10. PE Garrou, et. el., 42nd Electronics Components Tech. Conf. May 1992.
11. H Pranjoto and D Denton, Mat. Res. Soc. Proc., **203**, 1991, 295.
12. DJ Perettie, et. el.,Proc. of SPIE, San Jose, 1992, 1665.

# PART V

## Analytical Techniques and Other Topics

# IMAGING VLSI CROSS SECTIONS BY ATOMIC FORCE MICROSCOPY

Gabi Neubauer*, M. Lawrence, A. Dass* and Thad J. Johnson**
* Intel Corporation, 2200 Mission College Blvd., P.O. Box 58119, SC2-24, Santa Clara, CA 95052-8119
** Department of Materials Science and Engineering at Massachusetts Institute of Technology, 77 Massachusetts Ave, Cambridge, MA 02139

## ABSTRACT

Imaging is an integral part of VLSI technology development and quality control in device manufacturing. We report a novel application of Atomic Force Microscopy to image VLSI cross sections of metallographically polished samples. The major advantage of this technique over conventional imaging techniques, such as Scanning or Transmission Electron Microscopy, is the higher resolution achievable in combination with higher throughput and an easy access to quantitative data, such as line widths or re-entrant angles. We observe a very good correlation of AFM VLSI cross section images, acquired in air, with those acquired by SEM and TEM.

## INTRODUCTION

Imaging cross sections of VLSI devices is an essential part of VLSI process development. Common techniques used are Scanning Electron Microscopy (SEM) and Transmission Electron Microscopy (TEM). Turnaround times to accomplish an analysis with these techniques can vary substantially from several hours up to days, depending on the amount of sample preparation involved: SEM usually requires polishing and coating of the sample, whereas TEM specimen preparation, which needs electron transparent samples, is much more elaborate including extensive polishing and thinning of the sample. High vacuum conditions are necessary for both techniques.

We report here the first time application of Atomic Force Microscopy (AFM) to the imaging of VLSI cross-sections. AFM[1] is a descendant of Scanning Tunneling Microscopy (STM)[2]. Both STM and AFM are part of the rapidly growing field of Scanning Probe Microscopy (SPM), which constructs three-dimensional contour maps of surfaces with up to atomic resolution by probing a variety of different material properties.

Imaging by STM requires a conducting or semiconducting sample, whereas the AFM is a scanned-probe device that does not need a conducting specimen. The AFM seemed to us most suitable for imaging VLSI cross-sections, since VLSI devices consist of alternating conducting and insulating layers.

Most SPM work on semiconductors reported has been done by Scanning Tunneling Microscopy where several main topics have been addressed. Cleaved III-V heterostructures have been studied to learn about growth properties at the atomic scale [3,4], multiquantum wells [5], or double heterostructures for laser diodes [6]. On silicon, STM work has concentrated on dopant profiling by tunneling spectroscopy and I-V characteristics on cross sectioned p-n junctions [7] as well as junctions on the face of the substrate [8,9]. Few fully processed device structures have been studied by STM: Chemical etch delineation and subsequent topographic STM imaging was applied to a cleaved MOS structure [10]. Scanning Capacitance Microscopy, which measures C-V characteristics in an altered STM setup was used to image a cross sectioned MOSFET structure [11]. The only AFM application reported was done with a newly designed offspring of the AFM, the Capacitance Force Microscopy, on p-n test structures on Si [12,13]. There has been no attempt to image a fully processed VLSI device by STM or AFM.

## EXPERIMENTAL

The key component of our experiments is a Digital Instruments Nanoscope III AFM, which operates in air and is equipped with an x,y micro-mechanical stage for ease of sample positioning before and during imaging. Figure 1 illustrates the principle of AFM operation: The AFM probe is a very fine Si-nitride tip (a pyramid with a base area of 2 μm x 2 μm and an opening angle of about 60°), grown at the end of a microfabricated quartz lever (length 100 - 200 μm). The tip is in contact with the sample which is moved in x and y direction by piezoelectic scanners. The lever acts as a spring, keeping the tip pressed against the surface. In place of tunneling current in the STM, the AFM records contours of force - usually repulsive force. Lever deflections measure these forces and are detected optically. A feedback network assures imaging under constant repulsive force, small enough (~10nN) to prevent damage to the surface. A large dynamic range of magnifications of several orders of magnitude is accessible in all scanning probe microscopes, depending on the area covered by the scanners. The typical range of magnification which we apply to VLSI cross section imaging is about 1200 to 300,000 times, which corresponds to a point to point resolution of 175 nm to 8 Angstrom, respectively.

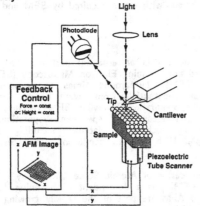

<u>Figure 1:</u>

Principal Operation of Atomic Force Microscopy.

Sample preparation for AFM imaging was similar to standard metallographic SEM cross section preparation. A glass cover slide was attached to the wafer piece containing the desired device, using a thin film of epoxy glue, to protect the top layers of interest. Grinding and polishing involved using increasingly higher grades of polishing papers, finishing eventually with 0.3 μm aluminum oxide or 0.5 μm diamond particle paper. A final polishing step with Syton, an alkaline slurry of colloidal silica, for 30 seconds before imaging, was found to give best results. Supporting electron microscopy data presented here, were acquired with a JEOL Field Emission 6400F SEM and a JEOL 1200EX TEM, respectively.

## RESULTS

In the experiments presented here, we study the cross section of a fully processed VLSI device, using AFM, SEM, and TEM. An AFM cross sectional image of device conductor and dielectric layers, is shown in Figure 2, in greyscale display, covering an area of 10μm x 10μm. Bright areas correspond to elevated features (such as the glue line which is highest in the whole image), darker areas represent valleys. The range of darkest to brightest spot in the AFM images covers about 150 nm change in topographical height, thus enabling us to identify all layers of the device clearly. The topography difference between layers is due to different mechanical polishing rates of the layers (see also section "Discussion" below).

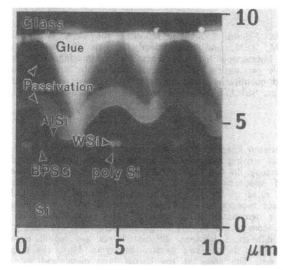

**Figure 2:**

AFM image of the device cross section with scan size 10 μm x 10 μm. All layers are clearly identified. Point to point resolution is 20 nm. The width of the polycide stack is measured to be 700 nm, and the re-entrant angle to 63°, respectively.

SEM and TEM cross section images are shown for comparison in Figure 3. The backscattered electron SEM image in Figure 3a was acquired on the same cross section as the AFM images, but most likely at other polycide stacks along the section. The TEM cross section image in Figure 3b was obtained from a different specimen of a similar wafer.

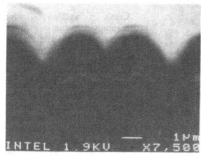

**Figure 3: a)**

SEM image of the same cross section as in Fig. 2, but at another spot. For scale see 1 μm marker.

**Figure 3: b)**

TEM cross section image of a similar device from a similar wafer. For magnification see 500 nm marker.

As we can see by comparing Figures 2 and 3, all details of the process layers are clearly reflected in the AFM images. Qualitative and quantitative information about the layer structure is easily obtained. In Figure 2, both the width of the polycide stack and the re-entrant angle can be measured directly on the screen to be 700 nm and 63°, respectively.

In all AFM images some polishing streaks are left visible. They are mostly due to the fact that during the final polishing steps the polishing direction had been maintained at top to bottom, resulting in some smearing of harder areas, such as the WSi layer into the poly2 layer. This will be improved in future studies.

## DISCUSSION

As we have seen above, Figures 2 and 3, reflecting the same structural information obtained by AFM, TEM, and SEM, clearly indicate the successful application of Atomic Force Microscopy to VLSI cross section imaging. After a standard metallographic sample preparation AFM images, acquired in air, offer an easy access to qualitative and quantitative structural information. Several important differences of scanning probe vs. SEM/TEM imaging are addressed in the following.

### Imaging Mechanism

The primary difference between imaging in electron microscopy and AFM is the probe used for imaging. Signals arriving from electron-specimen interactions are used for imaging in electron microscopy. In the case of force microscopy, the repulsive force interaction of tip and specimen causes the probe to track the surface topography. Different grinding rates of the different materials, present in the VLSI cross section, result in a surface topography and height differences, which are easily detected in the AFM. This is demonstrated in Figure 4 by a cross section line vertical to the layer structure, constructed by the image software from the data of Figure 2. Significant height differences between different layers are found, even between two such similar materials as the two passivation layers. Here, the step height is measured to 13.8 nm. The steps at the barrier layer/metal and gate-oxide/Si interfaces measure 45 and 3.9 nm, respectively.

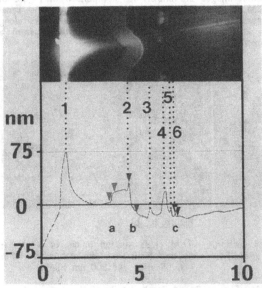

**Figure 4:**

Cross section line through the layer structure, constructed by the image software from the data in Figure 2. The cursors label different step heights at the interfaces of a) the two passivation layers, b) barrier layer/metal, c) gate-oxide/Si. Numbers refer to 1-glue, 2-TiN, 3-Ti, 4-WSi, 5-ONO, 6-gate oxide.

We could successfully delineate junctions by chemical staining, using a 1:100:25 mixture of hydrofluoric, nitric acid and water for 9 seconds [10]. An AFM cross section image, exhibiting doped regions in Si, is shown in Figure 5. The junction depth, i.e., its extension into the Si substrate, can be measured to be 163 nm. Due to the preferential etching of the doped region, it is depressed with respect to the substrate, as indicated by the darker area in the image. The measured height difference between the doped region and the substrate was 7 nm. In another experiment, also materials such as TiN and Ti (used in devices as a diffusion barrier), which have almost identical mechanical polishing rates, could be successfully distinguished by including an $Ar^+$ ion sputter clean as the final polishing step.

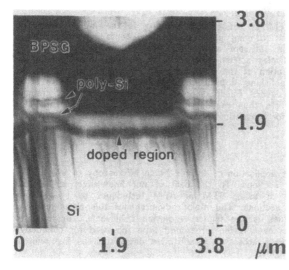

**Figure 5:**

AFM cross section image, exhibiting doped regions in Si between polycide stacks, after staining.

## Scan Sizes (Magnification)

As mentioned in the experimental section, scanning probe microscopes can cover a large range of magnifications. Higher magnifications than in the figures presented so far, can easily be accomplished by imaging still smaller areas. An example is shown in Figure 6 with a close up look at the lower part of a polycide stack (part of poly2, ONO, poly1, gate-ox, Si substrate), in a scan of 400 nm x 400 nm. The thickness of the poly1 layer is measured to be 140 nm. The magnification of this image corresponds to about 300,000 times. The point to point resolution is about 8 Angstrom and exceeds resolutions obtained by standard SEM imaging. With our current sample preparation, we do not expect to obtain atomic resolution images comparable to High Resolution TEM images, since substantial damage occurs to the top atomic layers from the polishing procedure. Other sample preparation techniques will be evaluated in the future.

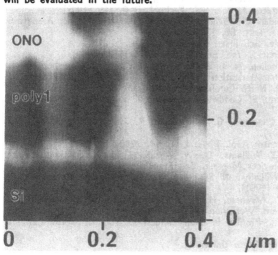

**Figure 6:**

High magnification AFM image of the lower part of a polycide stack, of size 400 nm x 400 nm.

## Data Acquisition and Imaging Environment

Another difference of AFM vs. SEM/TEM imaging is the image acquisition itself. In all state of the art scanning probe applications, data are acquired and stored digitally. The analysis software allows for easy quantification of crucial dimensions directly on the display screen. The accuracy of such measurements is, aside from an accurate calibration of the piezo scanners, mostly depending on the degree of sample tilt during polishing, a problem which is an inherent feature of all cross section imaging techniques. Finally, for AFM imaging, samples remain uncoated and are imaged in air, thus enabling much shorter turnaround times for AFM image acquisition than for conventional imaging under vacuum.

## SUMMARY

We present here the first application of Atomic Force Microscopy to image fully processed VLSI device cross sections. To the best of our knowledge, so far, no attempts had been reported of using STM or AFM techniques to image fully processed VLSI device cross sections. The major advantage of this technique over conventional imaging techniques is the higher resolution achievable, compared to SEM, in combination with an even faster turnaround time (no need for conductive coating or vacuum) and an easy access to quantitative data, such as line widths, step coverage or re-entrant angles.

## ACKNOWLEDGEMENTS

We wish to thank Bryan Tracy and Brad Sun for fruitful discussions and our staff in the metallography laboratory for assistance in sample preparation, Carmen Matos, Leslie Serrano, Richard Yoshimoto, Bertha Jones, Eleanor Foskett, and Ernestine Amaral. T.J.J. thanks Paul Davies and Bob McDonald for initiating and the Intel Coop program for funding this summer study project at Intel Santa Clara.

## REFERENCES

1. G. Binnig, C.F. Quate, Ch. Gerber; Phys. Rev. Lett. **56**, 930 (1986).
2. G. Binnig, H. Rohrer, Ch. Gerber, E. Weibel; Phys. Rev. Lett. **49**, 57 (1982).
3. H. Salemink, O. Albrektsen; J. Vac. Sci. Technol. B **9**, 779 (1991).
4. D. L. Abraham, A. Veider, Ch. Schonenberger, H. P. Meier, D. J. Arendt, S. F. Alvarado; Appl. Phys. Lett. **56**, 1564 (1990).
5. T. Kato, I. Tanaka; Rev. Sci. Instr. **61**, 1664 (1990).
6. M. Tanimoto, Y. Nakano; J. Vac. Sci. Technol. A **8**, 553 (1990).
7. S. Kordic, E. J. van Loenen, A. J. Walker; Proc. 1st Intern. Workshop on the Measurement and Characterization of Ultra-Shallow Doping Profiles in Semiconductors, Vol. **II**, North Carolina, April 1991.
8. R. Chapman, M. Kellam, S. Goodwin-Johansson, J. Russ, G. E. McGuire, K. Kjoller; ibid.
9. J. V. LaBrasca, R. C. Chapman, G. E. McGuire, R. J. Nemanich; J. Vac. Sci. Technol. B **9**, 752 (1991).
10. T. Takigami, M. Tanimoto; Appl. Phys. Lett. **58**, 2288 (1991).
11. J. A. Slinkman, C. C. Williams, D. W. Abraham, H. K. Wickramasinghe; International Electron Devices Meeting, San Francisco, CA, 73 (1990).
12. D. W. Abraham, C. Williams, J. Slinkman, H. K. Wickramasinghe; J. Vac. Sci. Technol. B **9**, 703 (1991).
13. M. P. O'Boyle, D. W. Abraham, H. K. Wickramasinghe, J. Slinkman, C. C. Williams; Proc. 1st Intern. Workshop on the Measurement and Characterization of Ultra-Shallow Doping Profiles in Semiconductors, Vol. **II**, North Carolina, April 1991.
14. P. Russel, North Carolina State University.

# THE EVALUATION OF THE TRUE TEST TEMPERATURE DURING WAFER-LEVEL ELECTROMIGRATION TESTS

G. L. BALDINI, A. SCORZONI, F. TAMARRI and D. TROMBETTI
CNR - Istituto LAMEL via Castagnoli 1, 40126 Bologna (Italy)

## ABSTRACT

The correct evaluation of the test temperature is an extremely important task when performing electromigration (EM) experiments on metal lines. In fact, reliable values of activation energy for the EM mechanism can be obtained from experimental data if and only if the true value of the temperature of each test structure during the tests is known. On the other hand true temperature measurements involve more than one practical and theoretical difficulties, ranging from heating due to Joule effect to bad thermal coupling between the test devices and the temperature sensors.

In this work a method is presented for true temperature evaluation of the test stripes during electromigration median-time-to-failure tests performed at the wafer level. In particular the problem of the thermal coupling between two or more samples tested at the same time is examined.

The experimental setup for simultaneous wafer-level testing of more than one stripe, requiring a probe-card and a switching matrix, is described, along with the development of a procedure for the calculation of the true temperature of each stripe, which takes into account Joule self-heating and mutual heating of the samples.

Results obtained on Al-Si and Al-Si/TiN/Ti metallizations are shown and design guidelines to avoid thermal coupling of the samples, thus allowing more reliable temperature measurements, are provided.

## INTRODUCTION

The problem of temperature control always represents a crucial point in experiments dealing with thermally activated processes, requiring temperature–accelerated tests at various temperatures in order to achieve the activation energy of the phenomenon.

Regarding EM, an additional thermal contribution comes from the Joule heating of the samples, since current density acceleration is also required when investigating mass transport phenomena related to current flow through metal lines. Therefore, different contributions must be taken into account to properly describe the thermal situation of a lot of metal lines, expecially in the particular case of wafer-level testing of more than one sample at the same time:

* "bulk" heating, due to the sample holder;

* Joule heating, due to the high current flowing through the lines;

* mutual heating, due to thermal coupling between the samples.

Moreover, the system cannot be considered as stationary, since during the EM test the electrical resistance of the samples varies (increases, in most cases), giving rise to variation of Joule heating. Finally, bulk temperature fluctuations must be taken into account, due to non-ideal temperature control of the hot-stage and perturbation in the surrounding environment. None of these contributions, although each of them plays a different role in the general thermal situation, may be neglected for the correct evaluation of the test temperature during the EM experiments. The detrimental effects of wrong test temperature estimation has been clearly shown in a work by Schafft et al. [1], comparing the results of EM experiments carried out in different laboratories, in which different techniques for temperature measurement were adopted.

**Mat. Res. Soc. Symp. Proc. Vol. 265.** ©1992 Materials Research Society

An effective approach to true temperature evaluation of a metal line is to consider for each sample a thermal parameter $R_{th}$ taking into account thermal exchanges between the sample, the substrate and the surrounding environment [2] (this corresponds, in principle, to the standard procedure suggested in [1] to estimate the Joule-induced temperature rise of a metal line). The so-called "thermal resistance" $R_{th}$ can be used to determine the temperature $T$ of the sample if the power $P$ dissipated on it is known, by means of the relationship:

$$T = T_{ch} + R_{th}P. \tag{1}$$

where $T_{ch}$ is the temperature of the sample holder.

In the case that $n$ metal lines are to be measured, the above equation must be written for each of the test structures and must be completed adding the heat contribution due to the power dissipated in the neighboring samples:

$$T_i = T_{ch} + \sum_{j=1}^{n} \Theta_{ij} P_j. \tag{2}$$

The $\Theta_{ij}$ terms are the thermal coupling coefficients between the $i$-th line and the other ones [3]. For $i = j$ the term $\Theta_{ii}$ corresponds to the above cited thermal resistance $R_{th}$ of the line.

In this work we study the application of eq. (2) in the preliminary phase of wafer-level EM experiments, directed to compare the results obtained, in terms of activation energy, from Median Time to Failure (MTF) measurements and Relative Resistance Change measurements. In particular we consider the determination of the $\Theta_{ij}$ elements of the *thermal coupling matrix* and we draw some conclusions on its effectiveness in our experimental case.

## EXPERIMENTAL

Electromigration measurement techniques, and in particular resistometric ones, are usually applied at the wafer level by testing one sample at a time. This procedure is clearly not suitable for MTF tests, which take a very long time and should be performed on a lot of sample at the same time.

Our aim was to carry out a compared study applying at the wafer level both the resistometric and the MTF technique. To this purpose we set up a measurement bench provided with a probe card suitable for contacting ten devices at the same time and two switch-control-units (HP3488A) to realize all needed routings between the current sources, the voltmeters and the test samples. Setting a suitable configuration of the switches in the HP3488A units, the samples to be tested are connected in series and two current sources are used to power the test-structures or to provide a low current for resistance measurement of the unstressed segments, as it will be explained in the following.

Each of the test structure we used includes a $1\mu m$ wide, $1000\mu m$ long metal line, which is the electromigrating element, a $4\mu m$ wide, $4000\mu m$ long, "Greek shaped" stripe, acting as a reference element for temperature fluctuation filtering [4] and two $2\mu m$ wide, $1000\mu m$ long stripes, placed nearby the test line and suitable for detecting EM-induced lateral metal extrusions. The metal is Al-1%Si, $1\mu m$ thick, magnetron sputtered on a $0.7\mu m$ thick field oxide layer with an interposing TiN/Ti barrier layer. Some wafers were passivated with a $0.7\mu m$ thick PSG coating. A schematic of the whole structure is shown in figure 1. This test structure, which will be referred to in the following as "type-A", was specially designed for these experiments, but preliminary tests were carried out on other samples (referred to as "type-B"), available in our laboratory and modified for our aims. They were $4\mu m$ wide, $320\mu m$ long stripes, with the same metallization scheme, but without the TiN/Ti barrier. Table I summarizes the characteristics of the two types of samples.

The experimental procedure is intended to obtain a *thermal coupling matrix*, that is a set of $\Theta_{ij}$ elements to be used during the EM life-tests to evaluate the true test temperature through eq. (2). Three distinct steps must be performed: first, the resistance vs.

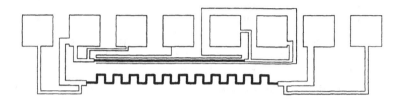

Figure 1: *Schematic of the test structure (type-A samples)*

| Sample | Metal | $W$ [$\mu m$] | $L$ [$\mu m$] | $J$ [$A/cm^2$] | $\Delta T$ [°$C$] |
|--------|-------------|-----|------|--------------------|----|
| A | Al-Si/TiN/Ti | 1 | 1000 | $4.9 \times 10^6$ | 27 |
| B | Al-Si | 4 | 320 | $3.0 \times 10^6$ | 25 |

Table I: *Experimental conditions. The last column reports the typical Joule-induced temperature rise when the sample holder is at $T_{ch} = 95°\,C$.*

temperature relationship must be determined for each sample, carrying out an appropriate temperature ramp. A linear fitting of the experimental points by means of the equation

$$R_i = R_{0i}(1 + \alpha_{0i}T_i) \qquad (3)$$

yields the $R_{0i}$ and $\alpha_{0i}$ parameters.

Second, one metal line at a time is subjected to high current flow, thus causing Joule self-heating of the sample and mutual heating of the neighboring stripes. The temperature increase of the $i$-th stripe due to the current flowing through the $j$-th one is determined as

$$\Delta T_{ij} = \frac{1}{\alpha_{0i}R_{0i}} (R_{ih} - R_{il}) \qquad (4)$$

where $R_{il}$ is the resistance of the $i$-th stripe when none of the sample is bearing the high stress current $I_h$ and $R_{ih}$ is the resistance of the $i$-th stripe when the high current flows through the $j$-th one. A low current value is used for $R_{il}$ and $R_{ih}$ measurements.

Last, the $\Theta_{ij}$ elements of the *thermal coupling matrix* are calculated as

$$\Theta_{ij} = \frac{\Delta T_{ij}}{P_j} = \frac{\Delta T_{ij}}{V_j I_h} \qquad (5)$$

where $V_j$ is the voltage drop across the $j$-th stripe when it is subjected to the current $I_h$.

The weak point of this procedure is the assumption that the chuck temperature corresponds to the temperature $T_i$ of the stripe during the ramp we use to extract the $\alpha_{0i}$ and $R_{0i}$ parameters. This is not exact, since the metal is not at direct contact with the chuck. However, we can suppose the fluctuations of $T_i$ with respect to $T_{ch}$ to be randomly distributed, so that the uncertainty over the slope of the $R(T)$ curve, i.e. the $\alpha_{0i}R_{0i}$ product, can be considered very small, if a large number of experimental points are taken. Therefore, $\Delta T_{ij}$ and subsequent $\Theta_{ij}$ calculations are slightly influenced by the above cited incorrect hypothesis, since only the $\alpha_{0i}R_{0i}$ product appears in eq. (4). Also, the stripe temperature when none of the stripes is bearing the high current is not measured on the thermochuck, but it is indirectly derived through $R_{il}$, thus avoiding the time-wasting and less reliable procedure of chuck temperature adjusting described in [3].

| | | | | | | | | | |
|---|---|---|---|---|---|---|---|---|---|
| 378.3 | 4.0 | 2.3 | 1.7 | 0.8 | 0.7 | 1.3 | 2.0 | 4.3 | 5.1 |
| 4.7 | 394.1 | 4.9 | 3.3 | 2.1 | 1.8 | 2.8 | 4.1 | 6.4 | 4.2 |
| 2.9 | 4.5 | 380.7 | 5.6 | 3.4 | 3.0 | 4.6 | 5.7 | 4.9 | 2.8 |
| 1.9 | 2.4 | 5.1 | 383.5 | 5.5 | 4.8 | 6.1 | 4.3 | 3.5 | 1.9 |
| 1.4 | 1.4 | 3.2 | 5.7 | 371.6 | 6.2 | 4.8 | 2.9 | 2.6 | 1.5 |
| 1.3 | 1.3 | 3.0 | 4.9 | 6.6 | 371.1 | 5.5 | 3.2 | 2.7 | 1.5 |
| 1.9 | 2.3 | 4.5 | 6.4 | 5.1 | 5.8 | 381.1 | 5.1 | 3.8 | 2.2 |
| 2.8 | 4.0 | 6.0 | 5.0 | 3.3 | 3.5 | 5.5 | 381.0 | 5.8 | 3.1 |
| 4.3 | 5.5 | 4.5 | 3.3 | 2.3 | 2.3 | 3.3 | 5.3 | 382.1 | 5.1 |
| 5.7 | 4.0 | 2.8 | 2.3 | 1.4 | 1.5 | 2.0 | 3.1 | 5.7 | 380.7 |

Table II: *Thermal coupling matrix relative to ten type-B samples.*

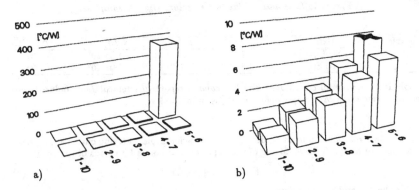

Figure 2: $\Theta_{ij}$ *distribution derived from column 5 of the thermal coupling matrix of table II. a) Full scale. b) Enlarged scale. The bar corresponding to $\Theta_{55}$ has been broken.*

## RESULTS

As stated above, a preliminary phase was carried out on $4\mu$m wide, $320\mu$m long Al-Si stripes. Ten stripes, arranged in two rows of five each, were measured at the same time. The interrow distance was about $1160\mu$m; the distance between two stripes within each row was also $1160\mu$m. In the resulting $10 \times 10$ *thermal coupling matrix* (see table II) the off-diagonal terms, representing the mutual heating parameters, are very small, but some of them are significantly higher. They correspond, as expected, to the samples near the stripe which was bearing the stress current. In figure 2 an histogram is shown in which each bar corresponds to the $\Theta_{ij}$ element for a fixed $j$, that is it represents the thermal coupling coefficients referred to the case of the high current $I_h$ flowing through the $j$-th stripe.

These samples showed another feature, which is clearly a function of the distance between the stripes on a row of five and between the two rows of samples: referring again to the histogram of figure 2, the thermal coupling resulted more effective, for example, between stripes 5 and 6 rather than between stripes 5 and 4. The reason is quite simple and lies in the fact that the temperature profile generated by a heating sample vanishes within a relatively short distance (obviously depending on the magnitude of the temperature peak, on the geometrical features of the structure and on the thermal characteristic of the materials). Considering a stripe as a series of elements, each acting as a heat source, it is clear that thermal coupling between the corresponding elements of stripes 5 and 6 is good, while only the nearest elements of stripes 5 and 4 can give a significant contribution

| 428.9 | 1.8 | 0.5 | 0.1 | 0.1 |
|---|---|---|---|---|
| 1.7 | 439.8 | 1.7 | 0.4 | 0.2 |
| 0.5 | 1.7 | 435.0 | 1.6 | 0.4 |
| 0.1 | 0.4 | 1.8 | 444.3 | 1.8 |
| 0.0 | 0.1 | 0.7 | 1.6 | 443.5 |

Table III: *Thermal coupling matrix relative to five type-A samples.*

to mutual heating.

In the second phase the tests were carried out measuring at the same time five type-A samples arranged in a single row. The distance between the metal lines in the row was again $1160\mu$m. Table III lists a typical $5 \times 5$ matrix of thermal coefficients obtained from these experiments. As can be seen, the off-diagonal terms are even smaller than in the case of type-B samples and the $\Theta_{ij}$ coefficients referring to the stripes far from the heating one are practically of the order of the experimental errors, showing that the thermal coupling between type-A samples is quite poor.

It must be remarked that a current density of the order of several MA/cm$^2$ was used in these tests (both on type-A and on type-B samples), causing a Joule heating of 25°C or more of the metal lines subjected to high current (see table I). Nevertheless, a quite moderate mutual heating could be detected in all cases.

This is a desirable condition for performing electromigration tests at the wafer level on a lot of samples at the same time. An immediate consequence is the fact that the temperature profile across the whole set of test-structures is quite uniform, while, in the case of a significant thermal coupling between the test-lines, a temperature peak would occur in the middle of the area in which the metal stripes lie, and the stripes located near the border would experience less severe stress conditions [3]. Moreover, at every stripe failure the temperature profile would be modified and it would be necessary to register the thermal history of each sample for the final data elaboration.

In our experimental setup we considered this possibility: the "Greek shaped" stripe appearing in fig. 1 was designed so that its characteristics, and in particular the heat generated, is the same of the test line. It is normally used as a thermal reference element [4], but it is wired so that, in the case the corresponding test stripe fails, it can be inserted to maintain the electrical continuity of the series of the test lines and to give approximately the same contribution previously provided by the failed stripe to the temperature profile across the set of test-structures. A rather complicated wiring is required to accomplish all these tasks, along with a greater number of switches and an obviously more complex software to take into account all possible configurations.

A "flat" temperature profile (or, better, a flat baseline with isolated peaks in correspondence of each stripe), with no thermal coupling of the samples, would allow for a quite simpler and more reliable experimental procedure, but, more important, in this case it would not be necessary, at the end of the experiment, to deduce an "average" temperature value for each stripe under test from the registered thermal history of each element. The latter is in any case an arbitrary procedure and might be a source of errors in subsequent calculations.

In our experiments the "flat" temperature profile across the set of test-structures is achieved with a good approximation. This is due to the high ratio of the $\Theta_{ii}$ terms to the mutual heating terms $\Theta_{ij}$, $i \neq j$. A relatively high thermal resistance has been achieved in our samples since they are narrow and not too long lines (almost completely following, in the case of type-A samples, the NIST recommendations for electromigration test-pattern design [5]), and therefore they have a limited area for thermal exchange with the substrate.

A great importance, anyway, must be given to the layout of the test structures on the wafer, since, as it appears from the data referring to type-B samples, thermal coupling is more effective between metal stripes on different rows rather than between the stripes on the same row.

## CONCLUSIONS

The problem of the true test temperature evaluation during electromigration tests was addressed in the case of wafer-level testing of a lot of samples at the same time. Added to the Joule self-heating of the samples, due to the high current flowing through the metal lines, mutual-heating terms should be generally considered to contribute to the temperature rise of the samples. Experiments have been carried out on two different types of samples, using a probe card for device contacting and a set of switch batteries to test five to ten stripes at the same time. A *thermal coupling matrix* was determined for each set of samples, showing, in the cases we examined, a limited thermal coupling between the test structures. This was essentially due to the layout of the metal lines, which are not too long, widely spaced in the wafer and, in the best case, arranged in a single row.

If thermal coupling of the samples is avoided, it is not required to register the thermal history of the tested lines during the EM life-tests, since the failure of a stripe does not influence the thermal situation of the others, nor it is necessary to replace a failed stripe with a heating element in order to maintain the temperature profile across the whole set of test structures. This ultimately results in a more reliable experimental procedure and data elaboration and, not irrelevant, in a simpler experimental setup.

## ACKNOWLEDGMENTS

The authors would like to thank Dr. C. Caprile of ST Microelectronics for providing the test structures, Dr. M. Impronta of CNR-LAMEL for his valuable help in developing the software, Dr. G. Brandoli and Dr. M. Giordano for performing most of the measurements.

## REFERENCES

[1] H. A. Schafft, T. C. Staton, J. Mandel and J. D. Shott, IEEE Trans. Electron Devices, **ED-34**, 673 (1987)

[2] A. Scorzoni, G. C. Cardinali, G. L. Baldini and G. Soncini, Microel. and Reliab., **30**, 123 (1990).

[3] N. Zamani, J. Dhiman and M. Buehler, Proc. 6th IEEE International Multilevel Interconnect Conference, 198 (1989).

[4] G. L. Baldini and A. Scorzoni, Proc. of the 1st Europ. Symp. on Reliab. of Electron Dev., Failure Phys. and Analysis (ESREF 90), Bari, 245 (1990); IEEE Trans. on Electron Devices, **ED-38**, 469 (1991).

[5] H. A. Schafft, IEEE Trans. Electron Devices, **ED-34**, 664 (1987).

# TEM INVESTIGATION OF IMPLANTED PHOTORESIST RESIDUES REMAINING AFTER OXYGEN PLASMA ASHING

S. YEGNASUBRAMANIAN, C. W. DRAPER and C. W. PEARCE[*]

AT&T Bell Laboratories, Princeton, New Jersey 08540
*AT&T Microelectronics, Allentown, Pennsylvania 18103

## ABSTRACT

This paper presents the results of a systematic TEM investigation to understand the chemical, morphological and microstructural makeup of photoresist residues following oxygen plasma ashing. The investigation was carried out on generic non-product control wafers and product wafers with a view to help aid in the identification of a suitable post-ashing cleaning process. Specimens in plan-view were prepared by mechanical grinding and Argon ion milling from the substrate end of the sample. Energy Dispersive X-ray Spectrometry (EDS) was used to obtain the composition and selected area electron diffraction (SAD) was used to obtain information on crystallinity. The residues were essentially amorphous and were found to be arsenic-rich in composition and exhibited a droplet-like morphology decorating areas where the photoresist was used as an implantation mask. In addition, hexagonally shaped crystals and spherical particulates of varying dimensions were seen in several regions and were found to be carbon rich.

## INTRODUCTION

The problems associated with stripping of photoresist that has been hardened by implanting ions such as arsenic or phosphorus is widely known in the microelectronics industry. High dose ion implantation affects the resist so much that a dark, hardened crust remains on the silicon wafer [1,2]. Removal of the hard-baked photoresist used as an implantation mask is an extremely difficult cleaning task. Some of the methods that address stripping tough photoresists currently were reviewed recently by Laura Peters [3]. Tenacious residues that will not be stripped in the cleaning cycles are of great concern for both performance and yield reasons. Improving the stripping process often enhances the yield and some of the approaches include low energy ashing, microwave plasma/reactive ion etch, hydrogen plasma to remove chlorinated residues, ultra/megasonics to accelerate residue removal rates, more powerful solvents to remove sidewall polymers etc. Applications involving multilevel metallization processes call for new stripping methods due to problems associated with corrosion.

Dry processes are widely used to volatize and remove the bulk of the resist followed by wet stripping techniques. However, in designs involving thinner gate oxides, radiation damage caused by dry strippers is a cause of great concern. Also, widely used process equipment and sequence of barrel-type plasma ashing followed by immersion in acid/peroxide wet chemical baths has appreciated disadvantages. Barrel ashers are known to cause high radiation damage as the wafers are exposed directly to the high energy plasma. High dose and high power implantation tend to lead to the deposition of a carbonized layer on the resist surface. Recent dry strippers available in the market seem to specifically address this issue of tackling high implant resists.

Studies to systematically understand the nature of the photoresist residues following ashing procedures have not been adequately explored. To the best knowledge of the authors, based on a preliminary literature search, a definitive study of the residues remaining after oxygen plasma ashing of arsenic implanted photoresist is not reported in the literature. Once the chemical, morphological and microstructural makeup of the residues are understood, the process engineer would be able to examine the wet chemical cleaning sequence for effectiveness against a known residue instead of relying on general cleaning procedures. In addition, the technologist would gain knowledge of the potential electrical activity of the contamination as well.

This paper summarizes the results of a systematic TEM investigation to characterize the stubborn residue that decorates source-drain photoresist masked areas following oxygen plasma ashing.

EXPERIMENTAL

The samples examined included blanks prepared from regular control silicon wafers with a ~ 20 nm thermal oxide, 1.4 μm of photoresist and a double implant of arsenic and phosphorus. Table 1 summarizes the implantation values.

Table 1. Ion Implantation of Photoresists

| Element | Dose ions.cm$^{-2}$ | Energy, keV | Beam Current, mA |
|---------|---------|---------|---------|
| As | 1 E 16 | 100 | 8 |
| P | 1 E 15 | 50 | 2.5 |

The photoresist and implantation process parameters were the standard used for the source-drain process sequence. Both window-patterned and fully-covered blanks were prepared. Two product wafers which had been ashed but not treated with post-ashing wet chemistry were examined following the preliminary survey of control wafers.

The specimens for TEM were prepared in plan view by thinning the sample from the bottom side of the wafer by mechanical grinding and argon ( 6 keV, 1 mA) ion milling. TEM micrographs and selected area electron diffraction patterns were recorded using a JEOL 2000FX TEM operated at 200 keV. The X-ray spectra were recorded using a TN 5500 pulse height analyzer system coupled to an ultrathin window HpGe light-element detector.

RESULTS AND DISCUSSION

Figures 1a - 1c show the TEM micrographs of typical residues found at various areas of the thin specimen of the control wafer. They exhibit a hillock or droplet-like morphology. They are roughly circular and range in diameter from tens of micrometers to submicrometer. The droplet is comprised of smaller droplets in some cases (Fig. 1a and 1b) and in others, the droplet is decorated by a ring of smaller droplets (Fig. 1c ). X-ray spectrometry of the dark regions of the droplets showed significant amounts of arsenic as seen in the spectrum of fig. 2. The oxygen and silicon lines in Fig. 2 are from the sacrificial oxide. Figures 3a and 3b show the micrographs of the patterned sample; in this case the residue seems to be predominantly left where the photoresist existed, but some smaller diameter residue is found within the windows. On the unpatterned controls the droplets seem to cover a much larger surface area when compared with that of the patterned wafers.

The residue is primarily composed of arsenic and was found to be an amorphous (noncrystalline) form of arsenic from electron diffraction experiments. Auger electron spectroscopy

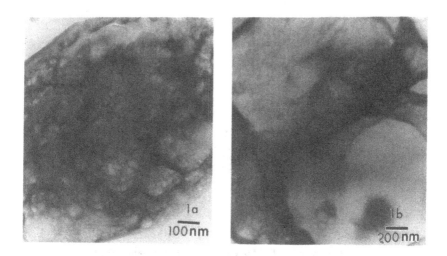

Fig. 1a,b & c  TEM of Residue from the Control Wafer

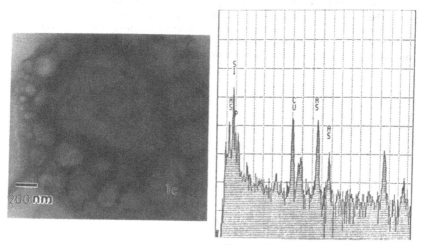

Fig. 2  X-Ray Spectrum of the Residue

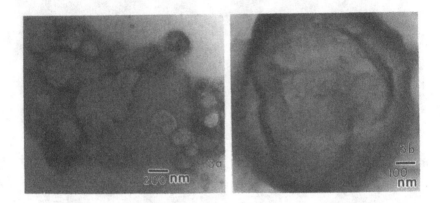

**Fig. 3a & 3b   TEM of Residue from the Product Wafer**

(AES) of the residue indicated that a small quantity of carbon, oxygen and silicon might reside in the residue but that the major component was arsenic [4].

Other residues that were observed in addition to the major arsenic-containing residues were : amorphous and crystalline carbon/graphite (Fig. 4a, 4b and 4c), (transition) metal-rich particulate debris (Fig. 5a and 5b), remnant hardened-photoresist flakes and spider webs (Fig. 6a and 6b) containing arsenic (dark features of Fig. 6). However these were relatively fewer in number density. Fig. 7 gives an idea of the extent of damage caused by the high-dose implant and the occurrence of a typical arsenic containing residue and a carbon particulate.

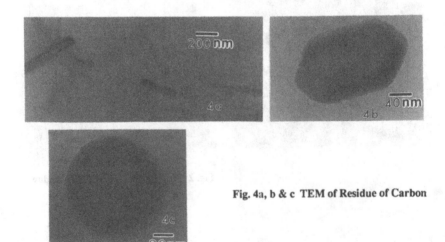

**Fig. 4a, b & c  TEM of Residue of Carbon**

Fig 5a   TEM of Metal
Particulate Debris

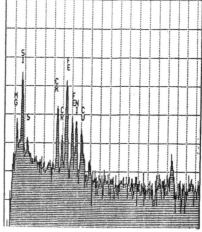

Fig. 5b  X-Ray Spectrum from the Debris

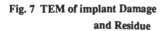

Fig. 6a & 6b   Spider Web-like Residue

Fig. 7  TEM of implant Damage
and Residue

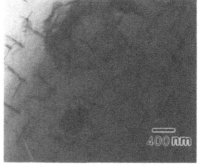

## SUMMARY

The problems associated with the stripping of photoresist that has been hardened by the implantation process are historic in nature and appreciated throughout the microelectronics industry. Likewise the widely used process equipment and sequence of barrel-type plasma ashing followed by immersion in acid/peroxide wet chemical baths has appreciated shortcomings. In this context it is rather a surprise that a more definitive understanding of the nature of the residue following ashing has not been pursued adequately in the literature. We have carried out a systematic TEM investigation of the residues remaining after oxygen plasma ashing of arsenic implanted photoresist to understand the microstructural, morphological and compositional makeup of these residues. The residues were found to be predominantly arsenic existing as amorphous arsenic. The observations were supported by AES results. Other residues that were observed include, carbon/graphite, metal-rich particulates, spider web-like features etc. but were significantly fewer in number density. The results help in the development of suitable cleaning alternatives in the removal of hard-baked photoresist in the industry.

## ACKNOWLEDGEMENTS

The authors thank R.F. Roberts for reviewing the manuscript.

## REFERENCES

[1]  D. Roche, J. F. Michaud and M. Bruel, Mat. Res. Soc. Symp. Proc., Vol. 45, 1985.

[2]  T. Venkatesan, S. R. Forrest, M. L. Kaplan, C. A. Murray, P. H. Schmidt and B. J. Wilkens, J. Appl. Phys. 54(6), June 1983.

[3]  Laura Peters, Semiconductor International, February 1992, p. 58.

[4]  P. J. Sakach, Private Communication.

# COMPUTER SIMULATION OF MATERIALS FAILURE PROCESSES IN A THERMAL OR RADIATIONAL ENVIRONMENT

S. SHIMAMURA
Department of Applied Science, Faculty of Engineering, Yamaguchi University, Ube 755, Japan

## ABSTRACT

The surfaces of various materials are often exposed to such environments as thermal cycling or irradiation. The failure processes caused by microcrackings in such an environment are studied by means of computer simulations. Crack growth is modeled as a series of processes of storage, release, and transfer of strain energies in the cubic lattice system. Surface crack pattern, crack penetration, and strain energy distribution are investigated for systems in different situations. The simulations reveal some correlations between crack pattern, crack penetration, and strain energy distribution. The results have some implications on the failure processes of materials in a thermal or radiational environment. The reliability of materials is discussed on the basis of the results of simulations.

## INTRODUCTION

Understanding the behavior of cracks that lead to materials failure is essential to the development of functional materials such as electronic materials [1] as well as structural materials. Studies in the past on materials failure have focused attention on the behavior of a few main cracks. However, the global evaluation of cracks is also required for discussing the reliability of materials. In recent years, there has been growing interest in the evaluation of cracks from this point of view [2-4].

Various state-of-the-art materials such as electronic materials are often exposed to such environments as thermal cycling or irradiation [5,6]. One of the characteristic features of these environments is the spatial randomness of their influences on materials. Thermal cycling generates thermal stresses at random on the surface of materials. Irradiation of X-ray, neutron and so on also generates spatially random stresses or damages on the surface of materials. These random stresses cause microcracks on the surface and finally lead to the failure of materials.

In this paper, we report the results of computer simulations of crack growth in a random environment as mentioned above. Initiation and propagation of cracks in such an environment can be regarded as a series of processes of storage, release, and transfer of strain energies in a material in terms of energetics. An irradiated material, for example, stores the strain energy at random as a result of radiation damages such as defects. Growth of generated microcracks is a release process of part of the strain energy, and is also a transfer process of the strain energy in the material. From this point of view, we have modeled crack growth using simple lattice systems and have investigated crack morphology by means of Monte Carlo simulations [7,8]. The results of the simulations have some implications on materials failure. We will here discuss the results of simulations in regard to the reliability of materials in a thermal or radiational environment.

## MODELING AND SIMULATION

Our model for materials failure is composed of the following processes. Consider a cubic lattice system; we call a cube the grain, and a side plane

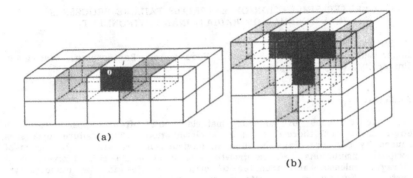

FIG. 1. Cubic lattice system for failure simulations. After a crack plane (shown by a dark plane) has been generated on the boundary between the i-th and j-th grains in (a), part of the strain energies of those grains is transferred equally to four shaded grains. The present model results in a distribution of strain energies as in (b) where shaded grains at the tips of a dark crack plane store more strain energies than neighboring unshaded grains.

of a cube, the grain boundary. We select a grain at random in the surface lattice layer to be exposed to thermal cycling or irradiation, give a "strain energy", $\Delta E$, to the grain, and repeat this storage process of strain energy. When the strain energies of adjacent two grains, say the i-th and the j-th grains, $E_i$ and $E_j$, satisfy the condition, $E_i \cdot E_j \geq E_t{}^2$, we generate a crack on the grain boundary of the two grains, where $E_t$ is a threshold energy for cracking. After cracking, the strain energies of the two grains both reduce to zero, and an energy, $E_r$, is released from the system. Then the rest of the strain energies of the two grains stored before cracking, $E_i + E_j - E_r$, is transferred equally to grains at the crack tips. Figure 1(a) shows this process. If any pair of grains at the crack tips satisfies the cracking condition as a result of this transfer process, further cracking follows. Otherwise the process of energy storage is repeated until the next crack occurs.

The processes mentioned above are performed so that cracks can penetrate from the surface to the interior of the system. That is, the processes of energy storage and transfer are performed for grains beneath the surface layer after a crack is generated in the surface layer. Figure 1(b) shows an example of crack plane and strain energy distribution resulting from the present model simulation. A more detailed explanation of the simulation was given in another paper [8].

Simulations have been performed in systems of 14400 (120 × 120) grains in the surface layer. For the boundary of the system, we have used periodic boundary conditions. The present model simulates the global behavior of cracks, that is, how cracks in a material in a random environment grow and how strain energies vary spatially in the material. Crack morphology and strain energy distribution revealed by the simulation, therefore, will give us some information on materials failure processes in a thermal or radiational environment.

## RESULTS AND DISCUSSION

In the present simulations, the system settles into a stationary state when the number of cracks, $N_c$, amounts to about 6000. In this state, the average strain energy per grain is almost constant because the strain energy stored by random inputs is released by crackings on the average.

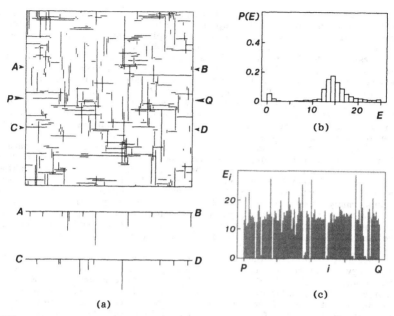

FIG. 2. Crack morphology and strain energy distribution for $\Delta E = 1$, $E_t = 25$ and $E_r = 1$: (a) surface crack pattern and crack penetration; (b) magnitude distribution of strain energies of grains; (c) spatial distribution of strain energies of grains.

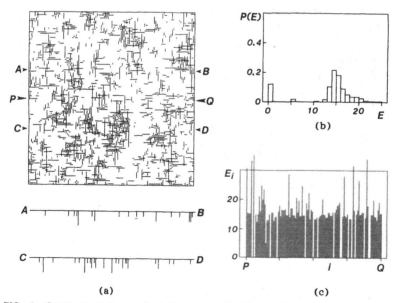

FIG. 3. Crack morphology and strain energy distribution for $\Delta E = 5$, $E_t = 25$, and $E_r = 1$. See also the caption of Fig. 2.

304

In other words, the distribution of strain energies in the system is stable. We then find a close relationship between crack morphology and strain energy distribution. We will report some typical results at $N_c = 6000$ and discuss the results in terms of materials reliability.

First we see the influence of the magnitude of $\Delta E$ on crack morphology and strain energy distribution. Figure 2 shows the surface crack pattern, the crack penetration, and the magnitude and spatial distributions of strain energies of grains for $\Delta E = 1$, $E_t = 25$, and $E_r = 1$. Figure 3 shows the results for $\Delta E = 5$, $E_t = 25$, and $E_r = 1$. A large $\Delta E$ results in a fine and nonuniform pattern of surface cracks and shallow penetration of the cracks. The characteristic features of the strain energy distributions are almost the same for $\Delta E = 1$ and $\Delta E = 5$.

A physical significance of a large $\Delta E$ is a rapid or large temperature difference in thermal cycling or thermal shock, and is an intense irradiation in various radiational situations; these situations induce large spatial fluctuations of strain energies on the surface of a material. Therefore the simulations suggest that severe thermal cycling or intense irradiation tends to cause many short cracks, but those cracks do not penetrate deeply into the material. In fact, some reported experiments on thermal shock [9,10] are consistent with our simulations [8]. The results of simulations will also give useful suggestions as to the failure of electronic materials exposed to thermal cycling or irradiation.

Secondly, we see the effect of the magnitude of $E_t$ on crack morphology. Figure 4 shows the crack pattern and the strain energy distributions for $\Delta E = 1$, $E_t = 5$, and $E_r = 1$. As compared with Fig.2 in which $E_t = 25$, a smaller $E_t$ results in a finer pattern of surface cracks and shallower penetration of the cracks. In other words, the system with a large threshold energy for cracking shows long and deep cracks. This

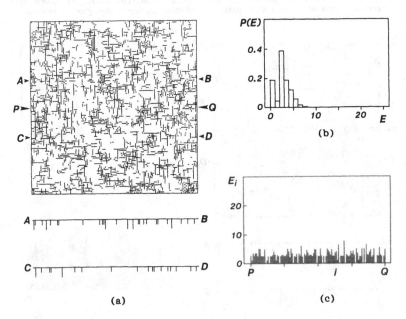

(a)                                    (c)

FIG. 4. Crack morphology and strain energy distribution for $\Delta E = 1$, $E_t = 5$, and $E_r = 1$. See also the caption of Fig. 2.

implies that a material with a strong grain-boundary strength is in danger of disastrous failure or fracture, though it is resistant to cracking.

Thirdly, we show the effect of the magnitude of $E_r$ on crack morphology and strain energy distribution. Figure 5 shows the crack pattern and the strain energy distributions for $\Delta E = 1$, $E_t = 25$, and $E_r = 25$. As compared with Fig.2 in which $E_r = 1$, a large $E_r$ results in an uniform and fine cracks on the surface and shallow penetration of the cracks. The system with a large release energy reveals characteristic features of strain energy distributions. That is, the strain energy distribution is uniformly wide in magnitude and randomly fluctuating in space. This gives us useful information on materials failure. It is desirable that a material contains microstructures that incorporate a release mechanism of strain energy. This is favorable for the material not only to prevent cracks from penetrating deeply, but also to resistant to fracture caused by subsequent random stresses or impact. A material with a large release energy tends to maintain a distribution of residual strain energies that is fluctuating in space, as shown in Fig.5(c). The large spatial fluctuations of strain energies are likely to arrest crack extension, being favorable to resistant to disastrous fracture.

The present model for failure processes is based on a series of processes of strain energy, therefore being applicable to various failure processes caused by random stresses. However, the model does not include parameters that are related with a specific material or environment. It is necessary to improve the model in order to investigate the aspects characteristic of a specific situation. In this sense, the global investigations in the present model simulations will be complementary with conventional approaches to the reliability of materials.

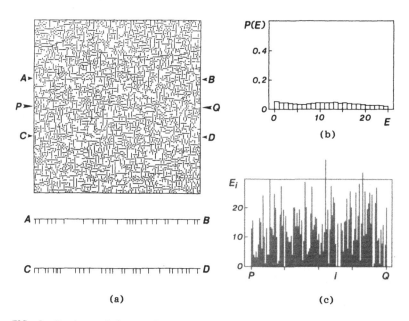

FIG. 5. Crack morphology and strain energy distribution for $\Delta E = 1$, $E_t = 25$, and $E_r = 25$. See also the caption of Fig. 2.

## ACKOWLEDGMENTS

This work was supported by a grant from the Iketani Science and Technology Foundation.

## REFERENCES

1. A. H. Burkhard, J. M. Kallis, L. B. Duncan, M. F. Kanninen, and D. O. Harris, in Advances in Fracture Research, Vol. 2, edited by K. Salama, K. Ravi-Chandar, D. M. R. Taplin, and P. Rama Rao (Pergamon Press, New York, 1989), p. 977.
2. Random Fluctuations and Pattern Growth: Experiments and Models, edited by H. E. Stanley and N. Ostrowsky (Kluwer Academic, Dordrecht, 1988), pp. 149-192.
3. Statistical Models for the Fracture of Disordered Media, edited by H. J. Herrmann and S. Roux (North-Holland, Amsterdam, 1990).
4. A. T. Skjeltorp and P. Meakin, Nature 335, 424 (1988).
5. Tai-Yan Kam, in Advances in Fracture Research, Vol. 4, edited by K. Salama, K. Ravi-Chandar, D. M. R. Taplin, and P. Rama Rao (Pergamon Press, New York, 1989), p. 2843.
6. J. Kameda and X. Mao, J. Mater. Sci. 27, 983 (1992).
7. S. Shimamura and K. Kuriyama, J. Mater. Sci. 26, 6027 (1991).
8. S. Shimamura and Y. Sotoike, J. Mater. Res. 7, in press (1992).
9. V. N. Gurarie and J. S. Williams, J. Mater. Res. 5, 1257 (1990).
10. E. H. Lutz, M. V. Swain, and N. Claussen, J. Am. Ceram. Soc. 74, 19 (1991).

# ANALYSIS OF STRESSES IN A MULTILEVEL CONFIGURATION BY USING A DESIGN OF EXPERIMENT METHOD

M. IGNAT, A. CHOUAF, J.M. TERRIEZ and M. MARTY*,

LTPCM - ENSEEG, BP.75 38402 St Martin d'Hères - France

* SGS-THOMSON Microelectronics Av. des Martyrs BP. 217 38019 Grenoble Cedex - France

## ABSTRACT

The effects of thermoelastic residual stresses on multilevelstructures was determined by finite element calculation. A design of experiment approach was used, with five geometrical parameters. The results are presented in the form of stress contours, shown as functions of pairs of parameters, with the other parameters held constant. This analysis shows the relative sensitivities to the different parameters with respect to potential damage and failure. Comparison of the results with previous observations of microstructures seems to validate this experimental design approach.

## INTRODUCTION

The design of present day integrated circuits introduces crucial reliability problems [References 1 to 4]. Among the mechanical problems arising when designing integrated circuits, the principal one comes from the matching of layers of different materials which have different mechanical, physical, and chemical properties. Simply taking into account the effects introduced by the thermal expansion mismatch of the materials, the strain incompatibilities can cause large stresses. When these stresses are high enough, depending on the material constituting the stressed layer, the stresses can relax by the activation of various deformation mechanisms, either immediately after the fabrication of the device , or when an external stress is applied. In particular, when multilayers with stiffer materials as tungsten combined with silicon oxide interlayers are submitted to cyclic stresses, crack opening and propagation in the brittle interlayer may be expected[5, 6, 7]. Moreover, since integrated circuits contain special geometrical structures (steps and kinks); the stresses are increased near the associated singularities, increasing the risk of irreversible mechanical damage, and the electrical failure of the device. For example, a standard BICMOS structure may show damaged zones near the edges of the metal line, and in the passivation layer (see Figure 1). Prior investigations have demonstrated that damage mechanisms can be correlated with the geometrical characteristics of a device. A decrease in the metal line spacing has been associated with an increase of the observed void density in the metal lines ; and an increase of the thickness of the metal lines has been associated with an increase in the probability of stress relaxation through the activation of crack opening processes in the passivation layer [8, 9, 10]. From these considerations it appears that establishing the mechanical design limits for a given structure would require microstructural analysis of a great number of samples, each one corresponding to one or several variations of the constitutive layers. The process would be long and tedious, and an preliminary optimization of the geometrical parameters of the device appears to be necessary, to reduce the number of samples required. We established this optimization for a standard BICMOS pattern, using an experimental design method. The variation of five geometrical parameters and their effects on stresses was analyzed. Patterns showing greater or less tendency to mechanical failure were determined and are discussed with respect to prior microstructural observations.

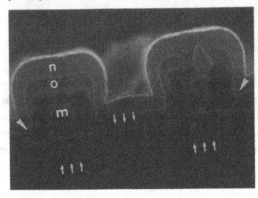

Figure 1: Typical BICMOS structure showing the evidence of damage. Little arrows point out interfacial decohesion, wider arrows: important damage at singular zones. The n and o letters correspond to the silicon nitride and silicon oxide passivation layers; m to the metal conducting lines.

## ANALYTICAL METHODS

### Experimental Design Method

The objective of the Experimental Design Method is to establish the mathematical equations which will relate the stresses in a multilevel device to its design rules, that is, the independent geometrical factors. The only analytical relations relating stresses to the geometry of deposited layers are for simple geometries (flat uniform layers); and no general relations are availble for the complex structures found in a multilevel system. We chose arbitrarily to use a full quadratic relation for interpolation among the results of finite element calculations. This sort of relation is able to represent linear relationships as certain types of non linear responses [11]. The general form of this relation is:

$$Y = b_0 + \sum_{i=1}^{n} b_i X_i + \sum_{i=1}^{n} b_{ii} X_i^2 + \sum_{j=2}^{n} \sum_{i=1}^{n-1} b_{ij} X_i X_j + \varepsilon \qquad (1)$$

Y is the value of the stress in a given media of the structure, n is the number of independent geometrical factors (in each media), the b terms are then the polynomial coefficients deduced from the finite element calculations and corresponding to a precise point in the structure, and the Xi terms are the geometrical parameters. For a typical multilevel structure, sketched in Figure 2, the parameters for the experimental design correspond to the geometrical properties, defining the top layers : thickness of the passivation layers as of the metal lines and width and space of the last ones. For a lack of space, we will not extend our analysis of equation (1) with respect to the experimental design, a DOEHLERT uniform network. A precise explanation of equation (1) and of the DOEHLERT uniform network are included in References [11] and [12]. The main purpose of a DOEHLERT uniform network is to generate a regular point distribution in parameter space. Then all the calculated points will remain in a hypersphere, and will be located at equal distances from the center. Because of the network uniformity, the precision of our interpolation will remain constant for all the calculated factors.

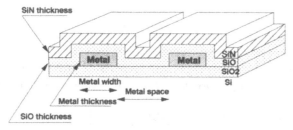

Figure 2: Schematic representation of the typical BICMOS structure.

The five geometrical parameters will interact and the stresses found for all 31 trials must be considered. The geometrical characteristics for each trial are given in Table I. Due to a lack of space, the design network used for the modelling and including the 31 trials is not reported here, but described in [12] and [13]. The main purpose of the experimental design we used was to develop a sequential approach for studying the second order response of surfaces. A set of selected points can be shifted by displacing the origin, in order to analyze adjacent points, and enlarge the analyzed surfac. In our case, the entire matrix includes 31 rows and 5 columns, each column corresponding to a geometrical parameter. These five parameters are: the thicknesses of the silicon oxide and silicon nitride passivation layers, which range from 0.35μm to O.8μm and 0.45μm to 0.95um respectively; the thickness and width of the metal lines, which vary between 1μm and 1.3μm, and 5.1μm and 1.9μm respectively; and the space between the metal lines, from 2.5μm to 5.5μm. The stress tensor was calculated for each row of the matrix.

Stress calculations

Stresses were calculated by a Finite Element Method, using a standard program. The structure was defined by isoparametric quadrilateral elements, with length to width ratio determined by the geometrical inputs for each layer.

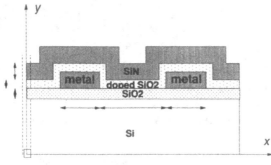

Figure 3: Representation of the boundary conditions used for our Finite Element Calculations.

The boundary conditions are shown in Figure 3 and the calculations were performed assuming Plane Stress conditions. The thermoelastic residual stresses were induced by cooling the structure from

773K (500°C) to 298K (25°C). We considered all the materials to be elastically isotropic and homogeneous ; their thermoelastic properties are reported elsewhere [14].

## RESULTS AND DISCUSSION

The results we obtained can be summarized in two and three dimensional plots. In the first case isostress contours are shown for a plane defined by axes which correspond to two of the geometrical parameters. For instance, in Figure 4, the contours of shear stresses in the passivation layers are plotted as a function of the thicknesses of the layers. In this case, the other three parameters are fixed.

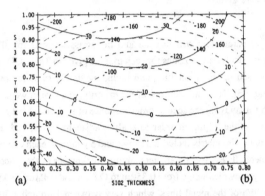

(a)                                SIO2_THICKNESS                                (b)

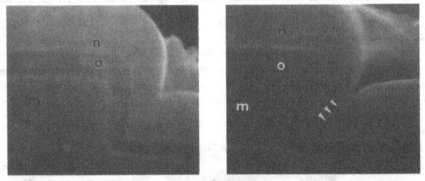

Figure 4: Diagram of isostress contours resulting from the Design of Experiment Plan: (a) and (b) are micrographs corresponding to the symbols in the diagram. The silicon oxide passivation layer is 0.35μm. (a) The thickness of the silicon nitride layer is a 0.785μm, no damage is observed. (b) When the thickness of the silicon nitride layer is increased to 0.96μm, damage is observed in the passivation layer.

In Figure 5, we show a three dimensional plot of a normal stress component with respect to two dimensional parameters, the metal width and metal thickness, with the other parameters held constant.

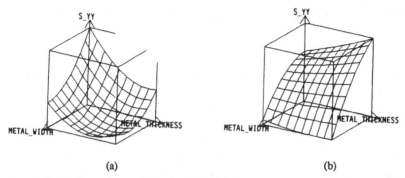

(a)                                                          (b)

Figure 5 : Normal stress contours resulting from the D.O.E plan reported as a function of Metal width and Metal thickness. (a) near to the middle of the metal line; (b) near to the middle of a lateral face of the matel line.

From such plots one can deduce parameters combinations which correspond to high enough stresses for a risk of failure; or choose dimensions which avoid high residual thermoelastic stresses. As pointed out previously, high stresses will tend to be relaxed by irreversible deformation mechanisms, such as interfacial crack opening. For example, the high stresses that can be induced in the silicon oxide layer, when the thickness of the silicon nitride layer is increased, may induce cracks. The contour plots of stresses in Figure 4 also show that for passivation layers of about 0.5μm thickness, the thermoelastic stresses are very low. Moreover, the calculations can be made for precise points, such as the ones corresponding to singular zones (edges of the metal line), where the stresses will be locally increased [15], and relaxation by crack opening is expected. Previous microstructural observations of these zones can be used to validate the method. For example, as shown in Figure 4a, the structure shows no cracks when the silicon nitride layer is 0.78μm thick, while cracks appear at singular zones when the passivation thickness is increased to 1μm (Figure 4b). With regard to the metal lines, we can also determine their optimum width and thickness for a fixed interline spacing, and fixed dimensions for the passivated layers. In each case, the analysis gives the stress as a function of the two interacting geometrical parameters, with the other three remaining fixed.

## CONCLUSIONS

The major stress effects resulting from the interactions among five geometrical factors in a multilevel device have been established, and optimization techniques demonstrated, with the use of an experimental design method.
- The results obtained by this approach are presented in contour plots. The contour plots show the stress as a function of pairs of geometrical parameters.
- The value of the method lies in its use to identify geometrical configurations which are potentially sensitive to mechanical failure.
- The correlation of the theoretical results with previous microstructural observations seems to show the validity of the method.

**ACKNOWLEDGMENTS**: AUthors thank Professor Paul A. Flinn from Stanford for helpful discussions and english improvement.

**REFERENCES**

1    J.R. Lloyd, F.G. Yost, P.S. Ho (editors) <u>Materials Reliability Issues in Microelectronics</u> (Published by the Mater. Res. Soc. vol. 225, 1991).

2    E. Suhir, R.C. Camarata, D.L. Chung, M. Jone (editors) <u>Mechanical Behaviour of Materials and structures in Microelectronics</u> (Published by the Mater. Res. Soc. Vol. 226, 1991).

3    P.A. Flinn in <u>Thin Films: Stresses and Mechanical Properties II</u>. Edited by M.F. Doerner, W.C Oliver, G.M. Pharr, F.R. Brotgen (Mater. Res. Soc. Proc. <u>188</u> Pittsburgh, P.A. 1990) pp.3-13.

4    M. Ignat, A. Chouaf, C. Bernard, J.M. Terriez in <u>Interfaces Between Polymers, Metals and Ceramics</u>. Edited by B.M. Dekoven, A.J. Gellman, R. Rosenberg. (Mater. Res. Soc. Proc. <u>153</u> Pittsburgh, P.A.1989) pp 357-362.

5    A. Chouaf, M. Ignat, J.M. Terriez, P.Normandon in <u>Electronic Packaging Materials Science</u> V. 203 MRS (1991) p 177-182.

6    M. Ignat, A. Chouaf, Ph. Normandon in Thin Films : <u>Stresses and Mechanical Properties II</u>, edited by M.F. Doerner, W.C Oliver, G.M. Pharr, F.R. Brotgen (Mater. Res. Soc. Proc. <u>188</u> Pittsburgh, P.A. 1990) pp 97-102.

7    M. Ignat, A. Chouaf, P. Normandon, P. Gergaud, J.J. Bacman in <u>Tungsten and other Advanced Metals for ULSI Applications in 1990</u>, edited by G.C Smith and R. Blumenthal (Mater. Res. Soc. Conf. Proc. Pittsburgh, PA, 1991) pp 305-310.

8    M.A. Korthomen, C.A. Paszkiet, C.Y. Li, J. Apply. Phys. $\underline{12}$, 69 (1991).

9    Q. Guo, C.S. Whitman, L.M Keer, Y.W. Chung, J. Appl. Phys. $\underline{11}$, 69 (1991).

10   S.K. Grothuis and W.H. Schroen in Transactions of the IEEE, IRPS.

11   G.E Box, N. Draper, "Empirical Model-Building and Responses Surfaces", edited by John Wiley Sons. New York 1987.

12   R. Phan Tan Luu, D. Feneville, D. Mathieu, "Méthodologie de la Recherche Expérimentale", I.P.S.O.I, Editeurs Paris 1983.

13   M. Marty, M. Ignat, A. Chouaf, J.M. Terriez in Transactions IEEE - IPFA (1991).

14   A.I. Sauter and W.D. Nix in <u>Finite Element Calculation of Thermal Stresses in Passivated and Un passivated Lines Banded to Substrates</u> (Mater. Res. Soc. Proc. <u>188</u>, 1990.

15   R.E. Jones in Transactions of the IEEE - IRPS p.5 (1987).

# THE IMPACT OF DEVICE ASYMMETRY ON THE ELECTRICAL AND RELIABILITY PROPERTIES OF FERROELECTRIC PZT FOR MEMORY APPLICATIONS

Jiyoung Kim, C. Sudhama, Vinay Chikarmane, Rajesh Khamankar, Jack Lee and Al Tasch, Microelectronics Research Center, The University of Texas at Austin, Austin, TX78712.

## ABSTRACT

The origin and the effects of asymmetrical electrical behavior in sputtered PZT (Zr/Ti=65/35) thin film capacitors with Pt electrodes have been studied. The asymmetry and constriction in the P-E hysteresis loops are understood to result from differences in mechanical stress at the top and bottom PZT/Pt interfaces because they experience different thermal cycles during fabrication. A method for correctly positioning asymmetric loops on the polarization axis is suggested. Both d.c. and a.c. electrical stressing (of either polarity) lead to hysteresis relaxation and symmetrization. A post-processing anneal leads to electrically symmetrical devices.

## INTRODUCTION

Thin films of ferroelectrics such as Lead Zirconate Titanate (PZT) have been extensively investigated for use in non-volatile and dynamic RAM cells. Ferroelectric materials are commonly characterized by a hysteresis loop in the polarization versus electric field plane, as shown in figure 1. Read/write operation involves A.C. cycling of the ferroelectric between the maximum polarization ($P_{max}$) and remanent polarization states ($P_r$) in DRAM cells (unipolar operation), and between $+P_r$ and $-P_r$ (positive and negative remanent polarization) states in at least one mode of (bipolar) NVRAM operation. Figure 1 shows $Q'_c$, the DRAM polarization, and the relevant NVRAM signals representing "1" and "0". The requirements for DRAM storage dielectric materials include high charge storage density ($Q'_c$) and low leakage current densities ($J_L$) [1]. For NVRAM cell dielectrics, high remanent polarization ($P_r$) and low coercive electric field ($E_c$) are needed. Additionally, reliability of materials is an important parameter in both types of memory. Ferroelectric materials exhibit a unique reliability-degradation mechanism: fatigue, which is the gradual loss or alteration of polarizability with a.c. stressing. Thus from a reliability perspective, it is important to understand the nature of fatigue.

Traditionally, hysteresis loops are depicted as symmetrical and smooth curves (see Figure 1). However, constricted and asymmetrical loops have also been reported in the literature [2,3]. The asymmetry has been attributed to radiation induced damage, mechanical clamping and doping effects. Our sputtered films exhibit significant asymmetry in the fresh device (prior to electrical stress) and are also constricted (as in Figure 2) in the region of zero electric field. It was observed that both a.c. and d.c. electrical stressing reduced the constriction and resulted in symmetrical loops in the stressed device. Similar effects were observed after a post-processing anneal of the film. This indicates the importance of the connection between electrical and mechanical properties in these piezoelectric materials. Understanding these dependencies can lead to greater insight into the fatigue mechanism.

In this paper we highlight the structural and electrical asymmetry of our ferroelectric films, and describe the effects of a.c. and d.c. stressing on the shape of the hysteresis loop. The results of a post-annealing experiment lead to a qualitative understanding of stress-effects.

## EXPERIMENTAL DETAILS

Test devices used in this study were metal-insulator-metal capacitors (with platinum used as electrode material and PZT film thickness of 4000Å) fabricated on $TiO_2/SiO_2/Si$ multi-layered substrates, as shown in Figure 3. $TiO_2$ is used as a barrier oxide for Si diffusion and to enhance the adhesion of Pt onto the substrate. For deposition of PZT films, we used reactive DC-magnetron sputtering from a multi-component Pb/Zr/Ti metal target in a pure oxygen ambient (the nominal Zr/Ti ratio is 65/35). A separate Pb metal target was sputtered

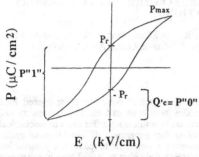

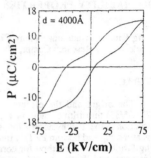

Figure 1. Typical hysteresis loop for a ferroelectric. $Q'_c$ (=$P_{max}$ - $P_r$) is the DRAM signal level. P"1" and P"0" represent switched charge densities for bit "1" and "0" respectively in the NVRAM read operation.

Figure 2. The asymmetrical quasi-static hysteresis loop (for example +$P_r$ = 3.8$\mu$C/cm$^2$ and -$P_r$ = 1.25$\mu$C/cm$^2$; +$Q'_c$ = 12.8$\mu$C/cm$^2$ and -$Q'_c$ = 15.3$\mu$C/cm$^2$) has been placed such that |+$P_{max}$| equals |-$P_{max}$|.

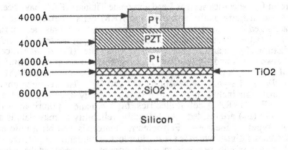

Figure 3. Schematic cross-section of test devices. The top electrode area is 87$\mu$m x 87$\mu$m.

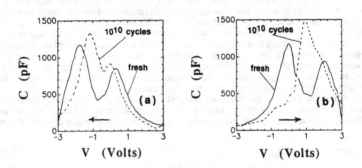

Figure 4. QSCV curves before (solid) and after (dashed) unipolar -3V dynamic stressing. (a) 3 to -3V sweep and (b) -3 to 3V sweep. Qualitatively similar effects are observed for d.c. stressing also.

simultaneously to compensate for Pb loss that occurs during high-temperature annealing (at ~650°C in nitrogen or oxygen for 1h). The top electrode was patterned using a lift-off process. Since the PZT films were annealed before deposition of top electrode, top and bottom electrode interfaces had different thermal histories. This asymmetry in the structure and processing of the device has important consequences. Some films also received a post-processing anneal (after top electrode deposition and patterning) for 1h at 400°C in $N_2$ ambient in order to study the effect of thermally-induced stress on the top electrode interface.

Various diagnostic tests were used to monitor the polarizability of the ferroelectric in fresh, stressed and post-annealed capacitors. The quasi-static CV technique involves the measurement of displacement current (which is proportional to the effective capacitance) for a voltage sweep with low, constant ramp-rate (0.5V/sec) using standard HP equipment. The C(V) curve for either sweep polarity is then integrated to yield P(V), which is the hysteresis loop. The RT66A ferroelectric tester was used to measure high-ramp rate (~5000V/sec) loops [4]. It was also used for the pulsed polarization test which involves the application of double bipolar pulses and the measurement of displaced charge density for the various transitions critical to NVRAM and DRAM operation. In general, the QSCV measurement yields polarization values higher than that from the other tests, possibly due to the frequency dependence of domain switching and mobile space charge effects [5]. The pulse polarization test, however, simulates memory cell operation more closely.

## RESULTS AND DISCUSSION

In contrast to the "ideal" loop depicted in Figure 1, a typical QSCV hysteresis loop from a sputtered film is constricted (as shown in Figure 2). Furthermore, it is visibly asymmetrical, in that the lobes for the two polarities on either side of the constriction are not identical in size or shape. The high ramp rate loops are qualitatively similar. Possible reasons for this asymmetry include structural asymmetry (the bottom platinum electrode is a macroscopic sheet while the top electrode is only about 100μX100μm in size) and asymmetry in thermal history (the bottom PZT/Pt interface was subjected to a high temperature annealing step while the top interface was not). When the capacitance-voltage curves are examined (Figure 4 (a), (b)), it is evident that for either sweep direction, there are two distinct peaks in the curve. It is understood that each peak represents a set of domains (with the position of the peak representing the value of the coercive voltage). This implies that within the film there are two regions of domains that switch independently; this might be related to the possibility of the presence of a sheet of space charge within the bulk of the film, as described in [3]. While there is no direct evidence for the presence of the charged layer, this model will be discussed further when the effects of electrical stress are described.

Differences in thermal history for the two PZT/Pt interfaces can have important consequences. When the PZT/Pt/TiO$_2$ structure is subjected to a high-temperature anneal (at ~600°C), the as-deposited amorphous phase transforms into the crystalline perovskite phase [6]. It is reasonable to expect the PZT layer to be stress-free at the high temperature; however, owing to a difference in the coefficients of thermal expansion in the two layers ($\alpha_{Pt} \sim 10 \times 10^{-6}$ $K^{-1}$, $\alpha_{PZT} \sim 1.5$ to $2 \times 10^{-6}$ $K^{-1}$ [7,8]), the cooling process will introduce compressive stress in the PZT layer (at the bottom Pt interface). At the Curie temperature (~350°C), the transition from paraelectric to ferroelectric phase causes a transition from cubic to tetragonal (or rhombohedral) structure in the material, and at the interface the c-axis possibly aligns itself in a direction that minimizes the compressive stress [9]. This may lead to the formation of an interfacial layer of domains isolated from the domains in the bulk of the film. Also, due to the columnar grain structure of the perovskite phase, no other regions of isolated domains are expected within a grain. Thus the two sets of domains are isolated from each other both physically and with respect to coercive fields (with the interfacial domains being clamped). This physical model seems to explain the asymmetry and the constriction in the hysteresis loops of the fresh devices.

Another question raised by the asymmetry in the hysteresis loop concerns the placing of the loop on the vertical (polarization) axis. When the capacitance curve is integrated to yield polarization values, there is an arbitrary constant that is indeterminate. Traditionally, this ambiguity is circumvented by equating the magnitude of $P_{max}$ for both polarities. For

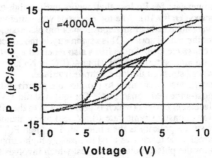

Figure 5. Asymmetrical 3V and 5V loops and a 10V loop. The 10V loop is on the verge of saturation (for both polarities) and is placed such that $|+P_{max}|$ equals $|-P_{max}|$. The 5V loop is placed to lie completely within the 10V loop.

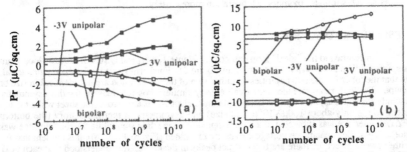

Figure 6. Polarization values as a function of number of stressing cycles under -3 and 3V unipolar and ±3 bipolar a.c. stressing. (a) $P_r$ increases prominently as the number of stressing cycles increases. (b) $P_{max}$ decreases for bipolar stressing, but increases for one polarity during unipolar stressing (of either polarity).

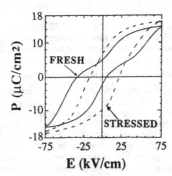

Figure 7. Hysteresis loops before and after $10^{10}$ cycles of -3V unipolar stressing. After fatigue, the hysteresis loop is more symmetrical and shows hysteresis relaxation.

asymmetrical devices like ours, this is probably inappropriate, because for the same magnitude of maximum electric field, the polarizability is arguably different for the two polarities. This implies that $|+P_{max}|$ doesn't necessarily equal $|-P_{max}|$. On the other hand, even for an asymmetrical device, a sufficiently large sweep amplitude drives the device into saturation for each polarity. In such a case, it is reasonable to expect that a domain that responds to one polarity also responds to the opposite polarity, implying that for a saturated loop, $|+P_{max}| = |-P_{max}|$ is valid. Figure 5 shows a 10V loop which is deep into saturation for the negative polarity, and on the verge of saturation for the positive polarity. This loop has been located consistent with $|+P_{max}| = |-P_{max}|$. The positioning of loops arising from smaller amplitude sweeps is not as simple. As explained by Miller et al., a hysteresis loop for a certain voltage sweep amplitude has to lie on or completely within the hysteresis loop arising from a sweep of larger amplitude [3]. Using this criterion, the 5V loop has been positioned within the 10V loop and the 3V loop within the 5V loop in Figure 5. It is interesting to note that in both cases $|+P_{max}|$ is considerably smaller than $|-P_{max}|$. All loops in this figure were obtained by the high ramp-rate test.

The correct positioning of the loop is crucial for estimating correctly the values of remanent polarization ($P_r$), but has no bearing on the differences in polarization between critical points on the loop. This means that as far as memory applications are concerned, the significant parameters can accurately be extracted from the loop without considering the positioning.

Included in Figure 4 are the quasi static C-V curves before and after $10^{10}$ cycles of unipolar -3V a.c. stressing (at 1MHz). As mentioned earlier, the fresh device has two distinct sets of domains, each with a different coercive voltage. It turns out that the peaks merge into a single peak during a.c. stressing. The hysteresis loop becomes more symmetrical as shown in Figure 7. The emergence of a single coercive voltage may be due to a complex process of pinning and de-pinning of different sets of domains. Alternatively, if a sheet of charge is understood to cause two coercive voltages in the fresh device, the merging of peaks during stress may be due to the neutralization or re-distribution of the space charge. It is interesting to note that d.c. stressing also results in the merging of the peaks and the emergence of a single peak in the C(V) curve (not shown in the figure). Moreover, the position of the new peak is identical for both polarities of d.c. stress.

Figures 6 (a) and (b) show the polarization values (derived from the pulsed polarization test) for devices subjected to various a.c. stresses. For each stress waveform, diagnostic pulse measurements were made for both polarities. Figure 6 (a) shows remanent polarization ($P_r$) as a function of number of cycles. Each stressing condition leads to the increase of $P_r$, which is often referred to as hysteresis relaxation (a loss of the constriction in the loop). Figure 6 (b) shows $P_{max}$ as a function of number of stressing cycles for different stress conditions. For unipolar stressing, there is an increase of $P_{max}$ for the polarity opposite to the stress polarity and a decrease of $P_{max}$ for the stress polarity, suggesting that there is an activation and pinning of domains for the two polarities respectively. Further, the increase of $P_{max}$ with negative unipolar stress is larger than that under positive stressing, due to the asymmetry of devices. As one may expect, bipolar stressing causes a decrease in $P_{max}$ for both polarities. As a result of changing $P_{max}$ and $P_r$ values, the hysteresis loop is more symmetric and relaxed after $10^{10}$ cycles of unipolar -3V AC stress, as shown in figure 7.

In order to test the hypothesis that asymmetric thermal history at the two electrode interfaces leads to asymmetric devices, a 400nm film was subjected to a post-processing anneal at 400°C (after the top electrode deposition and patterning). Figure 8 (a) shows hysteresis loops for two different devices on the same wafer, one without a post-anneal and one after a post-anneal. The loops have been normalized with respect to $P_{max}$ in order to highlight the symmetrization effect of the post anneal, and the decrease in the constriction. Figure 8 (b) on the other hand shows loops from a single device, before and after a post-processing anneal. The large increase in polarization after post-anneal is possibly due to stress-relief and the depinning of clamped domains. The behavior of post-annealed devices under a.c. stress needs to be examined in order to check whether the increase in polarization is accompanied by an increase in the fatigue rate. After a post-anneal there also is a symmetrization in the I-V characteristics. These effects bear evidence for the thermal-history hypothesis.

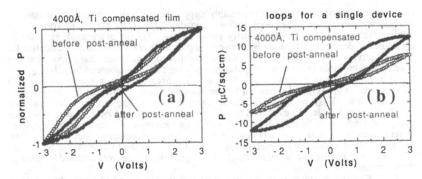

Figure 8 (a). Normalized polarization from two different devices: one without a post-processing anneal, and the other after a post-anneal (at 400°C for 1h in a $N_2$ ambient). The post-annealed film displays a symmetrical hysteresis loop and a decrease in the constriction. (b) Hysteresis loops from a single device before and after a post-anneal. There is a significant increase in polarizability after the anneal.

## CONCLUSIONS

Asymmetry in fresh sputtered PZT films is attributed to the asymmetry in the thermal-histories of the top and bottom electrodes. A method has been suggested for the positioning of the asymmetrical hysteresis loops on the polarization axis. Both d.c. and a.c. stressing tend to symmetrize the loop and also cause hysteresis relaxation. A post-processing anneal at 400°C after top electrode deposition tends to reduce the asymmetry and the constriction in the hysteresis loop.

## ACKNOWLEDGEMENTS

The authors wish to thank Prof. J. B. Goodenough of the University of Texas and Drs. Joe Mogab, Andy Campbell, Papu Maniar, Bob Jones and Reza Moazzami of Motorola Inc. for illuminating discussions and the use of their equipment.

This work was partially supported by SRC-SEMATECH contract # 88-MC-505 and by Texas Advanced Technology Program.

## REFERENCES

1. L. Parker and A. F. Tasch, IEEE Circuits and Devices Magazine, Jan. 1990, pp 17-26
2. Burfoot and Taylor, *Polar dielectrics and their applications*, (University of California Press, Berkeley and Los Angeles, 1979), p39
3. S. L. Miller et al., Journal of Applied Physics, **70** (5), 2849 (1991)
4. RT66A Ferroelectric Tester Operation Manual, Version 2.0, Radiant Technologies, Albuquerque, New Mexico.
5. J. Lee, et al., to be published in Ferroelectrics
6. V. Chikarmane et al., to be published in Jour. Vac. Sci. Tech. A, (1992)
7. B. Jaffe et al., *Piezoelectric Ceramics*, (Academy Press, London, 1971), p167
8. Y.S. Touloukian et al (ed.) *Thermophysical properties of matter Vol. 12*, (Plenum, New York, 1975) p 254
9. J. B. Goodenough, private communication.

# PLASMA ENHANCED LIQUID SOURCE-CVD OF Ta$_2$O$_5$ USING PENTA ETHOXY TANTALUM SOURCE AND ITS CHARACTERISTICS

P. A. Murawala, M. Sawai, T. Tatsuta and O. Tsuji
Research & Development Center, Samco International Incorporated,
33, Tanakamiya-Cho, Takeda, Fushimi-ku, Kyoto 612, Japan
and
Sz. Fujita and Sg. Fujita
Department of Electrical Engineering, Kyoto University,
Kyoto 606, Japan

## ABSTRACT

We report on plasma enhanced, liquid source, chemical vapor deposition (LS-CVD) of tantalum penta oxide (Ta$_2$O$_5$) material using a penta ethoxy tantalum [Ta(OC$_2$H$_5$)$_5$] liquid source. We have investigated several basic plasma deposition conditions such as - dependence of deposition rate and refractive index on the source tank temperature, carrier gas N$_2$ flow rate, reactive gas O$_2$ flow rate and substrate temperature. Structural properties investigated by $\theta$-$2\theta$ x-ray measurements showed amorphous nature of the films and Auger electron spectrosopy indicated carbon-contamination free growth of Ta$_2$O$_5$ films having proper stoichiometry (Ta/O = 0.4). In addition to this we have also performed electrical measurements on Au/Ta$_2$O$_5$/Si MOS structure which exhibit very well defined C-V characteristics with flat band voltage as low as +0.3V, low leakage current and high breakdown voltages. As a hitherto unreported step in Ta$_2$O$_5$ processing we also performed rapid thermal annealing at 700°C and 900°C for 5 minutes which showed much improved electrical properties. All results suggest growth of high quality Ta$_2$O$_5$ films from a carbon-based Ta liquid source, due to an effect of plasma enhanced deposition process.

## INTRODUCTION

Tantalum penta oxide is a promising material due to its high dielectric constant, making it a suitable alternative to conventional insulator materials like SiN and SiO$_2$ in I.C. technology. During the past decade many efforts have been made to deposit Ta$_2$O$_5$ by different techniques and from several source materials. Well known source materials of Ta in CVD are solid materials such as TaCl$_5$ and related halogens and liquid materials such as Ta(OCH$_3$)$_5$, Ta(OC$_2$H$_5$)$_5$ and related alkoxides. However, despite the good controllability of flow rate over a wide range using liquid sources, solid sources such as TaCl$_5$ are more widely used in order to avoid carbon contamination in the Ta$_2$O$_5$ film. In practice, the poor electrical properties of Ta$_2$O$_5$ deposited using liquid sources by thermal CVD or photo CVD have been attributed to carbon contamination. However, we consider that plasma CVD using liquid sources, which has been little investigated, is more effective in decomposing the carbon-related bonds in source materials, and contributes to lower carbon contamination, resulting in better electrical properties of the deposited films.

Shinriki et al [1] reported on LPCVD of Ta(OC$_2$H$_5$)$_5$ source followed by two step annealing of UV-O$_3$ and dry-O$_2$ at 800°C. UV-O$_3$ annealing resulted in reduction of leakage current and dry-O$_2$ annealing lead to reduction of defect density. Saitoh et al [2] reported on LPCVD of Ta(OC$_2$H$_5$)$_5$ and showed that leakage current in Au/Ta$_2$O$_5$/Si MOS diodes reduces considerably with dilute TiO$_2$ doping. Yamagishi et al [3] reported on photo-CVD of Ta$_2$O$_5$ from a Ta(OCH$_3$)$_5$ liquid source. They got poor data by thermal CVD technique as compared to photo-CVD. It seems that several workers have applied either photo technique or annealing technique, in order to improve their Ta$_2$O$_5$ film quality, grown from carbon-based liquid sources such as Ta(OCH$_3$)$_5$, Ta(O$_2$CH$_3$)$_5$.

In the present work, we have attempted to deposit Ta$_2$O$_5$ by plasma enhanced liquid-source chemical vapor deposition (LS-CVD) using penta ethoxy tantalum Ta(OC$_2$H$_5$)$_5$ source, and have investigated several basic deposition conditions. Structural properties have been investigated by x-ray and AES measurements. Some initial electrical measurements have been performed on

**Mat. Res. Soc. Symp. Proc. Vol. 265. ©1992 Materials Research Society**

Au/Ta$_2$O$_5$/Si MOS devices and the effect of rapid thermal annealing has been studied.

## PLASMA ENHANCED LS-CVD Of Ta$_2$O$_5$

Deposition of tantalum penta oxide was carried out in Samco's model PD240 by a using plasma enhanced LS-CVD technique. Fig.1 shows a schematic diagram of the experimental set-up used in this work, and table 1 shows the typical deposition conditions. The source material, Ta(OC$_2$H$_5$)$_5$, is liquid at room temperature, its melting point (M.P.) is 21°C, boiling point (B.P.) is 146°C and its vapor-pressure at 140°C is 0.1 mm [4].

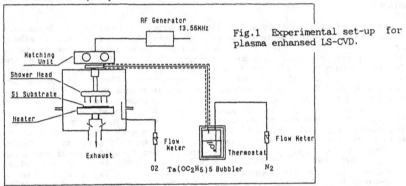

Fig.1 Experimental set-up for plasma enhansed LS-CVD.

Under the present optimised plasma condition, we investigated some of the basic deposition conditions.

## TABLE 1.TYPICAL DEPOSITION CONDITIONS

| | |
|---|---|
| Source | – Ta(OC$_2$H$_5$)$_5$ |
| Tank Temperature | – 160–190°C |
| Line Temperature | – 180°C |
| Substrate Temperature | – 475°C |
| Carrier-gas | – N$_2$ |
| Carrier-gas flow | – 50 sccm |
| Reactive gas | – O$_2$ |
| Reactive gas flow | – 50 sccm |

## CHARACTERISTICS OF PE-CVD Ta$_2$O$_5$ FILM

Fig.2 show, the dependence of the deposition rates (Rd) and refractive index (n) on the source temperature. The deposition rate increases in the temperature range of 160-190 °C to about 39 Å/min. Refractive index, n also showed similar dependence on the tank temperature. A study of the temperature dependence of vapor-pressure of Ta(OC$_2$H$_5$)$_5$ [4] indicates that a vapor pressure of about 2-3 mmHg is necessary for the deposition of high quality Ta$_2$O$_5$ thin films. It is obvious that at T$_T$=190°C, enough Ta source reaches to plasma chamber and deposition rate is enhanced.

## Table 2. : Deposition Rates

| Deposition Technique | Ts (OC) | Rd(Å/min.) | Ref. |
|---|---|---|---|
| LPCVD [Ta(OC$_2$H$_5$)$_5$] | 700 | 45 | Saitoh et al |
| Photo CVD [Ta(OCH$_3$)$_5$] | – | 70 (20) | Yamagishi et al |
| Photo CVD (TaCl$_5$) | 450 | 32 | Matsui et al |
| D.C.Sputtering (Ta target) | – | 16 | V.Teravaninthorn |
| ECR-CVD (TaCl$_5$) | 200 | 400 | Watanabe et al |
| PE-CVD [Ta(OC$_2$H$_5$)$_5$] | 475 | 64 | Present work |

Fig.3 shows the dependence of deposition rate and refractive index on carrier gas $N_2$ flow rate (through the source tank). The deposition rate and the refractive index both increase with the increase in $C-N_2$ flow. Deposition rates of up to about 64 Å/min. were achieved with a refractive index of n=2.25.

A comparative study of the reported deposition rates of $Ta_2O_5$ by different techniques and materials is shown in the table 2. Saitoh [2] has reported Rd=45 Å/min. by LPCVD, using $Ta(OC_2H_5)_5$ but substrate temperature was too high for use in device applications. Yamagishi et al [3] have achieved Rd=70 Å/min. by Photo-CVD using $Ta(OCH_3)_5$, however by thermal CVD they got only Rd=20 Å/min. Matsui et al [5] have reported Rd=32 Å/min by Photo-CVD using $TaCl_5$, Teravaninthorn [6] reported Rd=16 Å/min. by D.C. Sputtering (Ta target). In comparision to reported data, our Rd=64 AO/min. by PE LSCVD using $Ta(OC_2H_5)_5$ is much better and we hope that it will be improved by the use of photo-irradiation and/or UV-03 techniques. Of course, Rd=400 Å/min obtained by ECR is an exempt of a much faster rate [7].

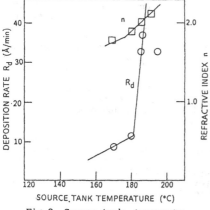

Fig.2 Source tank temperature dependence of the deposition rate Rd and refractive index n.

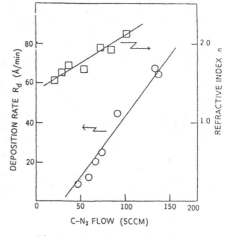

Fig.3 Carrier gas $N_2$ dependence of the deposition rate Rd and refractive index n.

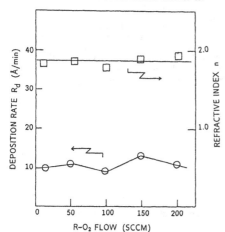

Fig.4 Reactive gas $O_2$ dependence of the deposition rate Rd and refractive index n.

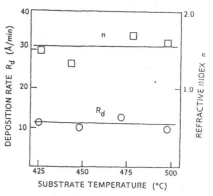

Fig.5 Substrate temperature dependence of the deposition rates Rd and refractive index n

Further investigations show that n and Rd are independent of the variation of the $O_2$ flow (10-200 sccm) and of substrate temperatures (425-500°C) as shown in the fig.4,5.

In order to achieve uniform deposition on 4"-6" wafers, we have investigated electrode distance dependance on deposition rate and we find as the electrode separation is increased from 20 mm to 70 mm, the uniformity improves from about 34% to about 10% on 4-6" wafers.

From the above results it appears that the deposition rate and refractive index strongly depends on the source tank temperature and carrier gas, $N_2$ flow rate, but is nearly independent of $O_2$ flow rate and substrate temperature. Deposition kinetics depend upon many factors such as i) flow rate of Ta source vapor ii) dissociation energy of Ta source iii) surface reaction and iv) absorption (diffusion) at the surface. In plasma CVD, chemical reactions result from electron-impact dissociation of precursor gases in the plasma glow discharge. As the rf power intensity increases, the number of ions also increases, which exhances plasma ion bombardment and results in dissociation of precursors. In the conventional rf discharge systems, sources like $Ta(OC_2H_5)_5$ dissociate due to an applied electric field, and the deposition occurs due to the relationship [8], given by

$$R_i = n_e k_i [P]$$

where, $R_i$ = deposition rate, $n_e$ = electron density, $k_i$ = rate constant for the dissociation reaction, [P] = concentration of reactant.
The rate constant $k_i$ is given by

$$k_i = \sqrt{(2/m_e)} \int E^{1/2} f(E) \sigma_i(E) dE$$

where, $m_e$ = electron mass, E = electron energy, f(E) = normalized electron distribution function and $\sigma_i(E)$ = cross-section for the reaction.

If the deposition process is dominated by an absorption process, then it should follow the Hertz-Knudsen relationship.

$$\Gamma_i = \beta_i P_i / \sqrt{2\pi M_i kT}$$

where, $\Gamma_i$ = Absorption rate, $\beta_i$ = Adhesive probability coefficient, $P_i$ = Pressure, $M_i$ = Mass of radical, T = Substrate Temperature and k = Boltzman constant, such that slope of $\ln \Gamma_i$ vs. $1/T_i$ is 0.5. However Fig.5 indicates negligible slope i.e. no temperature dependence. This means that surface absorption processes are not effective, perhaps due to the lower pressure of the Ta source under the present plasma deposition condition.

## STRUCTURAL PROPERTIES

## X-RAY ANALYSIS

In order to study the structural properties of the $Ta_2O_5$ films, we performed x-ray diffraction measurements in the $\theta-2\theta$ mode (fig.6) on a series of samples, grown under different conditions, by plasma enhanced LSCVD. Each of the samples showed identical x-ray spectra. The X-ray spectra show well-known sharp and intense peaks from Si substrate for $K\alpha_1$, $K\alpha_2$ and $K\beta$. In addition to these peakes, we observed a broad and low intensity peak at about 17 degrees. The nature of this x-ray spectrum indicates an amorsphous

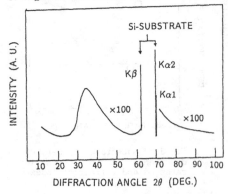

Fig.6 X-ray diffrection pattern ($\theta-2\theta$) mode of $Ta_2O_5$ on Si

structure but it also suggests the presence of a crystalline phase of the
Ta2O5 films. Amorphous Ta2O5 has also been reported by Shinriki et al [9].
They performed x-ray diffraction, SR-EXAFS and XPS measurement to determine
Ta2O5 crystal structure. From their results, they proposed an δ-Ta2O5
(hexagonal) structure of Ta2O5 and estimated the lattice constant assuming
that the crystal form is δ-Ta2O5 and the unit cell consists of one molecule.

## AUGER ELECTRON SPECTROSCOPY (AES)

In order to understand chemical compositions in the Ta2O5 films
deposited by plasma-enhanced LSCVD of Ta(OC2H5)5 liquid source, we have
performed AES measurements. Results shown in fig.7 are a plot of Ar-beam
sputtering time versus the atomic concentration percentage of the constituent
elements in the film. Analysis of this data indicates that; 1) at the surface
of the film, the Ta concentration is relatively low and O concentration is
higher as comparared to that in the bulk of the film. This is because, the
Ta2O5 film at surface is oxygen rich due to surface oxidation. 2)
Concentration of Ta and O in the bulk of the film is quite uniform,
suggesting uniform distribution of Ta and O atoms across the whole deposited
film. 3) The compositional atomic concentration of Ta and O in Ta2O5 film is
28% and 70% respectively, and therefore Ta/O ratio is 0.4 which very well
represents Ta2O5 stoichiometry of the film. 4) Carbon contamination which is
quite often found in films deposited from the Ta(OC2H5)5 source materials has
not been detected by AES measurement. It may be well below the detectable
limit of the AES equipment.

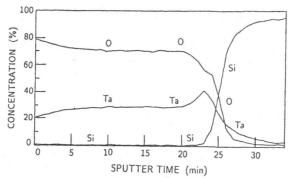

Fig.7 Auger Electron spectroscopy of PE-LS-CVD grown Ta2O5 film.

## ELECTRICAL PROPERTIES

Fig.8 shows a 1 MHZ C-V characteristics of Au/Ta2O5/n-Si MOS structure.
The C-V curve exhibits very well defined C-V characteristics obtained on
as-grown Ta2O5. We got similar results on Au/Ta2O5/p-Si MOS devices. The flat
band voltage (Vfb) is as low as about +0.3V. Tanimoto et al [10] also
reported C-V curves for Al/Ta2O5/Si MIS structure where Ta2O5 was grown by
photo-CVD using a TaCl5 solid source and ozone. Although they did obtain
typical C-V characteristics, flatband voltage was greatly shifted towards the
positive bias indicating some charge exists in Ta2O5 film.

Current-voltage measurements of Au/Ta2O5/Si MOS diode resulted in
leakage currents as low as about $6 \times 10^{-8} A/cm^2$ for 1 MV/cm electric fields,
which increased gradually with increasing applied fields with breakdown
occuring at 5 MV/cm. From a comparision of I-V data obtained for as-grown
samples prepared from different carbon-based liquid sources, we note that our
leakage currents are two orders of magnitude smaller then those obtained by
LPCVD. Our results suggest that up on the dissociation of Ta(OC2H5), carbon
is evacuated before it can be deposited in the Ta2O5 film, and a nearly
carbon free Ta2O5 can be deposited resulting in better I-V and C-V
characteristics.

324

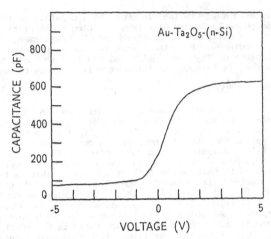

Fig.8 Well defined C-V characteristics of Au/Ta$_2$O$_5$/n-Si MOS device. Vfb = + 0.3V.

Rapid thermal annealing (RTP equipment manufactured by AST, Germany) of Ta$_2$O$_5$ performed at 700°C and 900°C for 5 minutes resulted much improved leakage currents. At a 1 MV/cm applied electric field, currents are 6 x 10$^{-8}$ A/cm$^2$, 1 x 10$^{-8}$ A/cm$^2$ and 1.5 x 10$^{-9}$ A/cm$^2$ respectively for as-grown; 700°C and 900°C annealed samples. It is important to note that breakdown does not occur even up to the investigated 10 MV/cm of electric field in annealed samples. Details of electrical properties will be reported elsewhere.

CONCLUSION

In conclusion, we have succeeded in growing high quality tantalum penta-oxide dielectric material from a carbon based liquid source Ta(OC$_2$H$_5$)$_5$ by using a plasma enhanced CVD technique. Structural properties indicate that the films are amorphous in nature, Ta and O atoms are uniformly distributed in the film and films are stoichiometrically good. Carbon contamination, which is usually considered as cause of poor electrical properties, has not been detected by AES. Our as-grown Ta$_2$O$_5$ shows very well defined C-V characteristics with Vfb close to zero and low leakage current as compared to those reported earlier by other CVD techniques for carbon based source materials. Hence plasma enhanced LSCVD grown film may be viewed as more suitable in I.C. technology.

REFRENCES

1. H. Shinriki, Y. Nishioka, Y. Ohji and K. Mukai IEEE Transaction on Electron Devices 36(2), pp 328-332, 1989.
2. M. Saitoh, T. Mori and Tamura, IEDM-86, IEEE, pp 680-683, (1986).
3. K. Yamagishi and Y. Tarui J.J. A. P., 25(4), pp. L306-L308, (1986).
4. TRI Chemical Laboratory Inc., Japan.
5. M. Matsui, S. Oka, K. Yamagishi, K. Kuroiwa and Y. Tarui, J.J.A.P. 27(4) pp. 506-511, (1988).
6. U.Tervaninthorn, Y. Miyahara and T. Morizumi J.J.A.P., 26(3), pp.347-351, (1987).
7. I. Watanabe and H. Yoshihara J.J.A.P. 24(6), pp.L411-L413, (1985).
8. Plasma deposited thin film by J.Mort and F.Jansen, CRC Press (1986).
9. H. Shinriki, N. Hiratani, A. Nakao and S.Tachi, Proc. Int. Conf. SSDM, Yokohama, pp. 198-200, (1991).
10. S. Tanimoto, M. Matsui, N. Aoyagi, K. Kamisako, K.KKuroiwa and Y.Tarui, Proc. Int. Conf.SSDM, Sendai, pp L195-L198, (1990).

# Author Index

326

# Subject Index